Fundamentals of Thermodynamics, Thermochemistry, and Gas Dynamics

Joseph A. Kunc
University of Southern California

Fundamentals of Thermodynamics, Thermochemistry, and Gas Dynamics
Joseph A. Kunc
University of Southern California

First published in 2020

Printed in the United States of America

Library of Congress Cataloging-in-Publication Data on file

ISBN 978-1-7330098-0-5

Managing Editor: Kimberly Hitchens, Booknook.biz
Typographer: Mike Dworski
Cover Designer: Shelley Savoy

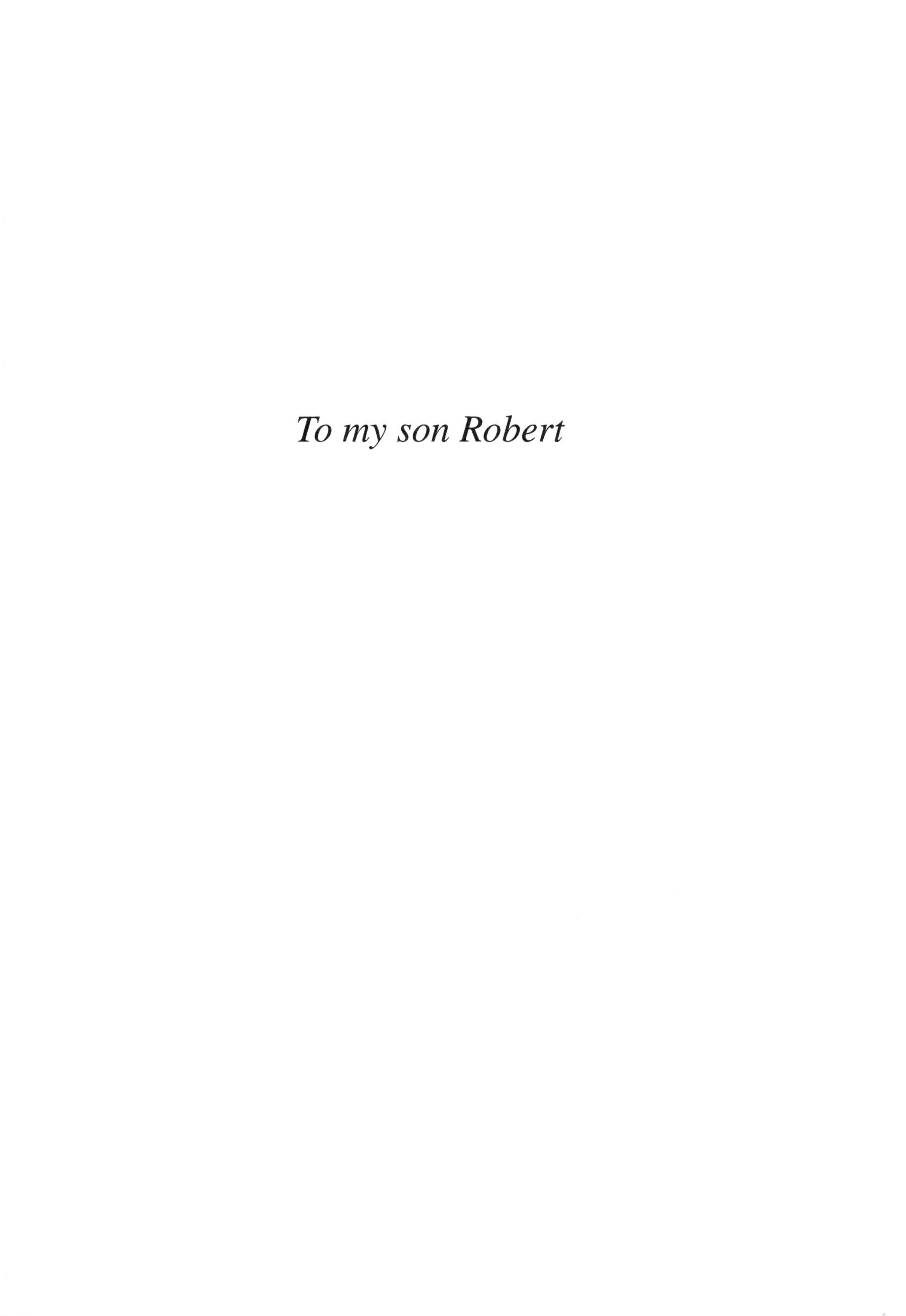

To my son Robert

Contents

CONTENTS

Preface

Classical Thermodynamics (also called Phenomenological Thermodynamics, or just Thermodynamics) is a general, mathematically comprehensive, and powerful theory of energy-processing phenomena that has contributed enormously to the progress of science and technology. Arthur Eddington described the Second Law of Thermodynamics as holding the supreme position among the laws of Nature, Albert Einstein famously said that thermodynamics is the only physical theory of universal content that will never be overthrown, and Roger Penrose called the theory a magnificent body of well-established physical understanding. The theory has been taught across the world but considered to be one of the most difficult both to learn and to teach. As stated, with a dose of humor, by Arnold Sommerfeld, one of the founders of modern physics: "Thermodynamics is a funny subject. The first time you go through it, you don't understand it at all. The second time you go through it, you think you understand it, except for one or two small points. The third time you go through it, you know you don't understand it, but by that time you are so used to it, it doesn't bother you anymore."

This book has evolved from my many years of research and teaching on molecular and radiative processes, statistical and quantum mechanics, thermodynamics, and non-equilibria in high-temperature gases and plasmas. The main goal of the book is to present, with rigor and clarity, the fundamental concepts and methods of three major disciplines (Thermodynamics and its derivatives Thermochemistry and Gas Dynamics) and to knit them together into a coherent and comprehensive stream of information based on a single set of scientific principles. The material can be covered in a fairly demanding upper-division academic course. The book will be useful to students, physicists, chemists, and engineers. Also, it meets the needs of the increasing number of researchers in inter-disciplinary and multi-disciplinary areas of science and technology.

Many professionals working in those areas have no extended training in thermal sciences but often face problems similar to those discussed here.

In preparing the text, a special effort was made to demonstrate the direct links of Gas Dynamics and Thermochemistry with Thermodynamics, and to emphasize the importance and generality of the thermodynamic method. The organizational structure of this text and its index, way of repetition of important definitions and descriptions of many concepts follow the suggestions of numerous students and faculty familiar with the text. (The style and the content of the book's covers also was suggested by students.)

Since the potential readers of the book may have different educational backgrounds, in the first two chapters we review basic concepts of thermal sciences, popular unit systems, calculus of multi-variable functions, and the role of the concepts in study of the properties of matter, energy, and entropy in both closed and open physical systems. A special emphasis is put on discussion of entropy flows because of its role in understanding complex thermodynamic systems. In Chapter 3, we review the fundamental constraints that are imposed by Nature on all real processes. In Chapter 4, we extend the discussion of properties of state and their role in thermodynamic theory. Chapters 5 and 6 concern the equations of state and heat capacities of matter, respectively, and their role in analyses of thermal phenomena. Chapter 7 is a review of common models of thermodynamic processes in a variety of applications. In Chapter 8, we discuss calculations and measurements of thermodynamic and thermochemical properties of various substances, and the practical use of databases such as the *NIST-JANAF Thermochemical Tables*. The general theory of multi-component and multi-phase systems is presented in Chapter 9. In Chapter 10, the thermal phenomena in a wide range of practical devices that operate along common thermodynamic cycles is discussed. Flows of compressible gases in open systems with and without gasdynamic shocks are covered in Chapter 11, and properties of chemical reactions and systems in chemical equilibrium are covered in Chapter 12. Chapter 13 deals with the production of entropy and energy availability (or exergy) in practical applications. Appendices A and B include some thermodynamic data on properties of single- and two-phase substances, respectively. Unit systems other than SI are discussed in the text because they dominate the vast volume of earlier literature and are still often used in today's publications. Scattered throughout the text are numerous illustrative problems with solutions intended to help readers increase their understanding of the concepts and methods studied.

I thank my friends at the University of Southern California, Dan Erwin, Mike Gruntman, Martin Gundersen, Bart Kosko, and Robin Shakeshaft, for their contributions to my professional life, including suggestions about the content of this book. I am also deeply indebted to several people who rendered important help and advice: Alex Dalgarno, Hans Griem, Michael Gryzinski, Evgueni E. Nikitin, Wolfgang Wiese, Gerhard Weissler, and Matthew Zgorzelski. Many thanks go to my undergraduate and graduate students, too numerous to name, from whom I have learned much and whose many comments and suggestions have been invaluable to the development of this text. Finally, I wish to thank Kimberly Hitchens, Shelley Savoy, and Mike Dworski for great help and sound advice during the production of the book.

Chapter 1

Introduction

1.1 Basic Concepts

The scientific discipline called **Thermodynamics** (also called **Classical Thermodynamics** and **Phenomenological Thermodynamics**) is a powerful theory of the relationships between heat, work, and the internal energy of physical systems. The theory originated from the following facts observable in all non-relativistic processes:

Fact 1: The amount of the matter (measured by its mass) present in a physical system has a significant influence on the system's properties. The mass of any system can be changed (increased or decreased) *only* as the result of a transfer of some matter between the system and its surroundings. In other words, matter cannot be produced or destroyed spontaneously. Thus, the *total* mass of every system that does not exchange matter with its surroundings remains constant regardless of the character of the processes taking place inside the system. This last statement is often called the **Principle of Conservation of Matter**. More discussion on the properties of matter in thermodynamic systems is given in Sections 2.1, 3.1, and 3.5.1.

Fact 2: Energy is the most common and the most important entity of the universe. It is carried everywhere by particles and photons. The *total* energy of any system can be changed (increased or decreased) *only* as a result of a transfer of some energy (in form of heat, work, and/or radiation) between the system and its surroundings. In other words, energy cannot be produced or destroyed spontaneously. One form of energy can be converted into another form, but the *total* energy of the system remains constant as long as the system does not exchange any energy with its surroundings. This last statement is often called the **Principle of Conservation of Energy**.

The amount of the energy stored in a physical system is usually meaningless in thermodynamic applications. What counts is the *change* of the energy since only such a change can produce a flow of heat, radiation, or work. More discussion on properties of energy is given in Sections 2.2, 3.2 and 3.5.2.

Fact 3: No real (that is, realizable in practice) process moving a physical system from one thermodynamic state to another thermodynamic state can *spontaneously* (that is, by an infinitesimally small change of the system's conditions at the latter state) reverse itself to restore both the system and its surroundings to their initial states. Thus, one can say that **all real processes occurring in the Universe are irreversible processes**. (Reversible processes are forbidden by Nature even though they do not violate Nature's Principle of Conservation of Matter and Principle of Conservation of Energy – see below.) Clearly, Nature always spontaneously and irretrievably dissipates some amount of 'something' during *every* real process in *every* system. We will learn later that this 'something' is the order in the distribution of microscopic (quantum-mechanical) properties of particles and photons, a phenomenon associated with an important thermodynamic property called **entropy**. More discussion on the entropic properties of thermodynamic processes is given in Sections 2.3, 3.3, and 3.5.3.

A **thermodynamic system** is defined as a region of space containing some amount of matter (particles) and radiation (photons). Examples of thermodynamic systems are: the air in a room, a car engine, a human cell, a human, the Sun, the Universe. The boundary of the region is often called the **control surface** of the system. The control surface can be real or imaginary (a great convenience in theoretical studies), fixed or movable, but it must be well defined during the entire thermodynamic process under consideration. Matter and energy (in form of work, heat, or radiation) may enter or leave the system through its control surface.

In thermodynamic studies, the Universe is usually divided into two parts: the **system** under consideration and the rest of the Universe which is called the **environment**. Therefore, any thermodynamic interaction between the system and the environment depends on the thermodynamic properties of both. The properties of the environment are always non-uniform through the huge space the environment occupies, so it is very difficult to describe the properties in physical terms. For example, the mean temperature of the universe is about 3 kelvins (K) while the temperatures of the direct surroundings of most thermodynamic systems of technological interest are between 200 K and 1000 K. Of course, the properties of a system's immediate vicinity are almost always much more important for processes taking place in the system than the properties of the remote parts of the environment. Therefore, it is practically a rule in thermodynamic studies to consider a system's immediate vicinity (the 'surroundings') as the 'environment'.

Studies of transfer of matter, heat, work, radiation, and entropy (see below) between the thermodynamic system and the environment often use a convenient concept of the **heat reservoir (heat bath, heat sink)**. The heat reservoir is a hypothetical body that can absorb or reject heat without undergoing meaningful changes in temperature, pressure, or density. Such a situation can be assumed when the amounts of matter, energy, and entropy transferred between the system and the reservoir are very much smaller than the amounts of matter, energy, and entropy of the reservoir, and when there are no processes inside the reservoir able to affect meaningfully its thermodynamic state.

Thermodynamic systems can be divided into open, closed, and isolated systems. An **open system** can exchange both matter and energy with its environment. The exchanged energy can be in the form of work, heat, or radiation. (In thermodynamic studies, transfer of radiation is often ignored or included in heat transfer.) A **closed system** can exchange energy but not matter with the environment. (Open systems are sometimes called the **control volumes**, and closed systems are called the **control masses**.) An **isolated system** exchanges neither matter nor energy with the environment. Thus, thermodynamic processes in isolated systems are not affected by processes that occur in the system's surroundings.

Any physical condition of a thermodynamic system at a given instant is called the **macrostate**, the **thermodynamic state**, or just the **state** of the system. When such a state is an equilibrium state, it can be described by the system's material properties (the amount of matter, its heat capacity), and a small number of macroscopic properties called the **thermodynamic variables** (the absolute temperature T, pressure P, volume V).

Any transformation of a physical system from one thermodynamic state to another thermodynamic state is called the **thermodynamic process, thermo- dynamic transition**, or **thermodynamic path**. The processes can be divided into two main categories: 1) **reversible transitions** which are often called the **quasi-static transitions**, or **quasi-equilibrium transitions**, and 2) **irre-**

versible transitions. A reversible transition is a theoretical idealization of real processes because it is a sequence of different quasi-equilibrium states, each being infinitesimally close to a state of an equilibrium. An irreversible transition is a sequence of non-equilibrium states. (As already mentioned, all thermodynamic transitions occurring in the world are irreversible processes.) If the sequence's initial and final states are equilibrium (quasi-equilibrium, local-equilibrium) states, then the *changes* of some important thermodynamic properties (the internal energy, enthalpy, entropy) of the process can be accurately calculated by the methods of Classical Thermodynamics. The methods give realistic descriptions of physical processes in a broad range of temperature and pressure that are common in practical thermal systems. However, study of the processes at very low temperatures or very high pressures requires consideration of some quantum-mechanical effects which are not discussed here.

A thermodynamic system can consist of a number of different **pure substances** that are often called the **components** of the system. A substance (H_2O, CH_4, Au, etc.) is called a pure substance when it consists of only one kind of molecule, and the structures of the molecule do not change during the process under investigation. For example, the substance having chemical symbol H_2O and heated at pressure of 1 bar from $T = 200\,K$ to $T = 1600\,K$ can be called 'component H_2O' because overwhelming majority of the H_2O molecules do not change their structures during the entire process even though the substance goes through several phases (ice (solid phase) $\rightarrow$ water (liquid phase) $\rightarrow$ vapor (gas phase)) during the process. However, if the process proceeds to higher temperatures, then dissociation of many of the H_2O molecules will produce new components such as H_2, O_2, O, O_3, OH, and H. As a result, the original single-component (pure substance) H_2O will become a **mixture** of many components.

A **phase** of a component is the collection of all regions of the system that are occupied by the component, have well-defined boundaries, and are homogeneous (in both physical structure and chemical composition) across their volumes. Components can be in the so-called *allotropic* forms, which are considered different phases of solid matter. (The different allotropic forms of a solid have different crystalline structures and, therefore, different physical (mechanical, electrical, thermal) properties.) An example of a multi-component, multi-phase thermodynamic system is a glass containing water, air, and three ice cubes floating in the water. Such a system consists of three components (the glass, the water, and the air) and four phases: solid glass, solid H_2O (the collection of the three ice cubes), liquid H_2O (the water), and gas (the air). In a rigorous analysis, the air should be considered a mixture of several components (nitrogen, oxygen, helium, etc.). However, mixtures common in applications (the air is such a mixture) are often treated in thermodynamic calculations as fictitious single-component substances of properties obtained from averaging the properties of all the components of the actual mixture. We will discuss such cases in Chapter 9.

Thermodynamic properties can be classified as either **state properties (thermodynamic functions of state)** or **path properties (thermodynamic functions of path)**. Thermodynamic functions of state are sometimes called the **thermodynamic potentials**. Examples of state properties are a system's internal energy, enthalpy, and entropy. The value of any **state property** of a physical system in a particular state of thermodynamic equilibrium can be obtained from the actual characteristics of this state (no knowledge of the 'history' of the creation of the equilibrium state is needed). Therefore, the **change of any state property during any process between any two equilibrium states is always the same regardless of the kind of the process, that is, regardless of the process's thermodynamic path. Thus, the change is just the difference between the values of the state properties at the final and initial states of the process. The formalism of Classical Thermodynamics can describe the properties of the two equilibrium states with great accuracy, but it is unable to accurately describe properties of non-equilibrium states. Therefore, Classical Thermodynamics can provide accurate information about changes of state properties only in processes that begin and end in equilibrium, quasi-equilibrium, or local-equilibrium states.** One should add that the formalism of Classical Thermodynamics treats the differentially close (with respect to thermodynamic variables pressure, temperature, volume) quasi-equilibrium and local-equilibrium states as if they were equilibrium states. Therefore, **every statement in this book that refers to an equilibrium state also applies to any differentially close quasi-equilibrium or local-equilibrium state.**

A **path property** is a thermodynamic property which depends on the way (thermodynamic path) in which a system has reached a state of interest. Examples of path properties are work and heat. Since every system can move between its initial state 1 and final state 2 in many different ways (that is, along many different thermodynamic paths), the change of every path property during the $1 \rightarrow 2$ transition depends on the character of the thermodynamic process moving the system between the two states. Thus, the work of the $1 \rightarrow 2$ transition at constant temperature will be different from the work of the process at constant pressure even though the initial states of the two transitions are the same and the final states of the two transitions are the same.

A thermodynamic process at a constant temperature is called an **isothermic (isothermal)** process. A process is called **isobaric** or **isochoric** if the system's pressure or volume, respectively, is constant during the process. A process is called **adiabatic** if no heat is exchanged between the system and its surroundings.

Thermodynamic properties can be either **extensive** or **intensive**. A property is called an **extensive property** when its value for the entire system is the sum of the property values for all the parts forming the system. System mass and energy are examples of extensive properties. Thermodynamic properties that do not depend on the size of the system are called the **intensive properties**. (System pressure and temperature are examples of intensive properties.) One should notice that all properties given per unit of volume of a homogeneous system in equilibrium are intensive properties. Identification of system properties as being extensive or intensive is often important because some thermodynamic laws apply to the properties of the former but not the latter category. Discussions of some thermodynamic laws do not require consideration of the (intensive or extensive) characters of the physical properties being described by the laws. In the cases where the character is important (see, for example, the Gibbs Phase Rule discussed in Section 9.1), the importance is usually emphasized in the laws' definitions. Sometimes, thermodynamic discussions ignore the different characters of the properties which are 'related' in an obvious and unique way. An example of such an informal (but usually not harmful) approach, common in studies of equilibrium systems, is the equivalent treatment of the system volume V (an extensive property) and the 'related' specific volume v (an intensive property).

1.2   States of Thermodynamic Systems

1.2.1   Non-Equilibrium

Matter inside of some volume of space can be in an equilibrium state or a non-equilibrium state. (The photons inside the volume can also be characterized by such concepts.) An amount of matter is said to be in a state of **macroscopic non-equilibrium** (the term 'macroscopic' is usually omitted) when the matter's

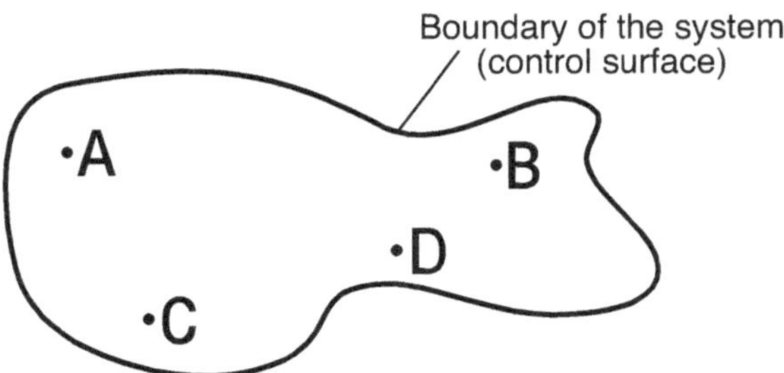

Fig. 1.1: A physical system divided into many very small regions A, B, C, D, ... On macroscopic scale, each of the regions ('cells') is small enough to claim its own (constant) values of thermodynamic properties because the changes of these properties across such small cells are negligible. When thermodynamic properties of all cells of the system are the same, the system is said to be in *(macroscopic) complete equilibrium*. If only a group of the cells (a part of the system) have the same properties, then such a part is said to be in *(macroscopic) local equilibrium*. A system is said to be in *(macroscopic) non-equilibrium* when at least some of its cells (or parts) have different thermodynamic properties.

macroscopic properties (pressure, temperature, or chemical composition) of various parts of the volume occupied by the matter differ from one another. Very small regions of the system volume are shown in Fig. 1.1 as 'points' A, B, C, D, ... and are called hereafter 'cells'. Each of these cells is small enough for the variations of the macroscopic properties across the cell to be ignored as meaningless for thermodynamic studies of the cell. At the same time, the cells must be large enough to be unaffected by the fluctuations of the microscopic (molecular) properties of the particles present in the cells.

A special kind of non-equilibrium is the **temperature steady-state**. A system is said to be in a temperature steady-state when the temperatures of its different cells (the cells A, B, C, D, ... in Fig. 1.1) differ from one another but the temperature of each of the cells does not vary with time ($T_A \neq T_B \neq T_C \neq \cdots$, but $dT_A/dt = dT_B/dt = dT_C/dt = \cdots = 0$). The same can be said about different groups (parts) of the system. In any system in temperature steady-state, there must be a steady transfer of heat between the system's cells (or parts) of different temperatures – such transfer is necessary to keep the temperatures time-independent. A good example of a system in a temperature steady-state is the gas in a working light bulb. The gas is hotter in the center of the bulb than at the bulb's wall. Therefore, the gas is in thermodynamic non-equilibrium.

However, it is in a temperature steady-state because the (different) temperatures of particular parts of the gas do not change with time. If the temperatures of some of the parts were varying with time, then one would face one of the following possibilities: 1) some parts of the bulb would become so hot that the bulb would melt; 2) some parts of the bulb would become so cold that the electronic excitation of the particles producing the visible radiation would become insignificant and, as a result, the bulb would stop producing the required amount of light; 3) the temperatures of the gas parts would increase and decrease with time in an oscillatory manner producing variations of the light intensity. None of these three situations can be called a temperature steady-state. However, the third situation can be called a temperature steady-state if the period of the temperature oscillations is much shorter than the time scale of the temperature variation important in applications of the bulb. This depends on the sensitivity of human eye to the rate of change of the intensity of the visible radiation produced by the bulb.

A thermodynamic system can be in **pressure steady-state** when the pressures in different cells (or parts) of the space occupied by the system differ from one another but the pressure in each of the cells (parts) does not change with time ($P_A \neq P_B \neq P_C \cdots$, but $dP_A/dt = dP_B/dt = dP_C/dt = \cdots = 0$). Thus, in a system in pressure steady-state, there must be a steady transport of matter between the cells (parts) of different pressures. The transport is necessary to keep the pressures constant with time.

Sometimes researchers use an informal terminology which refers to steady-state as 'equilibrium' while state of non-equilibrium is called 'instability.'

Since *isolated* thermodynamic systems do not interact with the environment, the processes inside the systems are controlled only by *natural* phenomena. Therefore, the systems can be used to study the role of such phenomena in real processes. For example, it is well known that every isolated system in a state of non-equilibrium will sooner or later reach a state of equilibrium. (The latter state is easily recognizable because there are no further physical and chemical *macroscopic* changes in the system once the equilibrium state has been reached.) Thus, one can say that **Nature itself always tries to bring thermodynamic systems to equilibrium ('Nature loves equilibrium').** Engineers are often able to successfully counteract this natural tendency by using various devices (engines, refrigerators, etc.) that are able to force some thermodynamic systems to move away from equilibrium by exchanging matter and/or energy between them and the environment, but such devices are not isolated systems.

1.2.2 Equilibrium

Consider a thermodynamic system shown in Fig. 1.1. The system is in a **complete macroscopic equilibrium** when: (1) the temperature of every cell of the system is the same and does not change with time, (2) the pressure of every cell of the system is the same and does not change with time, and (3) the system's chemical composition of all cells is the same and does not change with time. In other words, a system in complete macroscopic equilibrium is characterized by the following equalities: $T_A = T_B = T_C = \cdots = T = $ const (so there is no transport of heat inside the system), $P_A = P_B = P_C = \cdots = P = $ const (so there is no transport of matter inside the system), and the system has a fixed composition. If the matter of the system remains pure substance during the process of interest, then the state of complete macroscopic equilibrium is characterized only by requirements (1) and (2). The adjectives 'macroscopic' and 'complete' are usually omitted in thermodynamic discussions. Subsequently, the state of 'complete macroscopic equilibrium' is usually called 'complete equilibrium' or just 'equilibrium'.

1.2.3 Local Equilibrium

In some non-equilibrium systems one can identify some parts (groups of cells – see Fig. 1.1) where the spatial and temporal changes of thermodynamic properties are small enough to assume that they are constant within the parts. We then say that each such part is in its own state of the so-called **local equilibrium**. One often has such situations in open systems where the matter entering and leaving the system is in some state of local equilibrium. Then one can say that the flow of the matter between the system's inlet and its outlet is a thermodynamic process between two states of local equilibrium. Subsequently, one can study many aspects of such a flow using the formalism of Classical Thermodynamics since the formalism is quite successful in describing changes of state properties in processes between equilibrium (quasi-equilibrium, local-equilibrium) states – see below.

1.2.4 Phase Equilibrium

Another kind of 'equilibrium' can be seen in multi-phase systems where matter is in different (gaseous, liquid, or solid) phases, but the phases co-exist in a stable **phase equilibrium** where the temperatures and pressures of the phases remain equal and constant in time (but volumes of the phases can be changing). The additional requirement for existence of such phase equilibrium is that the phases' common pressure must be equal to the so-called **saturation pressure** of the substance at the phase equilibrium temperature (see Chapter 9).

1.3   Unit Systems

1.3.1   System International (SI)

The current scientific and technical literature is dominated by the International System of Units (SI). However, familiarity with other unit systems is strongly recommended since a vast volume of important work using units other than those of the SI system has already been published.

The SI is a modernized and extended version of the old metric systems MKS (with basic units meter (m), kilogram (kg), and second (s)) and MKSA (with basic units meter, kilogram, second, and ampere (A)).

The SI *basic units* of length, mass, time, absolute temperature, and electric current are meter (m), kilogram (kg), second (s), kelvin (K), and ampere (A), respectively. All the other ('coherent') units of the system are obtainable (without introducing additional numerical factors) from the basic units. The coherent units of physical properties can be derived from various relationships defining the properties. For example, the unit of speed ω can be found from the definition of speed (the derivative of the body displacement x with respect to time t),

$$\omega = \frac{dx}{dt}. \tag{1.1}$$

It follows that the SI unit of speed is m/s (the unit of displacement is meter while the unit of time is second). Other examples include units of acceleration a, force F, and energy E. The units can be derived from definitions of these entities (symbol M stands for mass);

$$a = \frac{d\omega}{dt} \rightarrow \left[\frac{\mathrm{m}}{\mathrm{s}^2}; \text{ the unit does not have any particular name} \right], \tag{1.2}$$

$$F = Ma \rightarrow \left[\mathrm{kg}\frac{\mathrm{m}}{\mathrm{s}^2}; \text{ the unit is called newton (N)} \right], \tag{1.3}$$

and

$$E = \frac{M\omega^2}{2} \rightarrow \left[\text{kg} \left(\frac{\text{m}}{\text{s}}\right)^2 ; \text{ the unit is called joule (J)} \right]. \qquad (1.4)$$

Similarly, the SI unit of pressure P can be derived from the definition

$$P = \frac{F}{S} \rightarrow \left[\frac{\text{kg}}{\text{s}^2 \cdot \text{m}} \text{ or } \frac{\text{newton}}{\text{m}^2}; \text{ the unit is called pascal (Pa)} \right], \qquad (1.5)$$

where F is the magnitude of force acting on the surface of area S.

A popular practical unit of pressure is

$$1\,\text{bar} = 10^5\,\text{Pa}. \qquad (1.6)$$

1.3.2 CGS Gaussian System

This unit system combines the electric units from the old 'electrostatic' CGS (centimeter, gram, second) system and the magnetic units from the old 'magnetostatic' CGS system. Combining these two systems resulted in the appearance of the speed of light in some relationships containing electric and magnetic entities.

In the CGS Gaussian system, the units of length, mass, time, and electric charge are centimeter (cm), gram (g), second (s), and statcoulomb ($\text{g}^{1/2}\text{cm}^{3/2}\text{s}^{-1}$, no particular abbreviation), respectively. Subsequently, force is measured in dynes (*dyne* $= \text{g·cm/s}^2 = 10^{-5}\,\text{N}$), and energy is measured in ergs (*erg* $= \text{g·cm}^2/\text{s}^2 = 10^{-7}\,\text{J}$).

In thermal sciences one usually includes kelvin (K) as the CGS Gaussian unit of absolute temperature.

The CGS Gaussian units are often used in atomic, molecular, and plasma physics literature.

1.3.3 Atomic Unit System

The basic units of the system are: e (electron charge) for electric charge, m_e (electron mass) for mass, and $\hbar$ for angular momentum ($\hbar = h/2\pi$, where h is Planck's constant). Subsequently, the other coherent units of the Atomic Unit System are:

the unit of length $(\hbar^2/m_e e^2) = 0.53 \times 10^{-10}\,\text{m}$ (the Bohr radius $a_\circ$),

the unit of speed $(e^2/\hbar) = 2.18 \times 10^6$ m/s,
the unit of energy $(e^2/a_\circ) = 4.36 \times 10^{-18}$ J,
the unit of time $(\hbar^3/m_e e^4) = 2.42 \times 10^{-17}$ s.

The Atomic Unit System is frequently used in atomic and molecular physics.

1.3.4   More Remarks about Units

Some relationships describing processes involving electric and magnetic quantities have different forms when written in different unit systems. One example of such a relationship is Coulomb's Law. According to the Law, the magnitude of the force F (in dynes) between two electric charges, q_1 and q_2, separated by a distance r is given in the CGS Gaussian units as

$$F = \frac{q_1 q_2}{r^2}, \tag{1.7}$$

where q_1 and q_2 are in statcoulombs ($g^{1/2}cm^{3/2}s^{-1}$) and r is in centimeters. Using the SI units the same Coulomb's Law gives the Coulomb force (in newtons) as

$$F = \frac{q_1 q_2}{4\pi\epsilon_\circ r^2}, \tag{1.8}$$

where q_1 and q_2 are in coulombs, r is in meters, and $\epsilon_\circ$ is a constant that is called the **permittivity of vacuum** and equal to 8.854×10^{-12} $C^2N^{-1}m^{-2}$ (C and N stand for coulomb and newton, respectively).

Another example is the definition of the Lorentz force (the force acting on a particle of electric charge q moving with velocity $\mathbf{v}$ across electric field $\mathbf{E}$ and magnetic field $\mathbf{H}$). In non-relativistic cases, the force is given as

$$\mathbf{F}_L = q\left(\mathbf{E} + \frac{1}{c}\mathbf{v} \times \mathbf{H}\right), \qquad \text{in CGS Gaussian units,} \qquad (1.9)$$

or

$$\mathbf{F}_L = q(\mathbf{E} + \mathbf{v} \times \mathbf{B}), \qquad \text{in SI units,} \qquad (1.10)$$

where c is the speed of electromagnetic waves in a vacuum, and $\mathbf{B}$ is the magnetic induction.

Yet other examples of relationships that have different mathematical forms in different unit systems are Maxwell's Equations which describe the dynamics of an electromagnetic field.

It is often easy to recognize what unit system is being used in formulas that include electric and magnetic fields because the permittivity ϵ_o and/or permeability μ_o appear explicitly in the formulas that use the SI system.

The unit systems discussed above also include some additional units such as the unit of angle radian (rad) and the unit of solid angle steradian (sr).

Several other non-coherent units are often used in science and engineering: unit of length called Angstrom (Å), unit of amount of matter called mol (mole), and unit of mass called the atomic mass unit (amu).

One Ångstrom is defined as

$$\text{Å} = 10^{-10}\,\text{m}. \tag{1.11}$$

One mol of a substance is defined as the number of grams equal to the atomic (molecular) mass number of the substance (see Section 2.1).

The atomic mass unit amu (also called a.m.u., not to be confused with the unit of mass (electron mass m_e) of the discussed-above Atomic Unit System) is equal to 1/12 of the mass m_C of a ^{12}C carbon atom (see Section (2.1)),

$$1\,\text{amu} = \frac{m_C}{12} = 1.6605 \times 10^{-27}\,\text{kg}. \tag{1.12}$$

Consequently, the mass m of any particle can be given as the product of the particle atomic (molecular) mass number A and one amu,

$$m = A \times \text{amu} \simeq A \times m_H \simeq A \times m_p, \tag{1.13}$$

because the mass of one amu is very close to the mass of a hydrogen atom (m_H) and to the mass of a proton (m_p).

Other non-coherent units of energy common in science and engineering are: eV (electronvolt), Ry (rydberg), cal (calorie), Cal (Calorie), and cm^{-1} (with no particular name).

One electronvolt is the energy gained by an electron traveling a distance of 1 cm in a constant electric field of 1 V/cm (volt per centimeter),

$$1\,\text{eV} = 1.602 \times 10^{-19}\,\text{J}. \tag{1.14}$$

The energy of one rydberg is 13.61 eV. Another unit of energy, common in atomic physics, is the hartree; 1 hartree = 2 rydbergs = 27.22 eV.

One calorie (cal, sometimes called the 'small calorie') is defined as the amount of the energy necessary to raise the temperature of 1 gram of water from 15 °C to 16 °C at pressure of 1 Atm (see below). The United States Bureau of Standards defined the calorie as equal to 4.184 joules. One Calorie (Cal, kcal, sometimes called the 'big calorie' or 'nutrition calorie') is equal to 1000 calories; the 'nutrition calories' are often listed on food products to indicate their dietary values.

The motion of a photon in vacuum can be characterized by the associated electromagnetic wave wavelength λ, or frequency v, because

$$\lambda v = c, \tag{1.15}$$

where $c = 3 \times 10^8$ m/s is the speed of electromagnetic waves in a vacuum. Subsequently, the energy of the photon is

$$E = hv = \frac{hc}{\lambda} = \frac{\text{const}}{\lambda}, \tag{1.16}$$

where $h = 6.62 \times 10^{-34}$ J·s is the Planck constant, and $1/\lambda$ (in m^{-1}) is called the wave number. Therefore, one can use the wave number as a measure of energy, as is often done in spectroscopy and the theory of radiation transfer. Using relationship (1.16) with $1/\lambda = 1$ m^{-1} one obtains the photon energy that corresponds to wavelength of 1 m,

$$E = hc(1 \text{ m}^{-1}) = 1.986 \times 10^{-25} \text{ J}. \tag{1.17}$$

A non-SI unit common in early literature is 1 kG (one-kilogram-force) defined as the force with which the earth's gravitational field attracts one kilogram of any substance at sea level. Since, according to the Newton Laws, the gravitational force acting on a mass M is equal to Mg (g is Earth's gravitational acceleration equal to 9.81 m/s^2), the force of magnitude of 1 kG is equal to 9.81 kg·m/s^2 = 9.81 N.

Two other non-coherent units of pressure are torr (1 mmHg) and atmosphere (Atm, atm). One torr is the pressure exerted at 0 °C by a column of mercury (Hg) one millimeter high when the gravitational acceleration is $g = 980.665$ cm/s^2. One atmosphere is equal to 760 torr, or

$$1 \text{ Atm} = 760 \text{ mmHg} = 1.013 \times 10^5 \text{ N/m}^2 = 1.013 \text{ bar}. \tag{1.18}$$

Note that the **pressure of one atmosphere is close to the pressure of one bar.**

Another popular unit of pressure is psi (pounds per square inch) defined as

$$1\,\text{psi} = 6.895\,\text{kPa} = 0.0689\,\text{bar}. \tag{1.19}$$

Kelvins (K) are used to measure the absolute temperatures of thermodynamic systems, while Celsius degrees (°C) and Fahrenheit degrees (°F) are more common in everyday life. If t_C and t_F are the temperatures of a system in Celsius degrees and in Fahrenheit degrees, respectively, then the system's absolute temperature T (in kelvins) is

$$T = t_C + 273.15 \tag{1.20}$$

and

$$T = \frac{5}{9}t_F + 255.38. \tag{1.21}$$

It follows from expression (1.20) that the *changes* of temperatures during a thermodynamic process are related as follows:

$$\Delta T = \Delta t_C, \tag{1.22}$$

where $\Delta T = T_2 - T_1$ and $\Delta t_C = (t_C)_2 - (t_C)_1$ are temperature *changes* in kelvins and Celsius degrees, respectively, and subscripts 1 and 2 denote the initial and final states of the process. This is why the values of some thermodynamic properties that are given per system's *temperature change* of one kelvin are equal to the values of these properties measured per temperature change of one degree Celsius; for example, the units of the specific heat capacities are $\text{J/kg·K} = \text{J/kg·(°C)}$.

1.4 Multi-Variable Functions

It is a common practice to use properties of differentials of multi-variable functions to study properties of several important thermodynamic functions. Therefore, we summarize in this section the mathematical properties of such differentials.

Consider an analytical function f of variables $x, y, z, \ldots$. The function's total elementary (differential) change df is a result of a (weighted) superposition of the differential changes of the function along the individual variables $x, y, z, \ldots$. This fact can be written in mathematical form as

$$df = g(x, y, z, ..)dx + r(x, y, z, ..)dy + t(x, y, z, ..)dz + \cdots, \tag{1.23}$$

where the coefficients $g(x, y, z, ..)$, $r(x, y, z, ..)$, $t(x, y, z, ..)$, $\ldots$ are some analytical functions ('statistical weights'). For example, it is well-known from Classical Mechanics that elementary work done by a force **F** along an elementary displacement $d\mathbf{s}$ in three-dimensional (x, y, z) space is

$$dL = \mathbf{F} \cdot d\mathbf{s} = F_x(x, y, z)dx + F_y(x, y, z)dy + F_z(x, y, z)dz, \tag{1.24}$$

where F_x, F_y and F_z are the x-, y-, and z-components, respectively, of the force vector **F**, and dx, dy and dz are the corresponding differential displacements.

The total differential (1.23) is called the **exact differential** when the partial derivatives of its coefficients have the following properties:

$$\frac{\partial g(x, y, z, ..)}{\partial y} = \frac{\partial r(x, y, z, ..)}{\partial x}, \tag{1.25}$$

$$\frac{\partial g(x, y, z, ..)}{\partial z} = \frac{\partial t(x, y, z, ..)}{\partial x}, \tag{1.26}$$

$$\frac{\partial r(x, y, z, ..)}{\partial z} = \frac{\partial t(x, y, z, ..)}{\partial y}, \tag{1.27}$$

$$\cdots\cdots\cdots\cdots\cdots\cdots\cdots\cdots\cdots\cdots\cdots\cdots, \tag{1.28}$$

or it is called the **inexact differential** (or **Pfaffian**) if one or more of equalities (1.25)–(1.28) are not true. The set of equalities (1.25)–(1.28) is often called *the cross derivative theorem*.

Multi-variable functions with *exact differentials* also have, in addition to the property shown in expressions (1.25)–(1.28), other useful properties. For example, the exact differentials of such functions can be written as

$$df = \frac{\partial f(x, y, z, ..)}{\partial x}dx + \frac{\partial f(x, y, z, ..)}{\partial y}dy + \frac{\partial f(x, y, z, ..)}{\partial z}dz + \ldots, \tag{1.29}$$

and, when there are only three variables, the following relationships between their differentials are true:

$$\left(\frac{\partial x}{\partial y}\right)_z \left(\frac{\partial y}{\partial x}\right)_z = 1, \tag{1.30}$$

$$\left(\frac{\partial x}{\partial y}\right)_z \left(\frac{\partial y}{\partial z}\right)_x \left(\frac{\partial z}{\partial x}\right)_y = -1, \tag{1.31}$$

$$\left(\frac{\partial f}{\partial x}\right)_z = \left(\frac{\partial f}{\partial y}\right)_z \left(\frac{\partial y}{\partial x}\right)_z, \tag{1.32}$$

$$\frac{\partial}{\partial y}\left[\left(\frac{\partial f}{\partial x}\right)_y\right]_x = \frac{\partial}{\partial x}\left[\left(\frac{\partial f}{\partial y}\right)_x\right]_y, \tag{1.33}$$

and

$$\frac{\partial^2 f}{\partial x \partial y} = \frac{\partial^2 f}{\partial y \partial x}, \tag{1.34}$$

where the subscripting indicates the variables which are held constant during the particular differentiating.

Both exact and inexact differentials play an important role in Classical Thermodynamics, but their properties are quite different. In order to show the difference, we study integral $\int df$ in two different situations (we limit the study to functions of two variables, x and y): (a) df_e is the exact differential of a function $f_e(x, y)$, and (b) df_i is the inexact differential of a function $f_i(x, y)$.

(a) df_e is the exact differential of function $f_e(x, y)$:

Consider a function $f_e(x, y) = x^3 y$ and its differential,

$$df_e = g_e(x, y)dx + r_e(x, y)dy, \tag{1.35}$$

where

$$g_e(x, y) = 3x^2 y \quad \text{and} \quad r_e(x, y) = x^3. \tag{1.36}$$

Since

$$\left[\frac{\partial g_e(x, y)}{\partial y}\right]_x = 3x^2 \quad \text{and} \quad \left[\frac{\partial r_e(x, y)}{\partial x}\right]_y = 3x^2, \tag{1.37}$$

the differential (1.35) is the exact differential of function $f_e(x, y)$ (see expression (1.25)).

Let us calculate the integral $\int df_e$ when variable x changes from $x_1 = 1$ to $x_2 = 2$ while variable y changes from $y_1 = 1$ to $y_2 = 2$. We choose two *different* paths for the function $f_e(x, y)$ between the states $x, y = 1, 1$ and $x, y = 2, 2$. The first path starts at $x = 1, y = 1$, goes to $x = 2, y = 1$, and ends at $x = 2, y = 2$.

The second path starts at $x = 1, y = 1$, goes to $x = 1, y = 2$, and ends at $x = 2, y = 2$.

In the first case $(x, y = 1, 1 \rightarrow 2, 1 \rightarrow 2, 2)$ one has

$$\int_{x_1,y_1}^{x_2,y_2} df_e = \left[\int_{1,1}^{2,1} g_e(x, y)dx + \int_{1,1}^{2,1} r_e(x, y)dy \right]$$
$$+ \left[\int_{2,1}^{2,2} g_e(x, y)dx + \int_{2,1}^{2,2} r_e(x, y)dy \right]$$
$$= \left[\int_{1}^{2} 3x^2 dx + 0 \right] + \left[0 + \int_{1}^{2} 8dy \right] = 7 + 8 = 15. \tag{1.38}$$

In the second case $(x, y = 1, 1 \rightarrow 1, 2 \rightarrow 2, 2)$ one obtains

$$\int_{x_1,y_1}^{x_2,y_2} df_e = \left[\int_{1,1}^{1,2} g_e(x, y)dx + \int_{1,1}^{1,2} r_e(x, y)dy \right]$$
$$+ \left[\int_{1,2}^{2,2} g_e(x, y)dx + \int_{1,2}^{2,2} r_e(x, y)dy \right]$$
$$= \left[0 + \int_{1}^{2} dy \right] + \left[\int_{1}^{2} 6x^2 dx + 0 \right] = 1 + 14 = 15. \tag{1.39}$$

The integral

$$\int_{x_1,y_1}^{x_2,y_2} df_e \tag{1.40}$$

can be calculated for any other path between (x_1, y_1) and (x_2, y_2) and the result will always be the same if the initial state (x_1, y_1) and the final state (x_2, y_2) of the function $f_e(x, y)$ are fixed. This is a very useful property of the *exact* differentials: **values of integrals (1.40) do not depend on the path of integration, they depend only on the values of the integration variables at the integration limits.** In other words,

$$\int_{x_1,y_1}^{x_2,y_2} df_e = f_e(x_2, y_2) - f_e(x_1, y_1). \tag{1.41}$$

Indeed, one can see that the use of function $f_e(x, y) = x^3 y$ in expression (1.41) with $x_1 = 1, y_1 = 1$ and $x_2 = 2, y_2 = 2$ gives

$$\int_{x_1,y_1}^{x_2,y_2} df_e = f_e(x_2, y_2) - f_e(x_1, y_1)$$
$$= \left[x^3 y \right]_{x=2,y=2} - \left[x^3 y \right]_{x=1,y=1} = 16 - 1 = 15, \tag{1.42}$$

the same result as that obtainable from integrals (1.38) and (1.39).

Another useful property of functions having exact differentials is that the integrals of such functions along any closed path ('thermodynamic cycle') vanish. This can be seen in relationship (1.41),

$$\oint df_e \equiv \int_{x_1,y_1}^{x_2=x_1,y_2=y_1} df_e = f_e(x_1, y_1) - f_e(x_1, y_1) = 0. \qquad (1.43)$$

Relationships (1.41) and (1.43) can be generalized on functions of more than two variables.

Functions that have exact differentials are called the functions of state because the value of any integral of type (1.40) depends only on the characteristics of the initial and final values ('states') of the functions regardless of the character of the function being integrated. Examples of important state properties are the internal energy, enthalpy, entropy, the Helmholtz free energy, and the Gibbs free energy of physical systems.

(b) df_i is the inexact differential of function $f_i(x, y)$:

Consider a function $f_i(x, y)$ having the following differential:

$$df_i = g_i(x, y)dx + r_i(x, y)dy, \qquad (1.44)$$

where

$$g_i(x, y) = xy \quad \text{and} \quad r_i(x, y) = x^2. \qquad (1.45)$$

Since

$$\left[\frac{\partial g_i(x, y)}{\partial y}\right]_x = x \quad \text{while} \quad \left[\frac{\partial r_i(x, y)}{\partial x}\right]_y = 2x, \qquad (1.46)$$

the differential (1.44) is an *inexact* differential.

Let us calculate the integral $\int df_i$ in the interval between $x_1 = 1, y_1 = 1$ and $x_2 = 2, y_2 = 2$ for two *different* paths of integration. The first path starts at $x = 1, y = 1$, goes to $x = 2, y = 1$, and ends at $x = 2, y = 2$. The second path starts at $x = 1, y = 1$, goes to $x = 1, y = 2$, and ends at $x = 2, y = 2$.

In the first case $(x, y = 1, 1 \rightarrow 2, 1 \rightarrow 2, 2)$ one has

$$
\int_{x_1,y_1}^{x_2,y_2} df_i = \left[\int_{1,1}^{2,1} g_i(x, y)dx + \int_{1,1}^{2,1} r_i(x, y)dy \right]
$$
$$
+ \left[\int_{2,1}^{2,2} g_i(x, y)dx + \int_{2,1}^{2,2} r_i(x, y)dy \right]
$$
$$
= \left[\int_{1}^{2} xdx + 0 \right] + \left[0 + \int_{1}^{2} 4dy \right] = 1.5 + 4 = 5.5. \tag{1.47}
$$

In the second case $(x, y = 1, 1 \rightarrow 1, 2 \rightarrow 2, 2)$ one has

$$
\int_{x_1,y_1}^{x_2,y_2} df_i = \left[\int_{1,1}^{1,2} g_i(x, y)dx + \int_{1,1}^{1,2} r_i(x, y)dy \right]
$$
$$
+ \left[\int_{1,2}^{2,2} g_i(x, y)dx + \int_{1,2}^{2,2} r_i(x, y)dy \right]
$$
$$
= \left[0 + \int_{1}^{2} dy \right] + \left[\int_{1}^{2} 2xdx + 0 \right] = 1 + 3 = 4. \tag{1.48}
$$

The (different) results (1.47) and (1.48) allow one to make the following statement: **an integral**

$$
\int_{x_1,y_1}^{x_2,y_2} df_i \tag{1.49}
$$

can be calculated only if the dependence of the function f_i **on its variables** x **and** y **between the initial and final states of the integral is known.** In other words,

$$
\int_{x_1,y_1}^{x_2,y_2} df_i \neq f_i(x_2, y_2) - f_i(x_1, y_1), \tag{1.50}
$$

and

$$
\oint df_i \neq 0. \tag{1.51}
$$

These statements can be generalized on functions of more than two variables.

Functions that have inexact differentials are called the functions of path because the value of any integral of type (1.49) depends not only on the characteristics of the initial and final values ('states') of the functions but also on the character of the function being integrated. Examples of thermodynamic functions of path are heat and work.

As can be seen in the above discussion, thermodynamic functions that have exact differentials can be handled mathematically with a significant easiness. Therefore, one would like to have a reliable method able to transform the inexact differentials of the functions into exact differentials. Such a method has been developed in the theory of differential equations, and its central feature is the so-called *Caratheodory theorem* that allows one to find the so-called *integrating factor* able to make the transformation. The product of the factor and the inexact differential of a function of path gives the exact differential of the function. For example, consider the following differential of function $f(x, y) = x^2 y^3$:

$$df = g(x, y)dx + r(x, y)dy, \tag{1.52}$$

where

$$g(x, y) = 2y \quad \text{and} \quad r(x, y) = 3x, \tag{1.53}$$

and

$$\left[\frac{\partial g(x, y)}{\partial y}\right]_x = 2 \quad \text{and} \quad \left[\frac{\partial r(x, y)}{\partial x}\right]_y = 3. \tag{1.54}$$

Thus, the differential (1.52) is an inexact differential. However, if it is multiplied by the integrating factor xy^2 (found from some other considerations), one obtains:

$$df' \equiv xy^2 df = g'(x, y)dx + r'(x, y)dy, \tag{1.55}$$

where

$$g'(x, y) = 2xy^3 \quad \text{and} \quad r'(x, y) = 3x^2 y^2. \tag{1.56}$$

Since

$$\left[\frac{\partial g'(x, y)}{\partial y}\right]_x = \left[\frac{\partial r'(x, y)}{\partial x}\right]_y = 6xy^2, \tag{1.57}$$

the differential df' is an exact differential.

Integrating factors able to transform inexact differentials into exact differentials are very useful and powerful tools in theoretical thermodynamic studies. However, in many practical cases, finding the factors can be difficult, and often impossible. It was shown by Constantin Caratheodory (1873–1950) that inexact differentials of heat and boundary work (both thermodynamic functions of path) of reversible (quasi-static) thermodynamic transitions have simple integrating factors (see Section 2.8.2).

The attractive features of exact differentials of multi-variable functions have no practical values in studies of thermodynamic transitions between non-equilibrium states. This is because Classical Thermodynamics is unable to accurately describe thermodynamic properties of such states. In other words, Classical Thermodynamics can fully exploit the attractive features of functions of state only when the initial and final states of the studied processes are equilibrium, quasi-equilibrium, or local-equilibrium states.

Chapter 2

Properties of Thermodynamic Systems

2.1 The Amount of Matter

A **pure substance** is a substance that does not change its chemical composition during a thermodynamic process. Pure substances are often called **components**. Components are either **elements** or **compounds**. Elements are pure substances made of atoms (e.g., gold Au, helium He, etc.), while compounds are pure substances made of molecules (e.g., water H_2O, methane CH_4, etc.). All elements are listed in the *Periodic Chart of Elements*. Each atom (element) in the chart is associated with two symbols: Z (the **atomic number**) and A (the **atomic mass number**). The atomic number is just the number of the electrons (equal to the number of protons) inside an atom. The atomic mass number is the number telling how many times the mass of the atom is greater than the mass of 1/12 of one atom of carbon isotope ^{12}C. The latter mass is equal to the so-called **amu (atomic mass unit)**, amu $= 1.6605 \times 10^{-27}$ kg. (This amu should not be confused with the unit of mass used in the Atomic Unit System where mass of an electron is taken as the basic unit of mass – see Section 1.3.3). The atomic mass numbers of atoms in a particular molecule are added to compute the molecule's **molecular mass number**.

Mass of a substance (in gaseous, liquid, or solid state) is often thought of as a measure of the amount of matter of the substance, and this amount is usually given in kilograms, grams, pounds, etc. Classical Thermodynamics and

Thermochemistry often use another measure of amount of substance. A popular unit of the measure is the so-called **mol** (or **mole**) (the terms 'gram-atom' and 'gram-molecule' were used in earlier literature of the subject for one mol of an element and one mol of a compound, respectively). **One mol of pure substance (element, compound), in gaseous, liquid or solid phase, is the amount of the substance whose mass in grams is equal to the substance atomic mass number (in the case of the element) or to the substance molecular mass number (in the case of the compound).** For example, according to the *Periodic Chart of Elements*, the atomic mass number of the element gold (chemical symbol Au) is $A = 196.966$ (the atomic number of gold is $Z = 79$), while the molecular mass number of the compound with chemical symbol H_2O is $2 \times 1.008 + 15.999 = 18.015$ (the atomic mass number of hydrogen is 1.008, and the atomic mass number of oxygen is 15.999). Thus, 196.966 grams of the element gold makes one mol of gold, and 18.015 grams of the compound H_2O makes one mol of the compound, irrespective of whether the substance is in gaseous ('vapor'), liquid ('water'), or solid ('ice') state.

A larger-than-mol unit of the amount of a substance is a **kilomol**,

$$1\,\text{kmol} = 1000\,\text{mols}. \tag{2.1}$$

A useful consequence of the concept of mol is the so-called **Avogadro's Law: one mol of any substance (in gaseous, liquid, or solid phase) contains the same number of particles.** This number is called **Avogadro's number**,

$$N_A = 6.022 \times 10^{23}\ \text{particles}. \tag{2.2}$$

Also see Problem 5.2 and Section 5.3.

Avogadro's number results from the fact that one mol of atomic mass units ($\text{amu} = 1.6605 \times 10^{-24}$ g) has the mass of one gram. Thus, denoting the number of particles in one mol of a substance by $\bar{N}$ (overbars are used throughout this book to indicate *molar* properties, that is, the values of the properties per *one* mol of the substance), one has

$$\bar{N} \times \text{amu} = 1\,\text{g} \quad \rightarrow \quad \bar{N} \equiv N_A = \frac{1\,\text{g}}{1\,\text{amu}} = 6.022 \times 10^{23}\ \text{particles}. \tag{2.3}$$

2.2 The Internal Energy

Theoretical and experimental studies of physical systems have shown that every system has an important property called the system **internal energy**. The internal energy, $\mathcal{E}$, is the sum of all energies of the matter present *inside* the system (or, more precisely, inside the control surface defining the system). Typically, the photons present in thermodynamic systems play an insignificant role in most applications and, therefore, are often neglected in thermodynamic studies. The possible role of the radiative energy in thermodynamic systems can be studied by methods of the Theory of Radiation Transfer.

The internal energy of a thermodynamic system can be written as a sum of the **kinetic energy** E_{kin}, **potential (gravitational) energy** E_{pot}, and **thermal energy** E_{th} of the matter inside the system,

$$\mathcal{E} = E_{\text{kin}} + E_{\text{pot}} + E_{\text{th}}. \tag{2.4}$$

A closed system where the kinetic and potential energies of the system's matter can be ignored (a situation common in many applications) is called the closed **nonflow** system. The internal energy of such a system is just its thermal energy E_{th}. It is a common practice in thermodynamic literature (this practice is also followed in this book) to denote the internal energy of closed nonflow systems by the letter U. Therefore, in such systems one has

$$U \equiv E_{\text{th}} \equiv \mathcal{E}. \tag{2.5}$$

As discussed below, kinetic energy E_{kin} and potential energy E_{pot} are always properties of state, but thermal energy E_{th} is a function of state only in processes between equilibrium (quasi-equilibrium, local-equilibrium) states. Therefore, the internal energy, $\mathcal{E} = E_{\text{kin}} + E_{\text{pot}} + E_{\text{th}}$, is a function of state only in the processes between equilibrium (quasi-equilibrium, local-equilibrium) states.

2.2.1 Kinetic and Potential Energies

The kinetic energy E_{kin} and potential (gravitational) energy E_{pot} of the matter inside a physical system are often called the matter's **motion properties**. The kinetic energy is a function of the speed ω of the matter's center-of-mass, and the potential energy depends on the center's elevation z. The **kinetic energy** of a mass M when its center of mass is moving with speed ω is

$$E_{\text{kin}} = M\omega^2/2, \tag{2.6}$$

and the **potential energy** of the mass with its center-of-mass at elevation z is

$$E_{\text{pot}} = Mgz, \tag{2.7}$$

where $g = 9.81\,\mathrm{m/s^2}$ is the gravitational acceleration.

The energy of rotational motion of the matter around its center of mass is usually negligible in most of thermodynamic devices.

The motion properties (E_{kin} and E_{pot}) of the matter moving as a whole inside a studied system are always properties of state regardless of the matter's *thermodynamic state*. This is because the values of the motion properties depend only on the actual speed ω, or elevation z, of the center of mass of the moving matter. Therefore, calculating the motion properties does not require knowledge of the character of the motion *before* reaching the speed ω and elevation z.

The change of the potential (gravitational) energy of matter in most thermodynamic applications is negligible since it is usually much smaller than the changes of the matter's thermal and kinetic energies – see Problem 3.1. The potential (electro-magnetic) energy of electrically-neutral matter placed in an electric field is usually negligible.

2.2.2   Thermal Energy

The system **thermal energy** E_{th} (usually denoted U in thermodynamic literature) is the energy of the system's matter measured in a coordinate frame moving with the matter's center of mass. In other words, the thermal energy is the sum of the translational energies of all free particles (atoms and molecules) present in the matter plus the sum of all energies 'stored' inside the particles (the energies are called the **intramolecular energies**) plus the sum of the energies of all interactions between the free particles (the energies are called the **intermolecular energies**).

The intramolecular energy stored inside each particle is the energy stored inside its nuclei plus the energies stored in the particle's internal motions (vibrational, rotational, and electronic). A very large fraction (more than 99 percent) of the intramolecular energy of every particle is stored in the particle's nucleus (nuclei), but it is practically impossible to release the nuclear energy in common thermodynamic applications. However, science and technology have developed many ways to access and manipulate the other intramolecular energies. Common examples of outcomes of such manipulations are combustion processes.

It was postulated by Rudolf Clausius (1822–1888) that the **thermal energy E_{th} of some amount of matter takes on a unique value for each state of thermodynamic equilibrium (quasi-equilibrium, local equilibrium) of the**

matter. This statement, originally called the *Fundamental Hypothesis*, is one of the central concepts of Classical Thermodynamics. It allows one to identify thermal energy of every system in such a state as a **property of state** (**function of state**). After the Clausius postulate was proposed, it was shown that the Fundamental Hypothesis also applies to other thermal properties such as enthalpy H, entropy S, the Helmholtz free energy A, and the Gibbs free energy G (see below).

We will see in Section 9.1 that only two *thermodynamic variables* (P and T, P and V, or V and T) and some *material parameters* (mass, heat capacities) are needed to describe the thermal properties of any equilibrium (quasi-equilibrium, local-equilibrium) state of a single-component and single-phase substance. A property of a multi-component, multi-phase substance in equilibrium can be calculated by superposing the properties of all the components and phases. Therefore, a large part of this book is devoted to studying the properties of single-phase, single-component systems.

2.3 Entropy

Detailed studies of entropic properties of thermodynamic processes require use of the formalism of Quantum Statistical Mechanics. However, the formalism of Classical Thermodynamics is often quite successful in describing the entropy changes in a broad variety of thermodynamic processes.

2.3.1 Molecular Interpretation of Entropy

In Classical Thermodynamics, a system's entropy S is a *macroscopic* (phenomenological) property characterized by the features that are discussed below. In Molecular Thermodynamics (a part of the Quantum Statistical Mechanics), the entropy represents the way in which the *microscopic* (quantum-mechanical) properties of the system's particles and photons are distributed. The connection between the macroscopic and microscopic representations of entropy can be made by introducing the concepts of macrostate and microstate.

A **macrostate** of a physical system at a given instant is its thermodynamic state characterized by values of a small number of the system's macroscopic properties (pressure, temperature, etc.) at this instant.

A **microstate** of a physical system at a given instant is its quantum state characterized by values of a huge number of the system's microscopic properties (energies, linear momenta, spins, etc.) of *all* particles and photons present in the system (photon-related effects are usually neglected in most thermodynamic studies). Statistical averaging of these quantum-mechanical properties gives values of a small number of the system's macroscopic thermodynamic properties such as pressure, temperature, etc. Since the system particles and photons interact in very many different ways and with high frequency, they can redistribute their positions, velocities, and other quantum-mechanical properties in a huge number of ways. Therefore, there is always a very large number of different microstates that can produce any particular macrostate of the system. This very large number is called the **thermodynamic probability** of the given macrostate, even though it is not 'probability' in a mathematical sense (the mathematical definition of probability allows only values between zero and one).

Thermodynamic probability is a very useful number. For example, Statistical Mechanics has derived the following simple expression for the total entropy of a macrostate of *ideal gas* (see Section 5.1) in a state of equilibrium:

$$S = k \ln \Omega, \tag{2.8}$$

where k is the Boltzmann constant, and Ω is the thermodynamic probability of the macrostate. Relationship (2.8) is the most famous relationship of Statistical Mechanics. It is engraved on Boltzmann's tombstone in a Vienna cemetery. If a particular macrostate could be produced by a single microstate ($\Omega = 1$), the system's entropy would have its lowest possible value, $S = 0$. On the other hand, when the number of microstates able to produce a particular macrostate is the largest possible for this macrostate, then the system's entropy S also has the largest value that is possible for the macrostate.

A system's thermodynamic path towards or away from a state of equilibrium follows the system's total entropy S: the entropy is increasing when the system is approaching a state of macroscopic equilibrium, but it is decreasing when the system is moving away from equilibrium. In other words, the number of microstates (the thermodynamic probability Ω) available to the system increases when the system is moving towards equilibrium, but the number decreases when the system is moving away from equilibrium.

2.3.2 Properties of Entropy

Property 1: Particles 'carry' entropy. Transfer of particles (matter) between a thermodynamic system and the environment changes the system's entropy because it changes the number of the particle-particle and particle-photon interactions and, as a result, the number of the microsates that can produce the macrostate of the system. The (partial) entropy 'carried' by the matter is called the **entropy of matter** S_m. The change of this entropy during a thermodynamic process between states 1 and 2 is

$$\Delta S_m = v_2 \bar{S}_{m,2} - v_1 \bar{S}_{m,1}, \tag{2.9}$$

where v_1 and v_2 are numbers of mols of the matter entering and leaving the system, respectively, and $\bar{S}_{m,1}$ and $\bar{S}_{m,2}$ are the molar entropies of the matter that enters and leaves the system, respectively.

Property 2: Photons 'carry' entropy. Transfer of photons (radiation) between a system and the environment changes the system's entropy because it changes the number of photon-particle and photon-photon interacions and, as a result, the number of the microsates that can produce the macrostate of the system. The (partial) entropy 'carried' by photons is called the **entropy of radiation** S_{ph}. The change of the radiation entropy of a system during a process between states 1 and 2 is

$$\Delta S_{ph} = N_2 \langle S \rangle_{ph,2} - N_1 \langle S \rangle_{ph,1}, \tag{2.10}$$

where N_1 and N_2 are numbers of photons entering and leaving the system, respectively, and $\langle S \rangle_{ph,1}$ and $\langle S \rangle_{ph,2}$ are the mean (per photon) radiation entropies entering and leaving the system, respectively.

Property 3: Heat 'carries' entropy. Transfer of heat between a thermodynamic system and the environment changes the system's entropy because it changes the energies of the system's particles and photons. This changes the intensity of their participation in creating the number of the microstates able to produce the system's macrostate. Rudolf Clausius (1822–1888) had shown that the amount of the **heat entropy** 'carried' by heat δQ exchanged between a system and its environment during an elementary (differential) transition at temperature T is (also see Section 3.7.1)

$$\delta S_h = \frac{\delta Q}{T}, \tag{2.11}$$

where we used (as in the rest of the book) symbol δY to denote the *inexact* differential and symbol dY to denote the *exact* differential of property Y; both δY and dY refer to the same mathematical concept of the calculus – see Section (2.8).

The change of heat entropy during a finite process between states 1 and 2 is

$$\Delta S_h = \int_1^2 \frac{\delta Q}{T}, \tag{2.12}$$

and, when the heat is exchanged with its environment in an isothermal manner (T = const), the change of the system heat entropy is

$$\Delta S_h = \frac{Q}{T}, \tag{2.13}$$

where Q is the heat exchanged between the system and its environment during the process.

Property 4: All real (that is, realizable in practice) processes always generate an additional (and very special) partial entropy called the natural entropy S_{nat}. This fact is emphasized in discussions of real processes by saying that 'all real processes always show some irreversibility effects', or 'real processes always have some irreversibilities', or **'real processes are always irreversible'** – see below for a description of the 'reversible' process. In other words, the **natural entropy is always added to (never removed from!) every system during every real process, regardless of whether the other partial entropies (S_m, S_{ph}, S_h) are being added to or removed from the system during the process.** Since particle-particle, photon-particle, and photon-photon interactions are present in all systems, the natural entropy is produced even in real processes in isolated systems which, by definition, do not exchange matter, heat, and radiation with the environment – see Property 8. This fascinating enigma (called the Second Law of Thermodynamics) demands that every process in the Universe must 'pay' some fee (in form of the natural entropy) for the privilege of being realizable in practice (!). In other words, Nature always and spontaneously produces some microstates in every real process regardless of the process character. The presence of these 'extra' microstates always 'pushes' the system undergoing a real process closer to equilibrium. The increase of system's total entropy by the always-present production of natural entropy can be counteracted by removing some matter, heat, and/or radiation from the system because each of these entities 'carries' its own (partial) entropy. Therefore, the *total* entropies of real systems other than the isolated systems can decrease during some thermodynamic processes.

An important result of the always-present production of natural entropy in *all* real processes is the so-called 'dissipation' of energy. The dissipation always moves some amount of the system's energy into its heat-designated part, regardless of the skills and intentions of the system designer. As a result, for example, the amount of work produced in every irreversible process is always smaller than the amount of the work expected by the process's designer who ignores irreversibility effects. It should be emphasized that the dissipation does not destroy any energy of the system, it only transforms some amount of it into heat. More discussion on dissipation of energy in irreversible processes is in Chapter 13.

In most applications, the main contributions to systems' entropies come from particle-particle interactions. Therefore, the contributions of photon-particle and photon-photon interactions to the entropies are usually ignored in thermodynamic studies.

Thermodynamic transitions where natural entropy is not produced are called **reversible processes**. These are *theoretical* models of real processes with insignificant irreversibility effects. The models are quite useful for thermodynamic studies since they significantly simplify calculations of properties of path such as work and heat transfers.

Any thermodynamic transition that resembles a sequence of different quasi-equilibrium states can be treated as a reversible process. Very slow (quasi-static) transitions are good examples of reversible processes because they are sequences of quasi-equilibrium states.

There are three kinds of reversible transitions: the externally reversible, internally reversible, and totally reversible transitions. A thermodynamic process is called *externally reversible* when no irreversibilities are present outside the system during the process, and it is called *internally reversible* when no irreversibilities are present inside the system. A process is called *totally reversible* when no irreversibilities are present inside and outside the system. It is a common practice in thermodynamic literature to use the term 'reversible process' without specifying whether the transition is externally, internally, or totally reversible. Since most thermodynamic studies focus on the internal properties of systems, the term 'reversible transition' usually means an internally reversible process.

The amount of natural entropy produced during a thermodynamic process is proportional to the first- and second-order gradients of temperature and pressure during the process. Therefore, it is often justified to treat the processes where the gradients are significant as irreversible transitions, and the processes where the gradients are small as reversible transitions. As mentioned earlier, the assumption that an actual thermodynamic process can be treated as a reversible process allows one to study its heat and work transfers in an easy way – see the discussions associated with relationships (2.71)–(2.72).

Property 5: The system entropy of any non-isolated thermodynamic system can increase or decrease: it increases when the system approaches a state of equilibrium, but it decreases when the system is moving away from such a state.

The change of entropy of a physical system during an elementary (differential) transition involving matter, radiation, and heat transfers between the system and the environment can be given as

$$\delta S = \delta S_m + \delta S_{ph} + \delta S_h + \delta S_{\text{nat}}, \tag{2.14}$$

where the entropy of matter δS_m, entropy of radiation δS_{ph}, and entropy of heat δS_h are added to or removed from the system with the matter, radiation, and heat, respectively, and δS_{nat} is the natural entropy produced in the system during the process. As before, we are using symbol δY to denote *inexact* differentials (symbol dY is used in the book to denote the *exact* differentials of property Y); both δY and dY refer to the same mathematical concept of the calculus – see Section (2.8).

In closed systems with no radiation transfer, relationship (2.14) becomes

$$\delta S = \delta S_h + \delta S_{\text{nat}}. \tag{2.15}$$

The change of entropy of a system during its transition from a state 1 to a state 2 can be given as

$$\Delta S = \int_1^2 \left(\frac{dS}{dt}\right) dt, \tag{2.16}$$

but use of the relationship is very limited since the dependencies of systems' entropies on time are usually difficult to establish. Fortunately, Clausius' studies of the so-called *Fundamental Hypothesis* and *Clausius Inequality* have shown that the *total* entropy of every system during any (reversible or not) process between equilibrium (quasi-equilibrium, local-equilibrium) states behaves like a function of state. Therefore, the changes of the system entropy during all processes between two such states are identical, but the contributions of some partial entropies to the changes can depend on the degrees of irreversibility of the processes. This can be seen while considering the change of the *total* entropy of a closed system undergoing a thermodynamic process between two states of equilibrium. The change is (see expression (2.15))

$$\Delta S = \Delta S_h = \int_1^2 \frac{\delta Q}{T} \qquad \text{reversible process,} \qquad (2.17)$$

and

$$\Delta S = \Delta S_h' + \Delta S_{nat} = \int_1^2 \frac{\delta Q'}{T} + \Delta S_{nat} \qquad \text{irreversible process,} \qquad (2.18)$$

where the amounts of the heat that the system allows to be taken in or released during the two processes are different. This demonstrates the effect of the energy 'dissipation' and the response of the system to the fact that its *total* entropy is a function of state regardless of the degree of irreversibility.

Property 6: The entropy of every system remaining in a state of equilibrium (quasi-equilibrium, local equilibrium) does not change. Thus, the following is true in all systems being in such states:

$$S = \text{const} \quad \rightarrow \quad \Delta S = 0. \qquad (2.19)$$

On the microscopic level, the *distributions* of the particles' quantum-mechanical properties stop changing when the system reaches a state of equilibrium. As a result, the number of microstates (the system's thermo-dynamic probability Ω) that can produce the system's actual macrostate stops changing too. This keeps the system's entropy constant as long as the sys-tem remains in the state of equilibrium. (The equilibrium-state distributions of quantum-mechanical properties of the particles are called the *Boltzmann distributions*.)

Property 7: The total entropy of every system in a state of equilibrium (quasi-equilibrium, local-equilibrium) depends only on the thermodynamic characteristics of the state; the entropy has no 'memory' of the way in which such a state has been created. This allows one to treat the system entropy as a function of state (thermodynamic potential) when the system moves between equilibrium (quasi-equilibrium, local-equilibrium) states.

In general, the partial entropies S_m, S_{ph}, and S_h, can be functions of path, that is, their changes can depend on the ways in which the entities carrying the entropies are transferred between the system and the environment. However, the **total entropy of the system always behaves like a property of state if the initial and final states of the entire process under consideration are equilibrium (quasi-equilibrium, local-equilibrium) states**. Thus, the inexact differential δS in relationship (2.14) can be replaced by exact differential dS when the studied process's initial and final states are states of equilibrium, quasi-equilibrium, or local equilibrium.

Property 8: Nature loves equilibrium. As mentioned earlier, the total entropy of every *isolated* system continuously and spontaneously increases during every real process until the system reaches some state of equilibrium. This entropy is just the system's natural entropy S_{nat} because there are no transfers of matter, heat, and radiation in isolated systems, The change of the system entropy during the process is

$$\Delta S = \Delta S_{nat} > 0 \qquad \text{real processes in isolated systems.} \qquad (2.20)$$

On the microscopic level, the natural entropy is generated by particle-particle, particle-photon and photon-photon interactions which continuously try to redistribute the quantum-mechanical properties of the particles in a way that brings the system closer to macroscopic equilibrium.

Relationship (2.20) is one of the statements of the **Second Law of Thermodynamics**. The Law was stated this way by Max Planck (1858–1947): 'If a system evolves in thermal and mechanical isolation from the ambient environment, then its entropy increases'. Originally, the Second Law was formulated in somewhat different form by Rudolph Clausius (1822–1888) and William Thompson (Lord Kelvin) (1824–1907).

Since real processes in *isolated* systems are not influenced by any processes other than the natural (spontaneous) phenomena, studying the properties of such systems allows one to learn about some fundamental laws of Nature.

Property 9: Work transfers change energies of thermodynamic systems but do not 'carry' entropy. Therefore, the term 'entropy of work' does not occur in this book.

Property 10: Entropy is an additive property: the entropy of a system consisting of several physical parts is equal to sum of these parts entropies.

2.4 Other Thermal Properties

In addition to thermal energy E_{th} and entropy S, other thermal properties important in thermodynamic studies are: enthalpy H, the Helmholtz free energy A, and the Gibbs free energy G. In a single-component, single-phase system in equilibrium, the properties are functions of only two **thermodynamic variables** such as the system pressure P and temperature T, or volume V and T, or P and V (see Section 9.1). It was shown, after Clausius' *Fundamental Hypothesis* was postulated, that it applies not only to a system's thermal energy E_{th} but also to enthalpy H, and entropy S. These thermal properties for many single-phase pure substances (components) in equilibrium have already been tabularized, usually as functions of P and T, in various thermodynamic and thermochemical databases (see Chapter 8). Therefore, a large portion of this book is devoted to studying single-component, single-phase systems. The properties for a multi-component, multi-phase substance in equilibrium can be calculated by superposition of the corresponding properties of all the phases and components (see Chapter 9).

2.5 Work, Heat, and Radiation

Thermodynamic systems usually exchange some energy with the environment. The energy transfer can be in the form of work, heat, and/or radiation. For example, the energy of gasoline being burned in a car engine moves from the engine to the rest of the world (the environment) in the following three ways: as the work that moves the car, as the heat that makes the engine hot, and as the radiation (in the invisible range of its frequency spectrum) that leaves the engine. (In Classical Thermodynamics, radiation energy is often neglected or considered a part of heat.)

The three ways of energy exchange differ significantly on both macroscopic and microscopic levels. On the macroscopic level, work is associated with a displacement of some physical quantity upon action by some forces. For example, an amount of matter can be displaced by an action of a mechanical force, while an electric force can be used to displace an electric charge.

The concept of heat is not as comprehensive as the concept of work. However, we know from experience that energy can be moved from one place to another without visible displacements of parts of the system under consideration. Indeed, no part of a working light bulb moves, yet some energy in the form of heat is leaving the bulb, and the amount of this energy is large enough to burn our skin when the bulb is touched.

Work and heat are two different *forms* of the same phenomenon – energy in motion. Therefore, units of work and heat are the same as units of energy. Work and heat are not 'stored' within the thermodynamic systems and the environment, they are just two different ways in which energy can move in space. The 'thing' that is 'stored' in every physical system (and in the environment) is its **thermal energy** E_{th} (often denoted U in thermodynamic literature). Usually it is easy to be aware of the presence of work and heat transfers but it is more difficult to realize that every amount of matter has some thermal energy.

As shown in Problem 2.1, the thermal energies stored in physical systems are huge. But it does not matter in applications how much thermal energy is stored in a particular system. What drives the behavior of every system undergoing a thermodynamic process is not the amount of the system's energy, but rather the *magnitude of change* of the energy during the process. A decrease of the system energy produces work, heat, and/or radiation, while an increase of the energy is caused by work done on the system and/or by heat and radiation absorbed by the system.

2.6 Mass-Energy

The total energy of a material object moving in space free of fields is the sum of its rest energy E_o and its kinetic energy E_k,

$$E_{tot} = E_o + E_k. \qquad (2.21)$$

(In thermodynamic terminology, $E_o \equiv E_{th}$, and $E_k \equiv E_{kin}$.) The Theory of Special Relativity has shown that

$$E_o = Mc^2, \qquad (2.22)$$

where M is the object's rest mass, and $c = 3 \times 10^8$ m/s is the speed of light in field-free space.

The kinetic energy of a material object having mass M and moving with speed ω in field-free space is

$$E_k = Mc^2 \left[\frac{1}{\sqrt{1 - (\omega/c)^2}} - 1 \right], \qquad (2.23)$$

where the object can have any speed slower than the speed of light.

The *total* energy of the object moving in field-free space can be written as

$$E_{\text{tot}} = \frac{Mc^2}{\sqrt{1-(\omega/c)^2}}.$$ (2.24)

When the speed of the moving object is significantly slower than the speed of light (then $(\omega/c)^2 \ll 1$), one can use in relationship (2.23) the following power series expansion ($a = (\omega/c)^2$):

$$\frac{1}{\sqrt{1-a}} = 1 + \frac{1}{2}a + \frac{3}{8}a^2 + \frac{5}{16}a^3 + \ldots,$$ (2.25)

so that

$$E_k = Mc^2 \left[\frac{1}{2}\frac{\omega^2}{c^2} + \frac{3}{8}\frac{\omega^4}{c^4} + \frac{5}{16}\frac{\omega^6}{c^6} + \ldots \right].$$ (2.26)

Thus, if the object speed is significantly slower than the speed of light (a non-relativistic case), relationship (2.26) can be written as

$$E_k \simeq \frac{M\omega^2}{2},$$ (2.27)

a result well known from Classical Mechanics.

Since meaningful relativistic effects are rare in thermodynamic applications, Classical Thermodynamics treats energy and mass in the same way as Classical Mechanics, that is, as two independent entities.

Problem 2.1: Calculate the total energy of a one-cent coin of rest mass $M_{\text{penny}} = 2.5$ grams.

Solution: Using expression (2.22) one has

$$(E_\circ)_{\text{penny}} = M_{\text{penny}}c^2 = (2.5 \times 10^{-3}\,\text{kg})(3 \times 10^8\,\text{m/s})^2 = 2.25 \times 10^{14}\,\text{J}.$$ (2.28)

The energy E_b released during the explosion of the atomic bomb dropped on Hiroshima in 1945 was about 0.5×10^{14} J (equivalent to the explosion of about 15 kilotons of TNT). Therefore,

$$\frac{(E_\circ)_{\text{penny}}}{E_b} = \frac{2.25 \times 10^{14}\,\text{J}}{0.5 \times 10^{14}\,\text{J}} = 4.5.$$ (2.29)

Thus, a release of the entire energy of a penny (more than 99% of this energy is stored in the nuclei of the coin's particles) would produce about 4.5 times more energy than the energy released during the explosion. However, modern technology does not know ways to force the penny's particles to release the huge amounts of their nuclear energies.

End of Problem 2.1.

2.7 Energy and Power in Life and Technology

As discussed in Section 1.3, the basic unit of energy, work, and heat in the SI (System International) unit system is one joule (1 J). Other common non-SI units of these entities are:

- calorie (cal) = 4.184 J;
- Calorie (Cal) = 1000 cal = 1 kilocalorie = 4184 J [Calorie is often used in the nutrition and diet industry; it is sometimes called the 'big calorie' or 'nutrition calorie'];
- kilowatt-hour (kWh) = 3.6×10^6 J [kWh is often used to measure electric energy];
- electronvolt (eV) = 1.6×10^{-19} J [eV is often used in molecular and plasma sciences].

It is not practical to compare the usefulness and quality of some energy-processing devices (engines, heat pumps, etc.) without saying how much of the energy is being processed by each of the devices during the same amount of time (say, per second or per hour). Therefore, it is necessary to introduce the concept of **power**. The power of an energy-processing system is defined as the amount of energy (mechanical, chemical, electrical, etc.) produced by the system or needed to operate the system in a unit of time.

The basic SI (System International) unit of power is one watt (1 W = J/s), that is, the power of a system that can produce one joule of energy during one second. A bigger SI unit of power is kilowatt, 1 kW = 1000 W. A common non-SI unit of power is horsepower (1 HP); 1 HP = 746 W.

Approximate energy of	
a small apple hitting the ground when dropped from a table	~ 1 J
a bullet leaving a rifle	2×10^3 J
a hamburger processed in a stomach	1.5×10^6 J
the daily minimum (from food) needed by an adult	10^7 J
burning of 1 gallon of natural gas	2×10^5 J
burning of 1 kg of coal	2×10^6 J
burning of 1 kg of gasoline	5×10^7 J
a TV set working for an hour	5.5×10^5 J
a refrigerator working for an hour	5.5×10^5 J
a vacuum cleaner working for an hour	1.5×10^6 J
a coffee maker working for an hour	2.5×10^6 J
a desktop computer working for an hour	7×10^5 J
fission of all nuclei in 1 kg of Uranium-235	8×10^{13} J
fusion of all nuclei in 1 kg of Deuterium	5×10^{14} J
used annually by the United States	2.5×10^{20} J
used annually by the entire World	10^{21} J
the entire Universe (mass-energy)	$\sim 10^{70}$ J

Approximate power of	
typical lightbulb	100 W
human heart	1.5 W
human brain	30 W
human working hard	150 W
human working easily	50 W
horse working hard	800 W
a TV set	150 W
a refrigerator	150 W
a vacuum cleaner	400 W
a coffee maker	700 W
a desktop computer	200 W
the four engines of Boeing 747	10^8 W

2.8 Changes of Properties of Thermodynamic Systems

2.8.1 Changes of State Properties

The motion properties (kinetic energy E_{kin} and potential energy E_{pot}) of some amount of matter moving in a system always behave like functions of state regardless of the thermal characteristics of the matter. According to the discussion in Section 2.2.1, the changes of the motion properties in any process between any state 1 and any state 2 are

$$\Delta E_{kin} = E_{kin,2}(M_2, \omega_2) - E_{kin,1}(M_1, \omega_1) = \frac{M_2 \omega_2^2}{2} - \frac{M_1 \omega_1^2}{2}, \qquad (2.30)$$

and

$$\Delta E_{pot} = E_{pot,2}(M_2, z_2) - E_{pot,1}(M_1, z_1) = M_2 g z_2 - M_1 g z_1, \qquad (2.31)$$

where ω_1 and ω_2 are speeds of the matter's center of mass at the beginning and at the end of the process, respectively; M_1 and M_2 are masses of the matter at the beginning and at the end of the process, respectively; z_1 and z_2 are the elevations of the center of mass in the beginning and at the end of the process, respectively, and g is the gravitational acceleration.

Thermal energy U, enthalpy H, entropy S, the Helmholtz free energy A, and the Gibbs free energy G of systems undergoing thermodynamic transitions between states of equilibrium also behave like properties of state. The change ΔF of a state property F ($F \equiv U, H, S, A, G$) during a thermodynamic transition between an equilibrium state 1 (where $F = F_1$) and an equilibrium state 2 (where $F = F_2$) can be found in one of the two following ways:

(a) F_1 and F_2 are unknown but thermodynamic variables of states 1 and 2 are known:

In this case, one can replace the actual process that moves the system from state 1 to state 2 by *any* (reversible or not) process because the function of state F has no memory of the way in which these states have been created. The most convenient replacement of this kind is a sequence of several *reversible* and *mathematically convenient* subprocesses (isobars, isochors, etc.) that connects states 1 and 2. Then, integrating dF between states 1 and 2 gives the property change, ΔF, in a single-component, single-phase system during the entire $1 \rightarrow 2$ transition,

$$\Delta F = \int_1^2 dF = f_F(P_1, T_1, P_2, T_2), \tag{2.32}$$

where P_1 and T_1 are the pressure and the temperature of state 1, respectively, and P_2 and T_2 are the pressure and the temperature of state 2, respectively. The details of such a procedure are discussed in Section 4.2, and use of relationship (2.32) in studies of multi-phase, multi-component systems is discussed in Chapter 9.

Alternatively, relationship (2.32) can be given as a function of another pair (P, V or T, V) of thermodynamic variables. As discussed in Section 9.1, properties of a single-component, single-phase substance in equilibrium can be given as functions of only two thermodynamic variables (P and T, or V and T, or P and V). We used variables P and T in relationship (2.32) because this choice is convenient in most thermodynamic studies. This is why the variables P and T are usually used in tabularization of thermodynamic databases.

(b) Values of F_1 and F_2 are already known:

In such a case, the change of a state property F during any (reversible or not) process moving a single-phase, single-component system between equilibrium states 1 and 2 is

$$\Delta F = F_2(P_2, T_2) - F_1(P_1, T_1). \tag{2.33}$$

Taking the above into account, one can write the following expressions for the changes of important properties of state during any transition between equilibrium states 1 and 2:

$$\Delta U = f_u(P_1, T_1, P_2, T_2) \qquad \text{(in case (a))} \tag{2.34}$$

or

$$\Delta U = U_2(P_2, T_2) - U_1(P_1, T_1) \qquad \text{(in case (b))}, \tag{2.35}$$

$$\Delta H = f_h(P_1, T_1, P_2, T_2) \tag{2.36}$$

or

$$\Delta H = H_2(P_2, T_2) - H_1(P_1, T_1), \tag{2.37}$$

$$\Delta S = f_s(P_1, T_1, P_2, T_2) \tag{2.38}$$

or

$$\Delta S = S_2(P_2, T_2) - S_1(P_1, T_1), \tag{2.39}$$

$$\Delta A = f_a(P_1, T_1, P_2, T_2) \tag{2.40}$$

or

$$\Delta A = A_2(P_2, T_2) - A_1(P_1, T_1), \tag{2.41}$$

and

$$\Delta G = f_g(P_1, T_1, P_2, T_2) \tag{2.42}$$

or

$$\Delta G = G_2(P_2, T_2) - G_1(P_1, T_1). \tag{2.43}$$

In addition, one should recall that the change of the *internal energy $\mathcal{E}$* during any process is (see expression (2.4))

$$\Delta \mathcal{E} = \Delta E_{\text{kin}} + \Delta E_{\text{pot}} + \Delta U. \tag{2.44}$$

The way in which the state properties change in processes between equilibrium states plays an important role in studies of thermodynamic cycles. (A **thermodynamic cycle** is any closed-path process that begins and ends in the *same* state of the system – see Chapter 10.) Thus, the changes of the state properties of any system undergoing a complete thermodynamic cycle are (see expression (1.43))

$$\Delta U_c = \oint dU = 0, \tag{2.45}$$

$$\Delta H_c = \oint dH = 0, \tag{2.46}$$

$$\Delta S_c = \oint dS = 0, \tag{2.47}$$

$$\Delta A_c = \oint dA = 0, \tag{2.48}$$

and

$$\Delta G_c = \oint dG = 0. \tag{2.49}$$

One can see that Classical Thermodynamics provides reliable tools (expressions (2.30)–(2.44)) to study changes of state properties of (reversible or not) transitions in all (open or closed) single-phase, single-component thermodynamic systems where the initial and final states of the transitions are equilibrium, quasi-equilibrium, or local-equilibrium states. These tools are further discussed in the following chapters.

The general formalism discussed in this section can be used to study processes in multi-phase, multi-component systems – see Chapter 9.

2.8.2 Transfers of Path Properties

Transfers of heat and work between thermodynamic systems and the environment are important properties of path. Therefore, they cannot be calculated in the way similar to that leading to expressions (2.34)–(2.43). Instead, heat Q and work L of a thermodynamic transition from a state 1 to a state 2 must be obtained from the following general relationships:

$$Q = \int_1^2 \delta Q \tag{2.50}$$

and

$$L = \int_1^2 \delta L, \tag{2.51}$$

where we used the symbol δ in the differentials δQ and δL in order to emphasize the fact that these are *inexact* differentials. The integrals (2.50) and (2.51) can be evaluated if the thermodynamic path that takes the system from its state 1 to its state 2 is known, and if properties of the system in the states are known. This complicates the issue because many thermodynamic paths are complex functions of several thermodynamic variables and time. The situation is much simpler when a thermodynamic transition can be treated as a reversible process. For example, when the work done on or done by the system is boundary work W, the assumption of the process reversibility significantly simplifies calculations – see below. No work of an irreversible process can be given by a simple relationship involving thermodynamic variables, but it can be obtained from the Energy Balance Equation for the process if the other parts of the equation are known. (The Energy Balance Equation is valid in all (reversible or not) processes.)

The total work of a process in a thermodynamic system can be given as

$$L = L_{PV} + K, \tag{2.52}$$

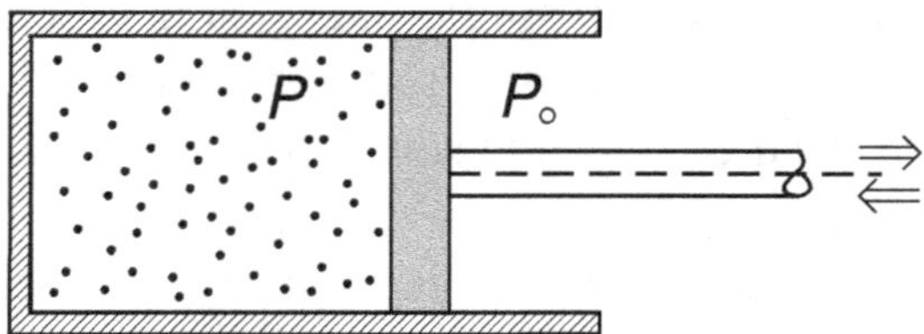

Fig. 2.1: A closed thermodynamic system in which a gas at pressure P is locked in by a movable piston that compresses or expands the gas. The outside pressure $P_\circ$ can change independently from the system's pressure P.

where L_{PV} is the work associated with the changes of the system's thermodynamic variables P and V during the process, and K is the shaft work or electrical work (or both) associated with some devices outside the system (see Figs. 2.6, 2.7, and 3.1). The P–V work can be given as

$$L_{PV} = \int_1^2 V\,dP \qquad \text{open systems,} \qquad (2.53)$$

and

$$L_{PV} = W = -\int_1^2 P\,dV \qquad \text{closed systems.} \qquad (2.54)$$

Relationships (2.53) and (2.54) are valid in reversible processes. In applications this restriction is routinely relaxed with little justification in many situations.

Work W in expression (2.54) (this is work associated with movement of the system's physical boundary) is called **boundary work** or **volumetric work**.

Work K in expression (2.52) and work L_{PV} in expression (2.53) are called **non-boundary works**, and work K is also often called **shaft work**.

Consider a gas inside a cylinder (the 'system') closed by a movable piston that has a cross section A (see Fig. 2.1). The pressure of the gas inside the cylinder is P and the pressure outside is $P_\circ$. Therefore, the gas inside the cylinder can increase (or decrease) its volume, producing (or receiving) during the process some boundary work W that depends on both the *system pressure P* and the *external (environmental)* pressure $P_\circ$. For example, the change of P can be caused by the change of the temperature of the gas, while the change of $P_\circ$ can be caused by a change of the environment's pressure that changes independently of the process in the system. Therefore, the resultant force **F** that is acting on the piston during the process depends, often in a complex time- and space-dependent way, on the pressures P and P_o. Then the boundary work transfer depends not only on the properties of the system (the pressure P) but also on the properties of the environment (the pressure P_o). However, if the expansion/compression is a reversible transition (a sequence of quasi-equilibrium states), then P and P_o are almost identical (differentially close to one another) during the entire process. Then, the boundary work of the reversible transition can be given as a function of the system's pressure P (or the external pressure P_o). Subsequently, the boundary work done on or done by the gas inside the cylinder during an elementary expansion/compression that moves the piston by a distance ds is

$$\delta W = -\mathbf{F}\cdot d\mathbf{s} = -(P\cdot A)\cdot ds = -P(A\cdot ds) = -PdV, \qquad (2.55)$$

where the minus sign has been introduced to secure, in agreement with the International Convention for the Sign of Energy (see below), a positive value of the boundary work for the system compression ($dV < 0$) or a negative value of the work for the system expansion ($dV > 0$). (It can be shown that relationship (2.55) is valid for systems of arbitrary geometrical shape.)

When the discussed processes are reversible processes with boundary works only (then $\delta L = \delta W$), one can transform integrals (2.50) and (2.51) into mathematically convenient expressions by using the so-called **integrating factors** associated with the inexact differentials δQ and δW. These factors are $1/T$ (for the differential δQ), and $-1/P$ (for the differential δW) – see Section 1.4. The products of the factors and the corresponding inexact differentials δQ and δW are the exact differentials $\delta Q/T$ and $-\delta W/P$, respectively. Thus, one can write for any reversible process that

$$\frac{\delta Q}{T} = dS \quad \rightarrow \quad \delta Q = TdS, \qquad (2.56)$$

and

$$-\frac{\delta W}{P} = dV \quad \rightarrow \quad \delta W = -PdV, \qquad (2.57)$$

where S and V are some functions of state. The property S is the **system entropy**, an important thermodynamic property that is discussed in many parts of this book.

Relationships (2.56) and (2.57) allow one to find the heat and the boundary work of any finite *reversible* process between an equilibrium state 1 and an

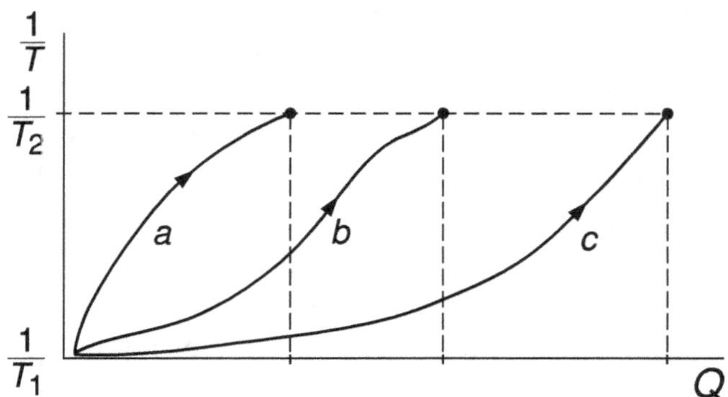

Fig. 2.2: Three (a,b,c) different thermodynamic paths of a physical system moving from an equilibrium state 1 (of temperature T_1) to an equilibrium state 2 (of temperature T_2) by adding some heat Q to the system. The value of the integral $\int \delta Q/T$ is the same for each of the three paths; the integral is represented by the area under the corresponding path.

equilibrium state 2,

$$Q = \int_{S_1}^{S_2} T \, dS, \tag{2.58}$$

and

$$W = - \int_{V_1}^{V_2} P \, dV. \tag{2.59}$$

The fact that $1/T$ and $-1/P$ are integrating factors for the heat and boundary work of reversible processes can be confirmed experimentally by bringing a closed system (that exchanges heat Q and boundary work W with the environment) from a fixed equilibrium state 1 to a fixed equilibrium state 2 in several different ways (each being as close to a reversible (quasi-static) process as possible) and plotting the results on $1/T$ vs. Q and $-1/P$ vs. W diagrams. Such a diagram for measured relationships $1/T$ vs. Q for three different thermodynamic paths is shown in Fig. 2.2. The area under each of the three paths is (almost) the same. (The term 'almost' is used here in order to emphasize the fact that it is practically impossible to execute a perfectly reversible process during a finite

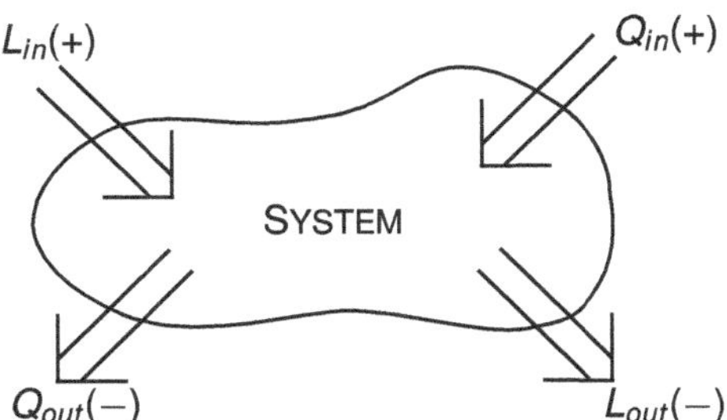

Fig. 2.3: The International Convention for the Sign of Energy: an energy transfer (in the form of work L or heat Q) between a system and the environment has a positive value when it is entering the system but a negative value when it is leaving the system.

amount of time.) This means that the term

$$\int_1^2 \frac{\delta Q}{T} = \Delta S \tag{2.60}$$

for each of the considered paths is the same – see expression (2.17). In other words, as long as the properties of the initial and final states of each path do not change, the change of the system entropy during each path is the same, that is, independent of the path's character. Thus, the system entropy S is a property of state. Similar statements can be made about the areas under the $-1/P$ vs. W curves taken during a corresponding experiment.

At this point, it is necessary to introduce a convention for signs (positive or negative) of heat Q and work L in balances of energy flows in thermodynamic systems. Two such conventions have been used in thermodynamic literature: 1) the heat and work that enter the system have positive ('+') values but they have negative ('−') values when they leave the system, and 2) the heat entering the system and the work done by the system have positive values, while the heat leaving the system and the work done on the system have negative values. In this book we use the first convention (called hereafter the **International Convention for the Sign of Energy**) because it has officially been recommended for use in scientific and technical literature by various international organizations. However, one should keep in mind that some authors are still using the second convention, and that it was used widely in earlier thermodynamic literature.

The International Convention for the Sign of Energy is demonstrated in Fig. 2.3. The Convention is easy to remember because it assumes that all forms of energy (work, heat, radiation) have positive numerical values when they increase the system internal energy but negative numerical values when they decrease the internal energy. For example, consider a closed thermodynamic system (where the changes of kinetic and potential (gravitational) energies of the matter moving inside the system can be ignored) producing 3 MJ of (boundary plus non-boundary) work L while absorbing 1 MJ of heat Q. Then, according to the Principle of Conservation of Energy and the International Convention for the Sign of Energy, the change of the system internal energy $\mathcal{E}$ during the process is (see Section 3.5.2)

$$\Delta\mathcal{E} = \Delta U = L + Q = (-3\,\text{MJ}) + (+1\,\text{MJ}) = -2\,\text{MJ}, \tag{2.61}$$

which means that the internal energy of the system has decreased during the process by 2 MJ.

The discussed above simple analytical expressions for heat and boundary work are valid only in reversible processes (see relationships (2.58) and (2.59)). However, **work (boundary, non-boundary) and heat of any (reversible, irreversible) process can also be obtained from solution of the Energy Balance Equation for the process.** (The Balance Equations for Matter, Energy, and Entropy are valid in all processes in all systems – see Section 3.5.)

Consider the processes shown in the T–S plane (the 'heat diagram') and in the P–V plane (the 'work diagram') – see Fig. 2.4. In the diagrams, a system moves thermodynamically from a *fixed* state A to a *fixed* state B along two different (independent) reversible thermodynamic paths, the paths 1 and 2. According to relationship (2.58), the heat Q_1 exchanged between the system and the environment during the A→B transition that follows path 1 is equal to the area of the T–S plane under the path (the area is filled with horizontal lines). The heat Q_2 of the A→B transition that follows path 2 is equal to the area under the path (the area is filled with vertical lines). These heat transfers can be given as

$$Q_1 = \int_{S_A}^{S_B} T_1(S)\,dS, \tag{2.62}$$

and

$$Q_2 = \int_{S_A}^{S_B} T_2(S)\,dS, \tag{2.63}$$

where S_A and S_B are entropies of the initial and final states of the transition, respectively, and functions $T_1(S)$ and $T_2(S)$ represent the system's temperature

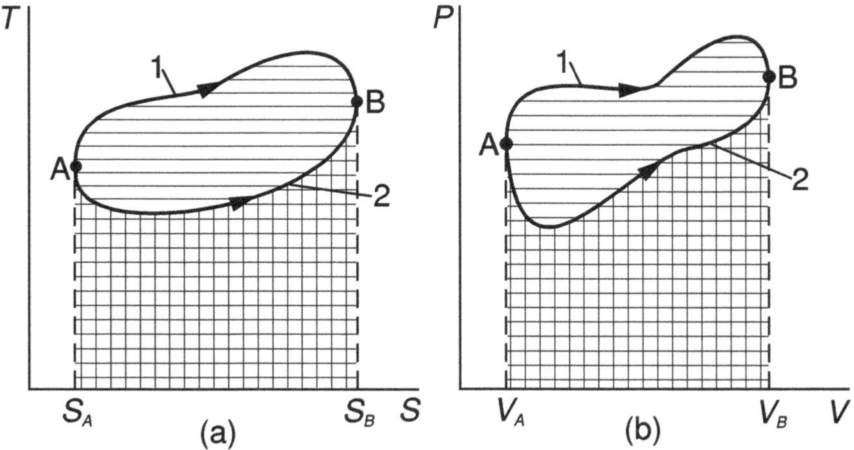

Fig. 2.4: Heats and boundary works of two different reversible processes between states A and B: (a) 'Heat diagram' (T is the temperature and S is the entropy of the system). Since heat is a function of path, the heat transfer during the process 1 differs from the heat transfer during the process 2 even though both processes start at the same state A and end at the same state B. (b) 'Work diagram' (P is the pressure and V is the volume of the system). Since boundary work is a function of path, the work transfers in processes 1 and 2 are different. The areas representing the heat and work of process 1 are marked with horizontal lines, and the areas representing the corresponding transfers in process 2 are marked with vertical lines.

dependence on entropy during paths 1 and 2, respectively. Since these paths are different, in the 'heat diagram' the area filled with horizontal lines differs from the area filled with the vertical lines,

$$Q_1 \neq Q_2. \tag{2.64}$$

Thus, **heat is a property of path in all reversible transitions in all systems** (in such transitions, it can be calculated accurately from expression (2.58)). The **heat of every irreversible transition also is a function of path** (heat transfers in irreversible processes will be discussed in the following chapters).

According to relationship (2.59), the boundary work W_1 transferred during the A$\rightarrow$B transition along the thermodynamic path 1 is equal, in the P–V plane, to the area under the path (the area is filled with horizontal lines in the 'work diagram' in Fig. 2.4), while the boundary work W_2 transferred during the A$\rightarrow$B transition along the thermodynamic path 2 is equal to the area under the path 2 (the area is filled in the diagram with vertical lines). In reversible transitions, the boundary works of the two processes are

$$W_1 = -\int_{V_A}^{V_B} P_1(V)\,dV, \tag{2.65}$$

and

$$W_2 = -\int_{V_A}^{V_B} P_2(V)\,dV, \tag{2.66}$$

where V_A and V_B are the volumes of the system in states A and B, respectively, and functions $P_1(V)$ and $P_2(V)$ represent the system's pressure changes during paths 1 and 2, respectively. Since these paths are different, one has

$$W_1 \neq W_2. \tag{2.67}$$

Thus, **boundary work is a property of path in all reversible transitions in all thermodynamic systems** and, in such transitions, it can be calculated accurately from expression (2.59). **Boundary work and non-boundary work of every irreversible transition also are functions of path** – the transfers will be discussed further in the following chapters.

Although integral (2.58) is well defined, it is usually difficult to calculate since the dependence of the system's temperature T on its entropy S is difficult to establish. Therefore, Classical Thermodynamics has proposed the following expression which is a good representation of heat transfers in *some* processes:

$$Q = \int_{T_A}^{T_B} C\,dT, \tag{2.68}$$

or, for elementary processes,

$$\delta Q = C\,dT, \tag{2.69}$$

where C is the so-called **heat capacity** of the *entire* matter of the system of interest, a property that is well-defined and can be accurately measured in some thermodynamic processes. Heat capacities will be discussed further in Chapter 6.

The boundary work transfers in closed systems undergoing some real (irreversible) transitions can be calculated quite easily. For example, in all *isochoric* transitions ($V = $ const, $dV = 0$), one has

$$W = 0. \tag{2.70}$$

Simple expressions for boundary work and heat transfers in some real processes can be proposed considering the fact that the degree of the irreversibility in the processes is proportional to the first- and second-order gradients $\partial T/\partial \mathbf{r}$, $\partial^2 T/\partial \mathbf{r}^2$, $\partial P/\partial \mathbf{r}$, and $\partial^2 P/\partial \mathbf{r}^2$ that are present in the processes. For instance, the constancy of pressure during an *isobaric* (P = const, dP = 0) transition in a closed system keeps the system in mechanical equilibrium but does not prevent the presence of a thermal non-equilibrium in the system. If, however, derivatives $\partial T/\partial \mathbf{r}$ and $\partial^2 T/\partial \mathbf{r}^2$ are not significant during the isobaric process, one can assume that the system's degree of irreversibility is small. In such a case, the boundary work is close to that of the corresponding reversible isobar, that is (see expression (2.59)),

$$W \simeq - \int_A^B PdV = -P(V_B - V_A). \tag{2.71}$$

Similarly, the constancy of temperature in an *isothermal* (T = const, dT = 0) transition in a closed system keeps the system in thermal equilibrium but does not prevent the presence of a mechanical non-equilibrium in the system. If derivatives $\partial P/\partial \mathbf{r}$ and $\partial^2 P/\partial \mathbf{r}^2$ during the isothermal process are not significant, one can assume that the degree of the system irreversibility is small. Then, the transfer of heat in the process is close to that of the corresponding reversible isotherm (see expression (2.58)),

$$Q \simeq \int_A^B TdS = T(S_B - S_A). \tag{2.72}$$

Problem 2.2: A substance has undergone, independently, three different *reversible* thermodynamic transitions in a closed thermodynamic system where changes of the kinetic and potential (gravitational) energies of the substance are negligible. The transitions start at the same equilibrium state A and end at the same equilibrium state B. They are shown in work diagram P–V in Fig. 2.5 as processes 1, 2, and 3. The product P^2V^3 is constant during the entire process 1; pressure P of the system is a linear function of the system's volume V during process 2; and process 3 consists of isobar 3*a* followed by isochor 3*b*. Find the boundary works done by or done on the system during each of the three transitions when the pressure of the initial state A is 1 bar and volumes of the system in states A and B are $0.1\,\text{m}^3$ and $0.4\,\text{m}^3$, respectively. What would the boundary work transfers in the processes be if they were reversible transitions moving in the opposite direction, that is, from state B to state A?

Solution:

Transition 1:

Since during the transition one has

$$P^2V^3 = \text{const} = P_A^2 V_A^3 = P_B^2 V_B^3, \tag{2.73}$$

the dependence of the pressure of the substance on its volume is

$$P_1(V) = P_A \left(\frac{V_A}{V}\right)^{3/2} = P_B \left(\frac{V_B}{V}\right)^{3/2}. \tag{2.74}$$

Thus, according to relationship (2.59), the boundary work of the transition is

$$W_1 = -\int_{V_A}^{V_B} P_1(V)dV = -\int_{V_A}^{V_B} P_A \left(\frac{V_A}{V}\right)^{3/2} dV$$

$$= -2P_A V_A \left[1 - \left(\frac{V_A}{V_B}\right)^{1/2}\right] = -10\,\text{kJ}. \tag{2.75}$$

Transition 2:

The linear dependence of the pressure of the substance on its volume during the transition is

$$P_2(V) = P_A + \frac{P_B - P_A}{V_B - V_A}(V - V_A), \tag{2.76}$$

where the final pressure P_B is the same as the final pressure of transition 1,

$$P_B = P_A \left(\frac{V_A}{V_B}\right)^{3/2} = 0.125\,\text{bar}. \tag{2.77}$$

Therefore, the boundary work of transition 2 is

$$W_2 = -\int_{V_A}^{V_B} P_2(V)dV = -\int_{V_A}^{V_B} \left[P_A + \frac{P_B - P_A}{V_B - V_A}(V - V_A)\right] dV$$

$$= -\frac{(P_A + P_B)(V_B - V_A)}{2} = -16.87\,\text{kJ}. \tag{2.78}$$

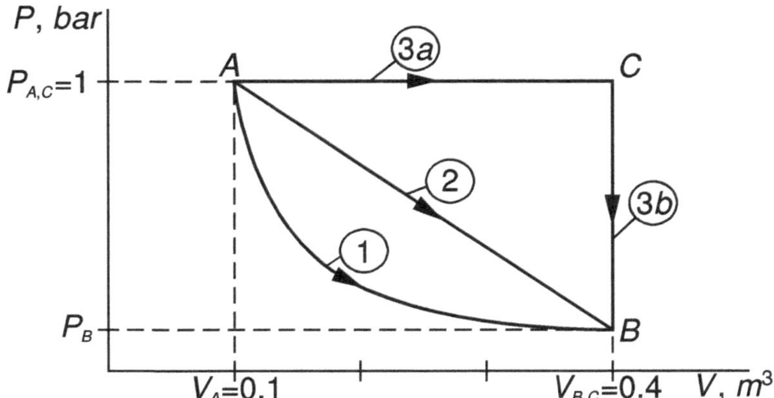

Fig. 2.5: Three $(1, 2, 3a + 3b)$ reversible thermodynamic transitions between state A and state B. The area under each line represents the boundary work of the corresponding transition.

Transition 3:

The boundary work of isochor $3b$ is zero because the volume of the system undergoing the isochoric process does not change. Thus, the boundary work of the entire transition 3 is

$$W_3 = W_{3a} = -\int_{V_A}^{V_C=V_B} P_A \, dV = -P_A(V_B - V_A) = -30 \, \text{kJ}. \tag{2.79}$$

One can see that even though the three discussed transitions start at the same state A and end at the same state B, the boundary works of the transitions are different. This agrees with the fact mentioned earlier that **every work transfer in every process in every system is always a function of path**.

The values of the work transfers in transitions 1, 2, and 3 are all negative numbers. This means, according to the International Convention for the Sign of Energy, that the transfers were done by the system on its environment, a result expected in expansion-type processes where $V_B > V_A$.

If transitions 1, 2, and 3 were moving the system from state B to state A, the values of the transitions' boundary works would be positive, that is, the works would be done on the system.

End of Problem 2.2.

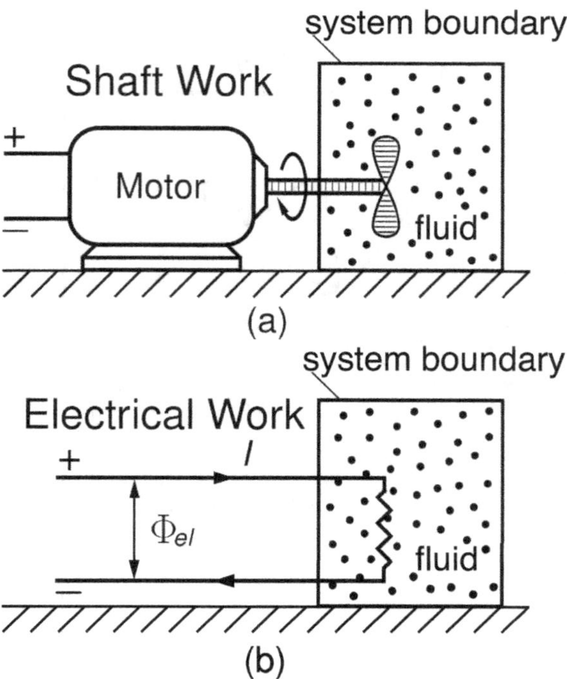

Fig. 2.6: (a) 'Shaft work': when a motor is connected by a shaft with an impeller located inside the system filled with a fluid, the work of the motor is transferred to the impeller causing its rotation; (b) 'Electrical work': the work done by an electric current, driven by an external voltage Φ_{el}, moving through the resistor immersed in the fluid.

We considered in Problem 2.2 only one kind of work – the boundary work. Common examples of other (non-boundary) work transfers are the **shaft work** and the **electrical work** – see Figs. 2.6a and 2.6b, respectively.

Boundary work is rarely present in open systems where matter flows through because physical boundaries of such systems usually do not move. The works pushing the flowing matter in and out of such a system's control surface along the direction of the flow, the 'flow works', are accounted for in the enthalpies of the matter entering and leaving the systems – see Section 3.5.2.

Non-boundary works of all thermodynamic processes (reversible or not) are, similarly to boundary works, functions of path. Also, every kind of work is an additive property. Thus, the net work done on or done by a system consisting of several parts is equal to the sum of the works done on or done by all of these parts. Similar remarks can be made about the additivity of heat transfers.

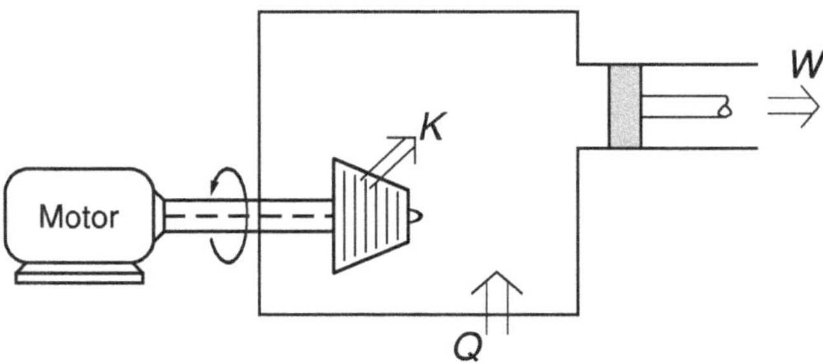

Fig. 2.7: An external motor drives a small turbine which supplies some non-boundary (shaft) work K to the closed thermodynamic system while some heat Q is added to the system. The motion of the piston transfers some boundary work W to the environment.

Problem 2.3: The total (net) work L (sum of all non-boundary and boundary works) done by a closed nonflow system during a thermodynamic process is 2 MJ. Find the process's boundary work W, non-boundary (shaft) work K, and heat Q when the following ratios are true (see Fig. 2.7):

$$\frac{\text{shaft work}}{\text{boundary work}} = \frac{K}{W} = -0.2 \quad \text{and} \quad \frac{\text{heat}}{\text{total work}} = \frac{Q}{L} = -0.4. \quad (2.80)$$

Is the process taking place in the system exothermic or endothermic?

Solution:

$$Q = -0.4L = -0.4(-2\,\text{MJ}) = +0.8\,\text{MJ}. \quad (2.81)$$

The internal energy of the nonflow system during the process can be obtained from the Principle of Conservation of Energy (see Section 3.5.2),

$$\Delta\mathcal{E} = \Delta U = Q + L = +0.8\,\text{MJ} + (-2\,\text{MJ}) = -1.2\,\text{MJ}. \quad (2.82)$$

Recall that the internal energy $\mathcal{E}$ is the sum of the thermal energy U, kinetic energy E_{kin}, and potential (gravitational) energy E_{pot} of the system's 'working' matter.

Since

$$\Delta\mathcal{E} = Q + L = Q + (W + K) = Q + (W - 0.2W) = Q + 0.8W, \quad (2.83)$$

one has

$$W = (\Delta\mathcal{E} - Q)/0.8 = [-1.2\,\text{MJ} - (+0.8\,\text{MJ})]/0.8 = -2.5\,\text{MJ}, \quad (2.84)$$

and

$$K = -0.2W = -0.2(-2.5\,\text{MJ}) = +0.5\,\text{MJ}. \tag{2.85}$$

Thus, this thermodynamic transition is, according to the International Convention for the Sign of Energy, an endothermic process ($Q > 0$), its boundary work $W = -2.5\,\text{MJ}$ was done by the system, and its non-boundary (shaft) work $K = +0.5\,\text{MJ}$ was done on the system.

End of Problem 2.3.

Chapter 3

Thermodynamic Principles

The most powerful tools for studying thermodynamic processes in a broad range of applications are the **Principle of Conservation of Matter**, the **Principle of Conservation of Energy**, and the **Principle of Production of Natural Entropy**. These three principles are studied below with the assumption that relativistic processes (such as some nuclear reactions) are not present in the thermodynamic systems under consideration so that mass (the amount of matter) and energy are independent entities.

3.1   Conservation of Matter

The Principle of Conservation of Matter states that **matter can neither be produced nor destroyed spontaneously**. This means that the change of the *total* mass of every (non-relativistic) system can be caused only by bringing some matter to or removing some matter from the system (possibly through many inlets and outlets). In other words, the Principle of Conservation of Matter tells us that the total mass M of any *closed* system does not change regardless of the kind of thermodynamic processes taking place in the system,

$$\Delta M = 0 \quad \rightarrow \quad M = \text{const} \qquad \text{closed systems.} \qquad (3.1)$$

The Principle of Conservation of Matter applies to all processes in all non-relativistic systems.

3.2   Conservation of Energy

The Principle of Conservation of Energy states that **one form of energy can be transformed into another form of energy** (for example, some chemical energy of gasoline can be transformed into heat during a combustion process), but **energy cannot be spontaneously produced or destroyed**. Thus, the change $\Delta\mathcal{E}$ of the *total* internal energy $\mathcal{E}$ of every physical system can be caused only by moving some energy (carried by radiation, heat, work, and/or matter) between the system and its environment. This means that in every *isolated* system one has

$$\Delta\mathcal{E} = 0 \quad \rightarrow \quad \mathcal{E} = \text{const} \qquad \text{isolated systems.} \qquad (3.2)$$

The Principle of Conservation of Energy is often called the First Law of Thermodynamics. It applies to all processes in all non-relativistic systems.

3.3   Production of Natural Entropy

As discussed in Section 2.3, a system's total entropy S can be changed by transfer of matter, heat, and/or radiation between the system and the environment. Consequently, scientists and engineers can have a substantial degree of control over the amount of entropy present in thermodynamic applications. However, one has no control over the fact that the irreversibilities present inside every real system undergoing a thermodynamic transition always produce some special kind of entropy (the natural entropy S_{nat}) even in processes in isolated systems where transfers of matter, heat, and radiation are absent. In other words, Nature always adds some 'extra' entropy (the natural entropy) to every system undergoing a real (that is, irreversible) thermodynamic transition. These statements form the **Principle of Production of Natural Entropy**. The principle results from the **Second Law of Thermodynamics**, and it is stated in many different (but equivalent) ways. For example, one of the popular statements of the Second Law says that heat cannot flow *spontaneously* from lower-temperature regions to higher-temperature regions of thermodynamic systems. The engineers know how to build devices, such as refrigerators, that can force flow of heat from lower-temperature regions to higher-temperature regions but such flows are not spontaneous. Another statement of the Second Law (**'Nature loves equilibrium'**) was already mentioned in Section 2.3.1.

The **natural entropy is always added to (never removed from!) every thermodynamic system undergoing a real (that is, irreversible) thermodynamic process**,

$$\Delta S_{nat} > 0 \qquad \text{all irreversible processes,} \qquad (3.3)$$

where ΔS_{nat} is the amount of the added natural entropy.

Since the natural entropy is the only partial entropy that changes during real processes in isolated systems which, by definition, do not interact with the environment, one can say that the **total entropy of every isolated system undergoing a real (that is, irreversible) process always increases**. In other words (see relationship (3.3)),

$$\Delta S = \Delta S_{nat} > 0 \qquad \text{irreversible processes in isolated systems.} \qquad (3.4)$$

This statement is sometimes used as one of the formulations of the Second Law of Thermodynamics.

When a system reaches a state of equilibrium, the production of the natural entropy stops, and the system remains in this equilibrium as long as it does not start to exchange any matter and energy with the environment. (Such an exchange would force a departure from equilibrium.) Thus, the **entropy of every system in equilibrium does not change** (see Section 2.3),

$$\Delta S = 0 \quad \rightarrow \quad S = \text{const}, \qquad \text{all systems in equilibrium.} \qquad (3.5)$$

One should notice that the **natural entropy is not produced in reversible transitions** because such transitions are sequences of (different) quasi-equilibrium states which are treated in Classical Thermodynamics in the same way as equilibrium states. This fact is behind the great appeal of reversible processes as theoretical models of some real thermodynamic processes.

3.4 Flow Rates and Fluxes

It is often important in studies of physical systems to know how quickly a physical property (some amount of matter, heat, etc.) is moving from one region of the system to another region, or how quickly the property enters or leaves the system during a thermodynamic process. These 'speeds' can be determined using the concepts of **flow rate** and **flux**.

Consider a flow of a physical property (for instance, some amount of matter) through a 'control area', which is an imaginary or real surface that allows the flow to pass through. The control area does not have to be normal to the direction of the flow. One can define the so-called **flow rate** of the property at the surface as the amount of the property passing through the area in some time interval (say, during one second). For example, the mass flow rate of water flowing out of a kitchen faucet during a typical dish washing is about 0.05 kg/s; in such a case, the control area A is the geometrical cross section of the faucet's outlet.

The flow rate of a property a moving through a control area A is usually denoted by symbol $\dot{a}_{th}$ or $\dot{a}$. Mathematically, the *theoretical* flow rate of a property a through an area A is defined as the following derivative:

$$\dot{a}_{th} = \frac{da}{dt} = \lim_{\Delta t \to 0} \left(\frac{\Delta a}{\Delta t} \right), \qquad (3.6)$$

where Δa is the amount of the property a that crosses the area A during time Δt. In practice, flow rates can be measured using relationship (3.6) while replacing Δt by some very small (but measurable) interval Δt_s. Then, the *practical* flow rate of the property can be given as

$$\dot{a} = \lim_{\Delta t \to \Delta t_s} \left(\frac{\Delta a}{\Delta t} \right). \qquad (3.7)$$

In most thermodynamic studies, one can replace the theoretical rates (3.6) by the (approximate) practical rates (3.7). Such a replacement introduces some small inaccuracy which is usually meaningless in practical applications.

When the flow rate $\dot{a}$ of a property a is constant with time, then the same amount of the property passes through the control area A during each second (minute, hour, etc.). A flow of a thermodynamic property a with a constant rate $\dot{a}$ is called **steady** flow of the property.

Examples of flow rates common in thermodynamic studies are

$$\dot{M} \qquad \text{mass flow rate [in kg/s]}, \qquad (3.8)$$

$$\dot{V} \qquad \text{volume flow rate [in m}^3\text{/s]}, \qquad (3.9)$$

$$\dot{L} \qquad \text{work flow rate [in J/s]}, \qquad (3.10)$$

and

$$\dot{Q} \qquad \text{heat flow rate [in J/s]}. \qquad (3.11)$$

In many cases, it is more convenient to deal with fluxes of physical properties than with the properties' flow rates. **Flux** of a property a is defined as the amount of the property passing during a unit of time through a unit of the control area D which is *normal* to the direction of the flow. Thus, the flux of a property a passing through the area D is

$$\dot{F}_a = \dot{a}/D, \tag{3.12}$$

where, as before, $\dot{a}$ is the flow rate of the property a entering the area. Subsequently, fluxes of the properties occurring in expressions (3.8)–(3.11) can be given, respectively, as

$$\dot{F}_M = \dot{M}/D \qquad \text{mass flux [in kg/m}^2\text{s]}, \tag{3.13}$$
$$\dot{F}_V = \dot{V}/D \qquad \text{volume flux [in m/s]}, \tag{3.14}$$
$$\dot{F}_L = \dot{L}/D \qquad \text{work flux [in J/m}^2\text{s or W/m}^2\text{]}, \tag{3.15}$$

and

$$\dot{F}_Q = \dot{Q}/D \qquad \text{heat flux [in J/m}^2\text{s or W/m}^2\text{]}. \tag{3.16}$$

3.5 Equations of Thermodynamic Principles

3.5.1 Balance Equations for Matter

Consider a thermodynamic transition of an *open* system from an arbitrary initial state to an arbitrary final state while some amount of matter is exchanged between the system and the environment through a **control surface** (an imaginary or real surface that defines the system under consideration – see Fig. 3.1). The space inside the control surface of an open system is called the **control volume**. The space taken by a closed system is called the **control mass**. According to the Principle of Conservation of Matter (see Section 3.1), the change of the total mass M of the matter inside the system's control surface during elementary time dt is (mass is always defined as positive property)

$$\frac{dM}{dt} = \sum_k \left(\frac{dM}{dt}\right)_k - \sum_j \left(\frac{dM}{dt}\right)_j, \tag{3.17}$$

where $(dM/dt)_k$ is the theoretical rate at which matter enters the system through its kth inlet, and $(dM/dt)_j$ is the theoretical rate at which matter leaves the system

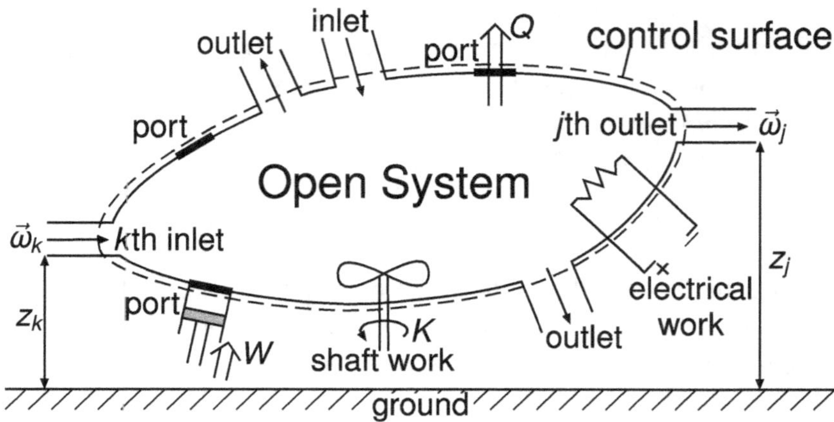

Fig. 3.1: An open thermodynamic system where matter enters through several inlets and leaves through several outlets while heat Q, boundary work W, and non-boundary work K are transferred to and out of the system through various ports. The inlets' and outlets' elevations, z_k and z_j, can be different.

through its jth outlet. The first sum on the right-hand side of expression (3.17) is taken over all inlets of the system, and the second sum is taken over all outlets of the system. Thus, the first sum represents the mass of all the matter that enters the control surface through all inlets of the system during time dt, and the second sum represents the mass of all the matter that leaves the control surface through all its outlets during this time.

Replacing in expression (3.17) theoretical rates $(dM/dt)_k$ and $(dM/dt)_j$ by the corresponding practical rates $\dot{M}_k$ and $\dot{M}_j$ gives

$$\frac{dM}{dt} \simeq \dot{M} = \sum_k \dot{M}_k - \sum_j \dot{M}_j. \tag{3.18}$$

Note that the replacement makes relationship (3.18) an approximate one, and that the accuracy of the approximation depends on the accuracy of the practical rates $\dot{M}_k$ and $\dot{M}_j$.

The rate of the mass flow through the system's kth inlet can be given as

$$\dot{M}_k = \rho_k A_k \omega_k, \tag{3.19}$$

where ρ_k and ω_k are mass density and speed of the matter flowing into the system through the kth inlet, respectively, and A_k is the area of the inlet's cross section. A similar relationship can be written for the matter flowing out of the system through its jth outlet.

When the thermodynamic process in the studied open system is a process that begins at time t_1 (when the system is in state 1) and ends at time t_2 (when the system is in state 2), relationship (3.18) can be integrated from t_1 to t_2. The most common choice of t_1 is the instant at which the matter enters the system, and the most common choice of t_2 is the instant at which the matter leaves the system. Then the change of the open system's mass during the process is

$$\Delta M = \sum_k M_k - \sum_j M_j, \tag{3.20}$$

where M_k is the mass of the matter that has entered the system through its kth inlet during time $\Delta t = t_2 - t_1$, and M_j is the mass of the matter that has left the system through its jth outlet during time Δt,

$$M_k = \int_{t_1}^{t_2} \dot{M}_k \, dt \qquad \text{and} \qquad M_j = \int_{t_1}^{t_2} \dot{M}_j \, dt. \tag{3.21}$$

Open systems often operate under steady conditions when

$$dM/dt = 0. \tag{3.22}$$

Then relationships (3.18) and (3.20) can be written as, respectively,

$$\sum_k \dot{M}_k = \sum_j \dot{M}_j, \tag{3.23}$$

and

$$\sum_k M_k = \sum_j M_j. \tag{3.24}$$

As mentioned earlier, the total amount of matter in closed and isolated systems remains constant at all times,

$$M = \text{const} \qquad \text{closed and isolated systems.} \tag{3.25}$$

3.5.2 Balance Equations for Energy

The Principle of Conservation of Energy (see Section 3.2) gives the following expression for the change of the internal energy $\mathcal{E}$ of the matter *inside* the control surface defining an open system that undergoes an elementary thermodynamic transition between any two states:

$$\frac{\delta\mathcal{E}}{dt} = \sum_k \left(\frac{\delta E}{dt}\right)_k - \sum_j \left|\left(\frac{\delta E}{dt}\right)_j\right| + \sum_k \left(\frac{\delta W_{\text{flow}}}{dt}\right)_k - \sum_j \left|\left(\frac{\delta W_{\text{flow}}}{dt}\right)_j\right|$$

$$+ \sum_c \left(\frac{\delta L}{dt}\right)_c - \sum_b \left|\left(\frac{\delta L}{dt}\right)_b\right| + \sum_l \left(\frac{\delta Q}{dt}\right)_l - \sum_p \left|\left(\frac{\delta Q}{dt}\right)_p\right|, \tag{3.26}$$

where we used the vertical bars (indicating the absolute values of the corresponding derivatives) in order to follow the International Convention for the Sign of Energy (see Section 2.8.2); $(\delta E/dt)_k$ is the (*theoretical*) rate of the entire (thermal+kinetic+potential) energy of the matter entering the system through its kth inlet, and $(\delta E/dt)_j$ is the rate of the entire energy of the matter leaving the system through the jth outlet; $(\delta W_{\text{flow}}/dt)_k$ is the rate of the so-called 'flow work' (see below) done on the flowing fluid in order to push it into the system through the kth inlet, and $(\delta W_{\text{flow}}/dt)_j$ is the rate of the flow work needed to push the flow out of the system through the jth outlet; $(\delta L/dt)_c$ is the rate of the (non-boundary plus boundary) work done on the system through its cth port, and $(\delta L/dt)_b$ is the rate of the (non-boundary plus boundary) work done by the system through its bth port; $(\delta Q/dt)_l$ is the rate of the heat entering the system through its lth port, and $(\delta Q/dt)_p$ is the rate of the heat leaving the system through its pth port.

One should keep in mind that displacements of physical boundaries of open systems (flows) are rare and, therefore, boundary work transfers are usually absent in thermodynamic studies. The most popular work transfers in open systems are 'shaft work' and 'electrical work' – see Figs. 2.6 and 2.6 – done on or done by the system through its ports.

The elementary **flow work** needed to push some amount of fluid a distance dx through the system's kth inlet is

$$\delta W_{\text{flow},k} = P_k A_k (dx)_k = P_k dV_k = P_k M_k dv_k, \tag{3.27}$$

where P_k is the fluid pressure at the kth inlet, A_k is the area of the cross section of the inlet ($P_k A_k$ is the force pushing the fluid through the kth inlet), V_k and

$\upsilon_k = V_k/M_k$ are volume and the specific volume, respectively, of the fluid being pushed through the inlet. (Note that the elementary work 3.27 is an inexact differential produced by an elementary force (inexact differential) acting over an elementary distance (exact differential).) Similarly, the flow work needed to push some matter of mass M_j out of the system through its jth outlet is (according to the International Convention for the Sign of Energy, this work is a negative number)

$$\delta W_{\text{flow},j} = -P_j M_j d\upsilon_j, \tag{3.28}$$

where P_j is the fluid pressure at the jth outlet, A_j is the area of the cross section of the outlet, and V_j and υ_j, are volume and the specific volume, respectively, of the fluid being pushed the distance $(dx)_j$.

The 'flow works' discussed above can be seen as 'P–V works' (see Section 2.8.2) *along* the directions of the flows that move through the system's inlets and outlets. However, they cannot be considered boundary works because they do not cause movements of the system's physical boundaries.

Replacing the theoretical rates in relationship (3.26) by the corresponding practical rates (see Section 3.4) one obtains

$$\frac{\delta\mathcal{E}}{dt} \simeq \left(\sum_k \dot{E}_k - \sum_j |\dot{E}_j|\right) + \left(\sum_k \dot{W}_{\text{flow},k} - \sum_j |\dot{W}_{\text{flow},j}|\right)$$
$$+ \left(\sum_c \dot{L}_c - \sum_b |\dot{L}_b|\right) + \left(\sum_l \dot{Q}_l - \sum_p |\dot{Q}_p|\right), \tag{3.29}$$

where

$$\dot{E}_k = \dot{E}_{\text{th},k} + \dot{E}_{\text{kin},k} + \dot{E}_{\text{pot},k} \tag{3.30}$$

is the practical rate of the energy carried by the matter entering the system through its kth inlet, and

$$\dot{E}_j = \dot{E}_{\text{th},j} + \dot{E}_{\text{kin},j} + \dot{E}_{\text{pot},j} \tag{3.31}$$

is the practical rate of the energy carried by the matter leaving the system through its jth outlet.

We did not include in sums (3.30) and (3.31) the possible energy of the matter that rotates around its center of mass because such rotations are usually absent in most thermodynamic applications.

Relationships (3.27)–(3.29) suggest introducing a new thermodynamic property called the **enthalpy**, which is defined as

$$H = E_{\text{th}} + PV, \tag{3.32}$$

where, as before, P and V are the pressure and volume of the matter, respectively. Subsequently, the rates of the enthalpy of the matter entering the system through the kth inlet and the enthalpy of the matter leaving the system through the jth outlet can be given as, respectively,

$$\dot{H}_k = \dot{E}_{\text{th},k} + P_k \dot{V}_k \qquad \text{and} \qquad \dot{H}_j = \dot{E}_{\text{th},j} + P_j \dot{V}_j, \tag{3.33}$$

and expression (3.29) can be written as

$$\frac{\delta \mathcal{E}}{dt} \simeq \sum_k (\dot{H}_k + \dot{E}_{\text{kin},k} + \dot{E}_{\text{pot},k}) - \sum_j (\dot{H}_j + \dot{E}_{\text{kin},j} + \dot{E}_{\text{pot},j})$$
$$+ \dot{L} + \dot{Q}, \tag{3.34}$$

where $\dot{L}$ and $\dot{Q}$ are rates of the total (net) work and the total (net) heat transfers in the system, respectively.

Relationship (3.34) is one of the most important in Classical Thermodynamics. It is often called the **First Law of Thermodynamics for open thermodynamic systems**. Integrating the relationship gives the change of internal energy of an open system during its operation that lasts $\Delta t = t_2 - t_1$ (as before, t_1 and t_2 are the instances at which the matter enters and leaves the system, respectively).

The practical rates $\dot{H}_k$, $\dot{H}_j$, $\dot{E}_{\text{kin},k}$, $\dot{E}_{\text{kin},j}$, $\dot{E}_{\text{pot},k}$ and $\dot{E}_{\text{pot},j}$ are often expressed in terms of their specific values. (A specific value of a system property is its value per unit of mass (or per unit of volume) of the system matter.) The specific properties are usually denoted in thermodynamic literature by lowercase letters of the alphabet. Thus, relationship (3.34) can be rewritten as

$$\frac{\delta \mathcal{E}}{dt} \simeq \sum_k \left[\dot{M}_k \left(h_k + \frac{\omega_k^2}{2} + g z_k \right) \right] - \sum_j \left[\dot{M}_j \left(h_j + \frac{\omega_j^2}{2} + g z_j \right) \right]$$
$$+ \dot{L} + \dot{Q}, \tag{3.35}$$

where h_k and ω_k are specific enthalpy (in J/kg) and speed (in m/s), respectively, of the fluid entering the system through its kth inlet; h_j and ω_j are specific enthalpy and speed, respectively, of the fluid leaving the system through its jth outlet; z_k and z_j (both in meters) are the elevations of the kth inlet and jth outlet, respectively; $g = 9.81$ m/s^2 is the gravitational acceleration; $\omega_k^2/2$ is the specific kinetic energy (in J/kg), and gz_k is the specific potential energy (in J/kg) of the fluid moving through the kth inlet; $\omega_j^2/2$ is the specific kinetic energy and gz_j is the specific potential energy of the fluid moving through the jth outlet of the system.

Introducing a state property called **specific stagnation enthalpy**,

$$h^s = h + \frac{\omega^2}{2}, \tag{3.36}$$

relationship (3.35) gives

$$\frac{\delta\mathcal{E}}{dt} = \sum_k \left[\dot{M}_k \left(h_k^s + gz_k \right) \right] - \sum_j \left[\dot{M}_j \left(h_j^s + gz_j \right) \right] + \dot{L} + \dot{Q}, \tag{3.37}$$

where h_k^s and h_j^s are specific stagnation enthalpies of the flowing matter entering the system through the kth inlet and leaving it through the jth outlet, respectively.

Relationship (3.35) can be rewritten as

$$\frac{\delta\mathcal{E}}{dt} \simeq -\Delta\dot{H} - \Delta\dot{E}_{\text{kin}} - \Delta\dot{E}_{\text{pot}} + \dot{L} + \dot{Q}, \tag{3.38}$$

where

$$\Delta\dot{H} = \sum_j \dot{M}_j h_j - \sum_k \dot{M}_k h_k \tag{3.39}$$

is the rate of change of the system enthalpy,

$$\Delta\dot{E}_{\text{kin}} = \sum_j \dot{M}_j \frac{\omega_j^2}{2} - \sum_k \dot{M}_k \frac{\omega_k^2}{2} \tag{3.40}$$

is the rate of change of the system kinetic energy, and

$$\Delta\dot{E}_{\text{pot}} = \sum_j \dot{M}_j gz_j - \sum_k \dot{M}_k gz_k \tag{3.41}$$

is the rate of change of the system potential energy during the process.

One should notice that the **properties of *steady flowing* matter may change with the flow position but not with time**. Therefore, the total energy, the control volume, and the mass of the matter inside the open system's control volume remain constant, and the boundary work is zero. As a result, in steady flow one has $\delta \mathcal{E}/dt = 0$ and $\dot{W} = 0$, the only work transfer is non-boundary work K (see discussion in Section 2.8.2, and the total power transfer is

$$\dot{L} = \dot{K}. \tag{3.42}$$

Thus, according to expression (3.38), the **Energy Balance Equation for a steady-flow open system is**

$$\Delta \dot{H} + \Delta \dot{E}_{kin} + \Delta \dot{E}_{pot} = \dot{K} + \dot{Q}, \tag{3.43}$$

or, when non-boundary work transfer is absent and changes of the flow kinetic and potential energies are negligible,

$$\Delta \dot{H} = \dot{Q}. \tag{3.44}$$

When an open steady-flow system has only one inlet and only one outlet (a situation common in thermodynamic applications), the Principle of Conservation of Matter allows one to write (see expression (3.23))

$$\dot{M}_1 = \dot{M}_2 = \dot{M}, \tag{3.45}$$

where $\dot{M}_1$ is the flow rate of the matter that enters the system through the inlet (the beginning of the process), and $\dot{M}_2$ is the flow rate of the matter that leaves the system through the outlet (the end of the process). The energy balance (3.43) for such a system can be rewritten as

$$\Delta \dot{H} + \Delta \dot{E}_{kin} + \Delta \dot{E}_{pot} = \dot{K} + \dot{Q}, \tag{3.46}$$

with

$$\Delta \dot{H} = \dot{M}(h_2 - h_1), \tag{3.47}$$

$$\Delta \dot{E}_{kin} = \dot{M}\left(\frac{\omega_2^2}{2} - \frac{\omega_1^2}{2}\right), \tag{3.48}$$

and

$$\Delta \dot{E}_{pot} = \dot{M}g\left(z_2 - z_1\right). \tag{3.49}$$

When flow of matter in and out of an open system is eliminated, it becomes a closed system. Then the system internal energy no longer depends on the thermal, kinetic, and potential energies of the matter *entering* and *leaving* the system but it still depends on the thermal, kinetic, and potential energies of the matter present, and possibly moving, inside the closed system. Subsequently, relationship (3.35) reduces to

$$\frac{\delta \mathcal{E}}{dt} = \dot{L} + \dot{Q}, \tag{3.50}$$

but is usually given in integrated form where the change of the closed system internal energy $\mathcal{E}$ during a thermodynamic transition from a state 1 to a state 2 is

$$\Delta \mathcal{E} = L + Q, \tag{3.51}$$

where, according to discussion in Section 2.8.2, the total work of the transition is

$$L = W + K, \tag{3.52}$$

where W is the boundary work and K is the non-boundary work of the process.

The matter moving *inside* a closed system has a thermal energy U (the letter U is frequently used in thermodynamic literature to denote system thermal energy), a kinetic energy E_{kin}, and a potential energy E_{pot}. (As already mentioned, the energy of the matter rotating around its center of mass is usually negligible in most thermodynamic applications.) Thus, the internal energy of the closed system at a given instant can be given as (see expression (2.4))

$$\mathcal{E} = U + E_{\text{kin}} + E_{\text{pot}}, \tag{3.53}$$

and the Energy Balance Equation of the system during its transition from a state 1 to a state 2 is

$$\Delta \mathcal{E} = \mathcal{E}_2 - \mathcal{E}_1 = \Delta U + \Delta E_{\text{kin}} + \Delta E_{\text{pot}} = W + K + Q, \tag{3.54}$$

where

$$\Delta U = U_2 - U_1, \tag{3.55}$$

$$\Delta E_{\text{kin}} = E_{\text{kin},2} - E_{\text{kin},1}, \tag{3.56}$$

and

$$\Delta E_{\text{pot}} = E_{\text{pot},2} - E_{\text{pot},1}. \tag{3.57}$$

Relationship (3.54) is the **First Law of Thermodynamics for processes in closed systems with no radiation transfer**.

In many thermodynamic applications, the matter locked up in closed systems does not experience meaningful changes in kinetic and potential energies. (As mentioned earlier, the closed systems where the movement of matter can be ignored are called 'nonflow systems'.) The internal energy of such a nonflow system is just its thermal energy U, and expression (3.51) can be written as (see expression (2.52))

$$\Delta U = L + Q = W + K + Q. \qquad (3.58)$$

This relationship is called the **First Law of Thermodynamics for processes in any closed system where changes of the kinetic and potential energies of the matter locked up in the system are negligible and where exchange of radiation between the system and the environment is negligible**.

Since no energy can be exchanged between an *isolated* system and the environment, the internal energy of every such system is constant during every process taking place in the system,

$$\Delta \mathcal{E} = 0 \quad \rightarrow \quad \mathcal{E} = \text{const} \qquad (3.59)$$

or, when movement of the matter inside the system can be ignored,

$$\Delta U = 0 \quad \rightarrow \quad U = \text{const.} \qquad (3.60)$$

Expression (3.54) does not show explicitly the dependence of $\Delta \mathcal{E}$ in processes in closed systems on enthalpy change ΔH. This dependence can be recovered by using the definition (3.32) and expression (3.58),

$$\Delta H = P\Delta V + V\Delta P + L + Q. \qquad (3.61)$$

When the considered thermodynamic transition is an *isobaric* process ($P = \text{const}, \Delta P = 0$) where the only work transfer is a *reversible* transfer of boundary work (then $L = W = -P\Delta V$ – see expressions (2.52) and (2.59), and discussion in Section 2.8.2), relationship (3.61) becomes

$$\Delta H = Q. \qquad (3.62)$$

3.5.3 Balance Equations for Entropy

Balance of entropy in an open thermodynamic system (with no radiation transfer) that has several inlets, outlets, and ports can be given as (see Section 2.3)

$$\frac{\delta S}{dt} = \sum_k \left(\frac{\delta S_m}{dt}\right)_k - \sum_j \left(\frac{\delta S_m}{dt}\right)_j$$
$$+ \sum_l \left(\frac{\delta S_h}{dt}\right)_l - \sum_p \left(\frac{\delta S_h}{dt}\right)_p + \frac{\delta S_{nat}}{dt}, \tag{3.63}$$

where $\delta S/dt$ is the theoretical rate of change of the system entropy, $(\delta S_m/dt)_k$ is the rate at which entropy of matter enters the system with the matter through the kth inlet, and $(\delta S_m/dt)_j$ is the rate at which entropy of matter leaves the system through the jth outlet; $(\delta S_h/dt)_l$ is the rate at which heat entropy enters the system (with heat) through the lth port, $(\delta S_h/dt)_p$ is the rate at which heat entropy leaves the system through the pth port, and $\delta S_{nat}/dt$ is the rate of production of the natural entropy inside the system that contains irreversibilities (this rate is always greater than zero in all real transitions regardless of their characters); sums $\sum_k$ and $\sum_j$ are taken over all the system's inlets and outlets through which matter enters and leaves the system, respectively, and sums $\sum_l$ and $\sum_p$ are taken over all the ports through which heat enters and leaves the system, respectively (see Fig. 3.1).

The reader should notice that we did not introduce any special convention for the signs of terms in relationship (3.63). This is because the entropies S_m, S_h, S_{nat}, and S are always defined as positive quantities (positive numbers) – see Section 2.3. Subsequently, the directions of the entropic flows are 'remembered' in the relationship by the plus signs (when entropy enters the system) or minus signs (when entropy leaves the system) located in front of the terms.

In general, not only the natural entropy S_{nat}, but also the partial entropies S_m and S_h can behave like functions of path in a process described by relationship (3.63). Note that the entropy S_h is always carried by heat which is a path property. Also, a partial entropy can enter and/or leave a thermodynamic system at the instants when the system is in some non-equilibrium states located between the initial and final states of the entire process under consideration. However, when the initial and final states of the entire process are equilibrium states (or, more precisely, local-equilibrium states, since we are discussing now an open system), then: 1) Classical Thermodynamics can provide an accurate description of thermodynamic properties of such states, and 2) the system's *total* entropy S behaves during the process like a function of state even when the transfers of the partial entropies behave like functions of path. Subsequently, the differential of the total entropy is an exact differential dS and relationship (3.63) can be rewritten as

$$\frac{dS}{dt} = \sum_k \left(\frac{\delta S_m}{dt}\right)_k - \sum_j \left(\frac{\delta S_m}{dt}\right)_j$$
$$+ \sum_l \left(\frac{\delta S_h}{dt}\right)_l - \sum_p \left(\frac{\delta S_h}{dt}\right)_p + \frac{\delta S_{nat}}{dt}, \qquad (3.64)$$

where the rates of the partial entropies associated with the heats passing the system ports at a given time are (see expression (2.11))

$$\left(\frac{\delta S_h}{dt}\right)_l = \frac{1}{T_l}\frac{\delta Q_l}{dt} \qquad (3.65)$$

and

$$\left(\frac{\delta S_h}{dt}\right)_p = \frac{1}{T_p}\frac{|\delta Q_p|}{dt}, \qquad (3.66)$$

where T_l and T_p are temperatures of the lth and pth ports at that time, and δQ_l and δQ_p are elementary heats that are passing through the ports at the moment. Note that Q_p is a negative number because heat Q_p leaves the system (see Section 2.8.2).

We derive below some expressions for the change of the system's total entropy during thermodynamic processes where the initial and final states are equilibrium, quasi-equilibrium or local-equilibrium states. Two cases are considered: 1) the values of S_1 (the entropy of the process's initial state) and S_2 (the entropy of the process's final state) are already known (from some measurements or calculations – see Chapter 8), and 2) the entropies S_1 and S_2 are not known but thermodynamic variables (pressure, temperature, etc.) of the states 1 and 2 are known.

In case 1, calculating the change of the system entropy is very easy because then (see Section 2.8.1 and expression (2.39))

$$\Delta S = S_2 - S_1. \qquad (3.67)$$

The situation is more complicated in case 2 where the entropies of the system in its initial and final states are not known. Then, the change of the system entropy during the transition can be given as a function f_s (see relationship (2.38)) obtainable from the following integral:

$$\Delta S = \int_{t_1}^{t_2} \frac{dS}{dt}\, dt, \tag{3.68}$$

where the integrand dS/dt is given in expression (3.64), and as before, t_1 and t_2 are the times at which the transition begins and ends, respectively.

Applying relationships (3.7) and (3.64) to an open-system transition between two states of local equilibrium one obtains

$$\frac{dS}{dt} \simeq \sum_k (\dot{S}_m)_k - \sum_j (\dot{S}_m)_j + \sum_l (\dot{S}_h)_l - \sum_p (\dot{S}_h)_p + \dot{S}_{\text{nat}}, \tag{3.69}$$

where all rates are practical rates at a given time t; $(\dot{S}_m)_k$ is the rate at which entropy of matter enters the system through its kth inlet, and $(\dot{S}_m)_j$ is the rate at which entropy of matter leaves the system through its jth outlet; $(\dot{S}_h)_l$ is the rate at which heat entropy enters the system through its lth port, $(\dot{S}_h)_p$ is the rate at which the heat entropy leaves the system through its pth port, and $\dot{S}_{\text{nat}}$ is the rate at which the natural entropy is produced in the system.

The rates $(\dot{S}_m)_k$ and $(\dot{S}_m)_j$ can be given as

$$(\dot{S}_m)_k = \dot{M}_k s_k \qquad \text{and} \qquad (\dot{S}_m)_j = \dot{M}_j s_j, \tag{3.70}$$

where $\dot{M}_k$ is the flow rate at which matter enters the system through its kth inlet, and $\dot{M}_j$ is the flow rate at which matter leaves the system through its jth outlet; s_k is the specific entropy of the matter entering the system through the kth inlet, and s_j is the specific entropy of the matter leaving the system through its jth outlet.

The rates $(\dot{S}_h)_l$ and $(\dot{S}_h)_p$ are (see expressions (3.65) and (3.66))

$$(\dot{S}_h)_l = \frac{\dot{Q}_l}{T_l} \qquad \text{and} \qquad (\dot{S}_h)_p = \frac{|\dot{Q}_p|}{T_p}, \tag{3.71}$$

where $\dot{Q}_i$ is the rate at which heat Q_i passes through the ith port of temperature T_i.

One should notice that in isentropic flow ($dS/dt = 0$), relationship (3.69) becomes

$$\sum_{k} \dot{M}_k s_k - \sum_{j} \dot{M}_j s_j + \sum_{l} \frac{\dot{Q}_l}{T_l} - \sum_{p} \frac{|\dot{Q}_p|}{T_p} + \dot{S}_{\text{nat}} = 0. \tag{3.72}$$

When a thermodynamic transition in an open system is a finite process that starts (at time t_1) in a local-equilibrium state 1 and ends (at time t_2) in a local-equilibrium state 2, the change of the system's entropy during the entire transition is

$$\Delta S = \sum_{k} \int_{t_1}^{t_2} \dot{M}_k s_k dt - \sum_{j} \int_{t_1}^{t_2} \dot{M}_j s_j dt$$

$$+ \sum_{l} \int_{t_1}^{t_2} \frac{\dot{Q}_l}{T_l} dt - \sum_{p} \int_{t_1}^{t_2} \frac{|\dot{Q}_p|}{T_p} dt + \int_{t_1}^{t_2} \dot{S}_{\text{nat}} dt. \tag{3.73}$$

When the system under consideration is a *closed* system (no transfer of matter between the system and the environment is possible), relationship (3.73) reduces to

$$\Delta S = \sum_{l} \int_{t_1}^{t_2} \frac{\dot{Q}_l}{T_l} dt - \sum_{p} \int_{t_1}^{t_2} \frac{|\dot{Q}_p|}{T_p} dt + \int_{t_1}^{t_2} \dot{S}_{\text{nat}} dt, \tag{3.74}$$

and in an *isolated* system, to

$$\Delta S = \Delta S_{\text{nat}} = \int_{t_1}^{t_2} \dot{S}_{\text{nat}} dt. \tag{3.75}$$

The amount of natural entropy produced during a transition in an *open* system can be obtained from expression (3.73) as

$$\Delta S_{\text{nat}} = \Delta S - \sum_{k} \int_{t_1}^{t_2} \dot{M}_k s_k dt + \sum_{j} \int_{t_1}^{t_2} \dot{M}_j s_j dt$$

$$- \sum_{l} \int_{t_1}^{t_2} \frac{\dot{Q}_l}{T_l} dt + \sum_{p} \int_{t_1}^{t_2} \frac{|\dot{Q}_p|}{T_p} dt, \tag{3.76}$$

while in a *closed* system the amount is

$$\Delta S_{\text{nat}} = \Delta S - \sum_l \int_{t_1}^{t_2} \frac{\dot{Q}_l}{T_l} dt + \sum_p \int_{t_1}^{t_2} \frac{|\dot{Q}_p|}{T_p} dt, \qquad (3.77)$$

and in an *isolated* system one has

$$\Delta S_{\text{nat}} = \Delta S. \qquad (3.78)$$

The Principle of Production of Natural Entropy requires that $\Delta S_{\text{nat}} > 0$ in all real (that is, irreversible) processes. Thus, such processes can be realized in practice only when (see expressions (3.76)–(3.78)):

1) open systems:

$$\Delta S - \sum_k \int_{t_1}^{t_2} \dot{M}_k s_k dt + \sum_j \int_{t_1}^{t_2} \dot{M}_j s_j dt$$

$$- \sum_l \int_{t_1}^{t_2} \frac{\dot{Q}_l}{T_l} dt + \sum_p \int_{t_1}^{t_2} \frac{|\dot{Q}_p|}{T_p} dt > 0; \qquad (3.79)$$

2) closed systems:

$$\Delta S - \sum_l \int_{t_1}^{t_2} \frac{\dot{Q}_l}{T_l} dt + \sum_p \int_{t_1}^{t_2} \frac{|\dot{Q}_p|}{T_p} dt > 0; \qquad (3.80)$$

3) isolated systems:

$$\Delta S > 0. \qquad (3.81)$$

Inequalities (3.79)–(3.81) result from the Second Law of Thermodynamics which requires that they must be true in all thermodynamic processes that are realizable in practice. If inequality (3.79), (3.80), or (3.81) for a studied process is weak, then production of the natural entropy in the process is small and, therefore, the process's dissipation effects are small. **Any thermodynamic process where production of natural entropy is insignificant can be treated as a reversible process**. More discussion on the entropic properties of irreversible processes can be found in Chapter 13.

3.5.4 Summary

Let us summarize the important general relationships that describe the transfers of matter, energy, and entropy in real thermodynamic processes in open, closed, and isolated systems:

(1) Conservation of matter is described by expressions (3.20) (open systems) and (3.25) (closed and isolated systems).

(2) Conservation of energy is described by expressions (3.35) (open systems), (3.51) (closed systems), and (3.59) (isolated systems).

(3) There is always a production of natural entropy in all real processes in all systems, a result of the presence of irreversibilities in such processes. This production is zero in reversible thermodynamic transitions which, therefore, are convenient theoretical models of some real processes.

(4) The amount of the natural entropy produced in real processes is given in expressions (3.76) (open systems), (3.77) (closed systems), and (3.78) (isolated systems).

(5) The amount of the natural entropy produced during irreversible processes is usually proportional to the first and second gradients of the thermodynamic variables during the processes. Therefore, the processes where the gradients are small can often be treated as reversible.

(6) The elementary heat entropy δS_h 'carried' by elementary heat δQ through a system's port of temperature T is $\delta S_h = \delta Q/T$ (see expression (2.11)).

(7) The change $\Delta S = S_2 - S_1$ of the system's total entropy in any real (irreversible) process between two equilibrium (quasi-equilibrium, local-equilibrium) states is the same as the change would be if the process was a reversible one. This is because the total entropy is a state property, that is, its change during any process between two such states is independent of the character of the process. (Also see discussion of relationships (2.17)–(2.18).) Similar remarks can be made about changes of thermal energy U and enthalpy H (both are properties of state) in thermodynamic processes between equilibrium (quasi-equilibrium, local-equilibrium) states.

(8) Many studies of thermodynamic systems use the Balance Equations for matter and energy but ignore the entropic inequalities discussed in this section. This is because the partial entropies occurring in the inequalities are difficult to calculate within the formalism of Classical Thermodynamics. Fortunately, the entropic requirements of the Second

Law of Thermodynamics are often 'accidentally' met in many applications. Consequently, studies of such processes while using only the Principles of Conservation of Matter and Conservation of Energy give acceptable pictures of physical reality (see Problems 3.2 and 3.3).

(9) Expressions (3.20), (3.29), (3.34), (3.43), (3.54), and (3.76)–(3.81) are general and very important. Some of them will be transformed in the following chapters into easy-to-calculate expressions convenient in studying a great variety of processes occurring in thermodynamic applications.

3.6 Perpetuum Mobile

It has been one of the longest dreams of humanity to build an engine, called 'Perpetuum Mobile' (the Latin name for a perpetual-motion engine), that would work 'for free'. Three kinds of such a marvel were proposed in the past (and are still occasionally proposed even today). *Perpetuum Mobile of the First Kind* is a machine that would work continuously and cyclically by creating its own energy. *Perpetuum Mobile of the Second Kind* is a machine that would work continuously and cyclically by using the internal energy of only one heat reservoir (see Section 10.2). *Perpetuum Mobile of the Third Kind* is a machine that would work continuously and cyclically between two heat reservoirs with no dissipative effects.

Our discussions in the previous sections have shown that Perpetuum Mobile of the First Kind is forbidden by the First Law of Thermodynamics (the Principle of Conservation of Energy), while Perpetuum Mobile of the Second Kind and Perpetuum Mobile of the Third Kind are allowed by the First Law of Thermodynamics but forbidden by the Second Law of Thermodynamics (the Principle of Production of Natural Entropy).

Problem 3.1: Water is pumped steadily from a pipe inlet i located at the ground to the pipe outlet o located 30 m above the ground. The radius of the pipe at the inlet is $r_i = 1$ inch. The temperature, pressure, and mass density of the water at the inlet are $t_i = 20\,°C$, $P_i = 1\,$bar, and $\rho_i = 999\,kg/m^3$, respectively. The water enters the inlet with speed $\omega_i = 1\,m/s$ and leaves the outlet with speed $\omega_o = 5\,m/s$. The temperature and pressure of the water at the outlet are $t_o = 30\,°C$, and $P_o = 0.9\,$bar, respectively. During the pumping, the water is heated with rate $\dot{Q} = 5\,\text{kJ/s}$. Calculate the power (in kW and in HP) required by the pump to maintain the steady flow of the water. Ignore radiation transfer between the pipe and the environment, and assume that the water at the inlet and at the outlet is in some states of local equilibrium.

Solution: The energy balance (3.46) for the steady pumping process (the pipe and the flowing water form an open thermodynamic system) can be written as

$$\dot{K} = \dot{M}\left[(h_\text{o} - h_i) + \frac{\omega_\text{o}^2 - \omega_i^2}{2} + g(z_\text{o} - z_i)\right] - \dot{Q}, \tag{3.82}$$

where h_i and h_o are the specific enthalpies of the water in the local equilibria at the pipe's inlet and outlet, respectively, z_i and z_o are the elevations of the inlet and outlet, respectively, and $g = 9.81\,\text{m/s}^2$ is the gravitational acceleration.

The mass flow rates of the steadily-pumped water at the inlet and at the outlet are the same (see expressions (3.19) and (3.23)),

$$\dot{M} = \dot{M}_i = \dot{M}_\text{o} = \pi r_i^2 \omega_i \rho_i = \pi (0.0254\,\text{m})^2 (1\,\text{m/s})(999\,\text{kg/m}^3)$$
$$= 2.02\,\text{kg/s}. \tag{3.83}$$

Since we assume in the calculations that the flowing water at the inlet and at the outlet is in states of local equilibrium, the water properties (density, enthalpy, etc.) at these locations can be taken from a database of thermodynamic properties of water in equilibrium. According to the *NIST-JANAF Thermochemical Tables* (see discussion in Chapter 8), the equilibrium values of the specific enthalpies of water at the inlet and at the outlet are: $h_i = 84.03\,\text{kJ/kg}$ at $t_i = 20\,°\text{C}$ and $P_i = 1\,\text{bar}$, and $h_\text{o} = 102.12\,\text{kJ/kg}$ at $t_\text{o} = 30\,°\text{C}$ and $P_\text{o} = 0.9\,\text{bar}$. Thus,

$$h_\text{o} - h_i = 102.12\,\text{kJ/kg} - 84.03\,\text{kJ/kg} = 18.10\,\text{kJ/kg}, \tag{3.84}$$

$$\frac{\omega_\text{o}^2 - \omega_i^2}{2} = \frac{5^2 - 1^2}{2}\,\text{m}^2/\text{s}^2 = 0.012\,\text{kJ/kg}, \tag{3.85}$$

and

$$g(z_\text{o} - z_i) = 9.81\,\text{m/s}^2(30 - 0)\,\text{m} = 0.29\,\text{kJ/kg}. \tag{3.86}$$

Expression (3.82) gives the following rate $\dot{K}$ at which work is being supplied to the system (note that the rate is just the power needed to maintain the pumping process),

$$\dot{K} = (2.02 \, \text{kg/s})(18.10 + 0.012 + 0.29) \, \text{kJ/kg} - 5 \, \text{kJ/s}$$
$$= 32.17 \, \text{kW} = 43.14 \, \text{HP}. \tag{3.87}$$

The power $\dot{K}$ needed to pump the water from the inlet to the outlet in a steady manner has a positive value ($+32.17\,\text{kW}$) which is in agreement with the International Convention for the Sign of Energy – the shaft work was done *on* the system.

One should notice that the changes of the flow kinetic energy (expression (3.85)) as well as the changes of potential energy (expression (3.86)) are much smaller than the changes of the flow's enthalpy (expression (3.84)). Such a situation is common in technological applications.

Since we have not analyzed production of the natural entropy in the system here, we take a risk that the theoretically-described process of water pumping may be unrealizable in practice.

End of Problem 3.1.

Problem 3.2: A closed thermodynamic system (with negligible radiation transfer) undergoes a thermodynamic transition from an equilibrium state 1 (of the internal energy $\mathcal{E}_1 = 20\,\text{MJ}$ and the total entropy $S_1 = 6\,\text{kJ/K}$) to an equilibrium state 2 (of the internal energy $\mathcal{E}_2 = 15\,\text{MJ}$ and the total entropy $S_2 = 1\,\text{kJ/K}$) while producing 2 MJ of heat and doing 3 MJ of work. Can such a process proceed isothermally at temperature $T = 1000\,\text{K}$? Note that use of the symbol $\mathcal{E}$ for the system's internal energy indicates that the matter inside the system may be moving.

Solution: The process can be realized in practice only if none of the three Principles of Thermodynamics (the Principle of Conservation of Matter, the Principle of Conservation of Energy, and the Principle of Production of Natural Entropy) is violated during the process.

1) conservation of matter: since the system under consideration is a closed system, the amount of matter (mass) of the system is 'automatically' conserved ($M = \text{const}$).

2) conservation of energy: the First Law of Thermodynamics for the closed system can be used to verify whether the system's total energy is indeed

conserved during the process. According to the Law (see expression (3.54)),

$$\Delta \mathcal{E} = \mathcal{E}_2 - \mathcal{E}_1 = 15\,\text{MJ} - 20\,\text{MJ} = -5\,\text{MJ} \qquad (3.88)$$

regardless of whether the process is isothermal or not and regardless of whether this is a nonflow system or not. Thus, the energy of the system is indeed conserved because (see expression (3.51))

$$\Delta \mathcal{E} = Q + L \qquad \text{or} \qquad -5\,\text{MJ} = -2\,\text{MJ} + (-3\,\text{MJ}), \qquad (3.89)$$

where the signs are chosen in accordance with the International Convention for the Sign of Energy.

3) production of natural entropy: according to expression (3.80), the Principle of Production of Natural Entropy is not violated during an isothermal ($T = $ const) process if the following is true during the process:

$$\Delta S_{\text{nat}} = \Delta S + \frac{|Q|}{T} > 0. \qquad (3.90)$$

This criterion would be violated during this process because

$$\Delta S = S_2 - S_1 = 1\,\text{kJ/K} - 6\,\text{kJ/K} = -5\,\text{kJ/K}, \qquad (3.91)$$

and, subsequently,

$$\Delta S_{\text{nat}} = -5\,\text{kJ/K} + \frac{2\,\text{MJ}}{1000\,\text{K}} = -3\,\text{kJ/K} < 0. \qquad (3.92)$$

Thus, the discussed isothermal process cannot be realized in practice even though it would violate neither the Principle of Conservation of Matter nor the Principle of Conservation of Energy.

End of Problem 3.2.

Problem 3.3: We would like to build a separation chamber (shown schematically in Fig. 3.2) that operates in a steady manner taking in a stream of ideal diatomic nitrogen gas at pressure $P_o = 1.2$ bar and temperature $t_o = 600\,°$C and dividing it isentropically ($S = $ const) into two streams, each having the mass flow rate equal to half of the mass flow rate of the stream entering the chamber. One of the outgoing streams has its outlet pressure $P_1 = 1.1$ bar and temperature

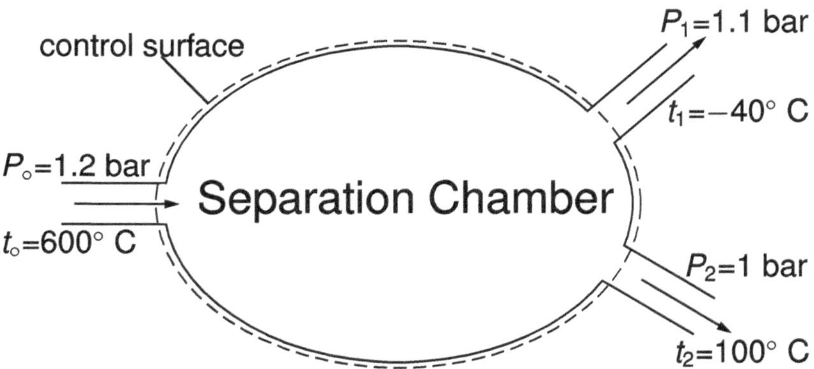

Fig. 3.2: A steady flow entering the separation chamber is divided inside the device into two streams that leave the device through two independent outlets.

$t_1 = -40\,^\circ\mathrm{C}$, and the other outgoing stream has its outlet pressure $P_2 = 1$ bar and temperature $t_2 = 100\,^\circ\mathrm{C}$. The gases at the inlet and at the outlets of the chamber are in some local-equilibrium states. Neither heat nor radiation are exchanged between the chamber and the environment, and no work is done on or done by the device. Can such an operation of the chamber be realized in practice?

Solution: The isentropic separation of the incoming stream into the two outgoing streams is possible if none of the three Principles of Thermodynamics (the Principle of Conservation of Matter, the Principle of Conservation of Energy, and the Principle of Production of Natural Entropy) is violated during the operation.

1) conservation of matter: since the separation chamber operates in a steady manner, one has (see expression (3.23))

$$\dot{M}_\mathrm{o} = \dot{M}_1 + \dot{M}_2. \tag{3.93}$$

where $\dot{M}_\mathrm{o}$ is the practical rate of the mass flow of the stream entering the chamber, and $\dot{M}_1$ and $\dot{M}_2$ are the practical rates of the mass flows through outlets 1 and 2, respectively.

Relationship (3.93) would indeed be true during the operation of the separation chamber because $\dot{M}_1 = \dot{M}_2 = \dot{M}_\mathrm{o}/2$. In other words, the operation would not violate the Principle of Conservation of Matter.

2) conservation of energy: the balance of the energy inside the system defined by the chamber's control surface can be formulated in a straightforward way using expression (3.46). However, even if the energy of the device is conserved, it does not guarantee that the steady operation of the chamber can be realized in practice. Therefore, we skip the analysis of the chamber energy and move to the entropic analysis of the device to see if the Principle of Production of Natural Entropy would be violated during operation of the chamber. If the Principle is not violated, then we will go back to check the validity of the Principle of Conservation of Energy.

3) production of the natural entropy: the process in the chamber is expected to be an isentropic process. Therefore, one can write (see expression (3.72))

$$\dot{M}_\circ s_\circ - (\dot{M}_1 s_1 + \dot{M}_2 s_2) + \dot{S}_{nat} = 0, \tag{3.94}$$

where $s_\circ$ is the specific entropy of the matter entering the system, s_1 and s_2 are the specific entropies of the matters leaving the system through outlets 1 and 2, respectively, and $\dot{S}_{nat}$ is the practical rate of production of the natural entropy in the chamber during its operation. Solution of equation (3.94) gives

$$\dot{S}_{nat} = -\dot{M}_\circ s_\circ + (\dot{M}_\circ/2)s_1 + (\dot{M}_\circ/2)s_2 = (\dot{M}_\circ/2)(-2s_\circ + s_1 + s_2). \tag{3.95}$$

Since the gases at the inlet and at the outlets are in local-equilibrium states, the specific entropies of ideal diatomic nitrogen at the locations are $s_\circ = 7.944\,\text{kJ/kg·K}$, $s_1 = 6.567\,\text{kJ/kg·K}$, and $s_2 = 7.081\,\text{kJ/kg·K}$. (These values were obtained using expression (8.33) and the corresponding values of the standard-state entropies listed in the *NIST-JANAF Thermochemical Tables* discussed in Chapter 8.) Thus,

$$(-2s_\circ + s_1 + s_2) = (-2 \cdot 7.944 + 6.567 + 7.081)\,\text{kJ/kg·K}$$
$$= -2.240\,\text{kJ/kg·K}, \tag{3.96}$$

and the natural entropy S_{nat} would be steadily removed from the chamber during its operation. This would violate the Principle of Production of Natural Entropy (the Second Law of Thermodynamics). Therefore, the stream separation discussed in this problem cannot be realized in practice even though the Principle of Conservation of Matter and (possibly) the Principle of Conservation of Energy are not violated during the separation process.

End of Problem 3.3.

3.7 Analytical Thermodynamics

3.7.1 SC–System

Let us apply the Principles of Classical Thermodynamics and the corresponding Balance Equations for the internal energy $\mathcal{E}$, enthalpy H, and entropy S to a special kind of closed thermodynamic system called hereafter the **SC–System (Simple Closed System)**. The SC–System is defined as follows:

1) it is a closed nonflow system where motion of matter is ignored ($\mathcal{E} = U$);
2) it has no internal boundaries (rigid, impermeable, or adiabatic);
3) it consists of a single-component substance in a single phase;
4) no radiative energy enters or leaves the system;
5) the only kind of work that can be done on or done by the system is boundary work;
6) some heat is exchanged between the system and the environment;
7) the thermodynamic process taking place in the system is a reversible process that does not produce natural entropy (see Section 2.3.2), and begins and ends in states of equilibrium.

The differential changes of important properties of the SC–System during a reversible process between two states of equilibrium are (see expressions (2.11), (2.56), and (2.57)):

$$\delta R = 0, \tag{3.97}$$

$$\delta W = -PdV, \tag{3.98}$$

$$\delta K = 0, \tag{3.99}$$

$$\delta S_{ph} = 0, \tag{3.100}$$

$$\delta S_h = \frac{\delta Q}{T}, \tag{3.101}$$

$$\delta S_{\text{nat}} = 0, \tag{3.102}$$

$$dS = \delta S_{ph} + \delta S_h + \delta S_{\text{nat}} = \delta S_h = \frac{\delta Q}{T}, \tag{3.103}$$

and, according to the First Law of Thermodynamics (see expression (3.58)),

$$dU = \delta Q + \delta W = TdS - PdV, \tag{3.104}$$

where, as before, δR and δQ are the radiative energy and heat exchanged between the system and the environment, respectively, δW is the boundary work done on or done by the system, and δK is the non-boundary work of process; dU, δS_{ph}, δS_h, δS_{nat} and dS are the changes of the system's thermal energy, radiation entropy, heat entropy, natural entropy, and the total entropy, respectively. As before, P, V, and T are the system's pressure, volume, and absolute temperature, respectively.

Equation (3.104) includes very simple terms which allows for easy description of the system thermodynamics. The process's boundary work δW is present in the equation but non-boundary work and radiation energy are not. This is because only boundary work of a reversible process can be given by the simple expression (3.98) that includes easy-to-measure and easy-to-deal-with thermodynamic variables such as pressure and volume. Elementary heat δQ also is given by the simple expression (3.101). The simplicity of these convenient choices is preventing using the SC–System to study heat and work transfers (*properties of path*) in real processes. However, the choices allow one to derive general relationships for changes of the thermodynamic *properties of state* that are valid in *any* (reversible or not) process between equilibrium states. The first step leading to doing so is the derivation of the Gibbs Equations – see below.

Taking the above into account, one can write for the SC–System,

$$\frac{\delta Q}{T} = \frac{dU}{T} + \frac{PdV}{T}, \tag{3.105}$$

$$dS = \frac{dU}{T} + \frac{PdV}{T}, \tag{3.106}$$

and

$$dS = \delta S_h = \frac{\delta Q}{T}. \tag{3.107}$$

As already mentioned (see Section 2.3.2, and the discussion associated with relationships (2.17) and (2.18)), studies of the *Clausius Hypothesis* and *Clausius Inequality* have shown that the **changes of a system's entropy during all processes between two fixed equilibrium (quasi-equilibrium, local-equilibrium) states are identical because the system entropy is a property of state.**

3.7.2 The Gibbs Equations

Relationship (3.104) gives the exact differential of the **thermal energy** of the SC–System undergoing an elementary thermodynamic transition,

$$dU = T\,dS - P\,dV. \tag{3.108}$$

Adding $d(PV) = V\,dP + P\,dV$ to each side of equation (3.108) gives

$$dH = T\,dS + V\,dP, \tag{3.109}$$

where the thermodynamic function of state H is called the system **enthalpy**,

$$H = U + PV \quad \rightarrow \quad dH = dU + P\,dV + V\,dP. \tag{3.110}$$

Subtracting $d(TS) = S\,dT + T\,dS$ from each side of equation (3.108) yields

$$dA = -S\,dT - P\,dV, \tag{3.111}$$

where the thermodynamic function of state A (the **Helmholtz function** or the **Helmholtz free energy**) is defined as

$$A = U - TS \quad \rightarrow \quad dA = dU - T\,dS - S\,dT. \tag{3.112}$$

Finally, subtracting $d(TS)$ from each side of equation (3.109) leads to

$$dG = -S\,dT + V\,dP, \tag{3.113}$$

where the thermodynamic function of state G (the **Gibbs function** or the **Gibbs free energy**) is defined as

$$G = H - TS \quad \rightarrow \quad dG = dH - T\,dS - S\,dT. \tag{3.114}$$

Clausius *Fundamental Hypothesis* allows one to treat the system thermal energy U and entropy S as properties of state in processes that begin and end in states of equilibrium. Therefore, the enthalpy H, the Helmholtz free energy A, and the Gibbs free energy G also are functions of state in such processes. It should be noticed that all differentials (including dP, dV and dT) on the right-hand sides of definitions (3.110), (3.112), and (3.114) are exact differentials.

Relationships (3.108)–(3.114) are called the **Gibbs Equations**. They have been derived for *reversible* thermodynamic processes in a simple SC–System that moves between equilibrium (quasi-equilibrium, local-equilibrium) states. But, as already mentioned, the changes of the state properties during the processes have no 'memory' of what has happened between the processes' initial and final states. In other words, the changes are not influenced by the way in which heat, boundary work, and non-boundary work are transferred between the system and the environment. Also, the changes of the state properties do not depend on the details of the system's structure. Therefore, the magnitudes of the changes in *any* (reversible or not) process between equilibrium (quasi-equilibrium, local-equilibrium) states of *any* system are the same as they are in the SC–System moving between the same initial and final states. This makes the Gibbs Equations a very important part of Classical Thermodynamics.

The Gibbs Equations can be summarized in the following compact form:

$$dU = dU_{\text{irr}} = dU_{\text{rev}} = \quad T\,dS - P\,dV, \tag{3.115}$$

$$dH = dH_{\text{irr}} = dH_{\text{rev}} = \quad T\,dS + V\,dP, \tag{3.116}$$

$$dA = dA_{\text{irr}} = dA_{\text{rev}} = -S\,dT - P\,dV, \tag{3.117}$$

$$dG = dG_{\text{irr}} = dG_{\text{rev}} = -S\,dT + V\,dP, \tag{3.118}$$

where the subscripts *rev* and *irr* indicate the character of the process. Again, U is the thermal energy of the system,

$$H = U + PV, \tag{3.119}$$

$$A = U - TS, \tag{3.120}$$

and

$$G = H - TS. \tag{3.121}$$

Since relationships (3.115)–(3.118) do not include heat and work transfers, they cannot provide any information about the role of such transfers in thermo-dynamic processes. One can only make the following general statement: heat and work transfers in a reversible process differ from the corresponding transfers in the irreversible process between the same initial and final states,

$$\delta Q_{\text{irr}} \neq \delta Q_{\text{rev}}, \quad \text{and} \quad \delta L_{\text{irr}} \neq \delta L_{\text{rev}}, \tag{3.122}$$

and this is a result of the energy dissipation present in every real (that is, irreversible) process (see Section 13.1).

One should recall that the heat and work transfers in any process can be related by the Principle of Conservation of Energy (the First Law of Thermodynamics). For example, the Balance Equation for Energy of a real process moving a nonflow closed system between two states of equilibrium while exchanging heat δQ_{irr} and work δL_{irr} with the environment can be written, *without the need to specify the degree of the process's irreversibility*, as (see expression (3.58))

$$dU_{\text{irr}} = \delta Q_{\text{irr}} + \delta L_{\text{irr}}, \tag{3.123}$$

where dU_{irr} can be obtained from the Gibbs Equation (3.115). Of course, in any reversible process one has

$$dU_{\text{rev}} = \delta Q_{\text{rev}} + \delta L_{\text{rev}} = dU_{\text{irr}}. \tag{3.124}$$

3.7.3 SCR–System

The Gibbs Equations (3.115)–(3.118) can be generalized on the **Simple Closed Reactive System (SCR–System)**. Such a single-phase system is similar to the SC–System but consists of more than one component, and the number of mols of the αth component in the mixture is ν_α. The generalization includes adding sums $\sum_\alpha \mu_\alpha \delta \nu_\alpha$ to the right-hand sides of the Gibbs Equations. The resulting **Generalized Gibbs Equations** can be written as follows:

$$dU = T\,dS - P\,dV + \sum_\alpha \mu_\alpha \delta \nu_\alpha, \tag{3.125}$$

$$dH = T\,dS + V\,dP + \sum_\alpha \mu_\alpha \delta \nu_\alpha, \tag{3.126}$$

$$dA = -S\,dT - P\,dV + \sum_\alpha \mu_\alpha \delta \nu_\alpha, \tag{3.127}$$

and

$$dG = -S\,dT + V\,dP + \sum_\alpha \mu_\alpha \delta \nu_\alpha, \tag{3.128}$$

where the sums account for the thermochemical effects produced in the reactive system by the mol number changes $\delta \nu_\alpha$, and μ_α is the so-called **molar chemical potential** of the αth component.

3.7.4 Thermodynamic Relationships

The Gibbs Equations (3.115)–(3.118) are widely used in thermodynamic studies. In the equations, the thermodynamic *functions of state* (the thermal energy U, enthalpy H, the Helmholtz free energy A, the Gibbs free energy G) can be given as the following functions (the choice of the variables of a particular function of state is dictated by mathematical convenience that will be discussed later):

$$U = U(S, V), \tag{3.129}$$

$$H = H(S, P), \tag{3.130}$$

$$A = A(T, V), \tag{3.131}$$

$$G = G(T, P). \tag{3.132}$$

Using relationship (1.29),

$$df = \frac{\partial f(x, y, z, \ldots)}{\partial x} dx + \frac{\partial f(x, y, z, \ldots)}{\partial y} dy$$
$$+ \frac{\partial f(x, y, z, \ldots)}{\partial z} dz + \ldots, \tag{3.133}$$

the exact differentials dU, dH, dA, and dG of some amount of matter undergoing any (reversible or not) elementary thermodynamic transition between two equilibrium states can be given as

$$dU = \left(\frac{\partial U}{\partial S}\right)_V dS + \left(\frac{\partial U}{\partial V}\right)_S dV, \tag{3.134}$$

$$dH = \left(\frac{\partial H}{\partial S}\right)_P dS + \left(\frac{\partial H}{\partial P}\right)_S dP, \tag{3.135}$$

$$dA = \left(\frac{\partial A}{\partial T}\right)_V dT + \left(\frac{\partial A}{\partial V}\right)_T dV, \tag{3.136}$$

$$dG = \left(\frac{\partial G}{\partial T}\right)_P dT + \left(\frac{\partial G}{\partial P}\right)_T dP, \tag{3.137}$$

where the subscripts denote the variables to be kept constant during the corresponding differentiation.

Relationships (3.134)–(3.137) and (3.115)–(3.118) lead to

$$\left(\frac{\partial U}{\partial S}\right)_V = T \quad \text{and} \quad \left(\frac{\partial H}{\partial S}\right)_P = T, \tag{3.138}$$

$$\left(\frac{\partial U}{\partial V}\right)_S = -P \quad \text{and} \quad \left(\frac{\partial H}{\partial P}\right)_S = V, \tag{3.139}$$

$$\left(\frac{\partial A}{\partial V}\right)_T = -P, \quad \text{and} \quad \left(\frac{\partial G}{\partial P}\right)_T = V, \tag{3.140}$$

$$\left(\frac{\partial A}{\partial T}\right)_V = -S \quad \text{and} \quad \left(\frac{\partial G}{\partial T}\right)_P = -S, \tag{3.141}$$

and, using expressions (1.25)–(1.28), to

$$\left(\frac{\partial T}{\partial V}\right)_S = -\left(\frac{\partial P}{\partial S}\right)_V, \tag{3.142}$$

$$\left(\frac{\partial T}{\partial P}\right)_S = \left(\frac{\partial V}{\partial S}\right)_P, \tag{3.143}$$

$$\left(\frac{\partial P}{\partial T}\right)_V = \left(\frac{\partial S}{\partial V}\right)_T, \tag{3.144}$$

$$\left(\frac{\partial V}{\partial T}\right)_P = -\left(\frac{\partial S}{\partial P}\right)_T. \tag{3.145}$$

The last four expressions are called the **Maxwell Relationships**.

Expressions (3.138)–(3.145) are useful in many thermodynamic studies since they allow one to find difficult-to-calculate (or difficult-to-measure) changes of some system properties in terms of other (easy-to-calculate or easy-to-measure) properties.

Relationships (3.129)–(3.145) can be generalized on the SCR–System discussed in the previous section. Then, however, the number of thermodynamic variables describing the system is larger than the number of the variables in relationships (3.129)–(3.145). The required additional variables are the mol numbers v_1, v_2, v_3,... of the components present in the system. Subsequently, the internal energy, enthalpy, the Helmholtz free energy, and the Gibbs free energy can be written, respectively, as

$$U = U(S, V, v_1, v_2, ...), \tag{3.146}$$

$$H = H(S, P, v_1, v_2, \ldots), \tag{3.147}$$

$$A = A(T, V, v_1, v_2, \ldots), \tag{3.148}$$

and

$$G = G(T, P, v_1, v_2, \ldots). \tag{3.149}$$

Use of expressions (3.146)–(3.149) in relationship (3.133) gives the following differential changes of functions of state U, H, A, and G in any transition between two states of equilibrium:

$$dU = \left(\frac{\partial U}{\partial S}\right)_{V,v_1,v_2,\ldots} dS + \left(\frac{\partial U}{\partial V}\right)_{S,v_1,v_2,\ldots} dV + \sum_\alpha \left(\frac{\partial U}{\partial v_\alpha}\right)_{S,V,v_1,v_2,\ldots} dv_\alpha,$$

$$dH = \left(\frac{\partial H}{\partial S}\right)_{P,v_1,v_2,\ldots} dS + \left(\frac{\partial H}{\partial P}\right)_{S,v_1,v_2,\ldots} dP + \sum_\alpha \left(\frac{\partial H}{\partial v_\alpha}\right)_{S,P,v_1,v_2,\ldots} dv_\alpha,$$

$$dA = \left(\frac{\partial A}{\partial T}\right)_{V,v_1,v_2,\ldots} dT + \left(\frac{\partial A}{\partial V}\right)_{T,v_1,v_2,\ldots} dV + \sum_\alpha \left(\frac{\partial A}{\partial v_\alpha}\right)_{V,T,v_1,v_2,\ldots} dv_\alpha,$$

$$dG = \left(\frac{\partial G}{\partial T}\right)_{P,v_1,v_2,\ldots} dT + \left(\frac{\partial G}{\partial P}\right)_{T,v_1,v_2,\ldots} dP + \sum_\alpha \left(\frac{\partial G}{\partial v_\alpha}\right)_{T,P,v_1,v_2,\ldots} dv_\alpha,$$

$$\left(\frac{\partial U}{\partial S}\right)_{V,v_1,v_2,\ldots} = T \quad \text{and} \quad \left(\frac{\partial H}{\partial S}\right)_{P,v_1,v_2,\ldots} = T, \tag{3.150}$$

$$\left(\frac{\partial U}{\partial V}\right)_{S,v_1,v_2,\ldots} = -P \quad \text{and} \quad \left(\frac{\partial H}{\partial P}\right)_{S,v_1,v_2,\ldots} = V, \tag{3.151}$$

$$\left(\frac{\partial A}{\partial V}\right)_{T,v_1,v_2,\ldots} = -P \quad \text{and} \quad \left(\frac{\partial G}{\partial P}\right)_{T,v_1,v_2,\ldots} = V, \tag{3.152}$$

$$\left(\frac{\partial A}{\partial T}\right)_{V,v_1,v_2,\ldots} = -S \quad \text{and} \quad \left(\frac{\partial G}{\partial T}\right)_{P,v_1,v_2,\ldots} = -S, \tag{3.153}$$

$$\left(\frac{\partial T}{\partial V}\right)_{S,v_1,v_2,\dots} = -\left(\frac{\partial P}{\partial S}\right)_{V,v_1,v_2,\dots}, \tag{3.154}$$

$$\left(\frac{\partial T}{\partial P}\right)_{S,v_1,v_2,\dots} = \left(\frac{\partial V}{\partial S}\right)_{P,v_1,v_2,\dots}, \tag{3.155}$$

$$\left(\frac{\partial P}{\partial T}\right)_{V,v_1,v_2,\dots} = \left(\frac{\partial S}{\partial V}\right)_{T,v_1,v_2,\dots}, \tag{3.156}$$

$$\left(\frac{\partial V}{\partial T}\right)_{P,v_1,v_2,\dots} = -\left(\frac{\partial S}{\partial P}\right)_{T,v_1,v_2,\dots}. \tag{3.157}$$

Problem 3.4: A closed nonflow thermodynamic system is undergoing a reversible (quasi-static) transition between an equilibrium state 1 (of volume $V_1 = 1\,\mathrm{m}^3$ and temperature $T_1 = 300\,\mathrm{K}$) and an equilibrium state 2 (of volume $V_2 = 0.5\,\mathrm{m}^3$ and temperature $T_2 = 700\,\mathrm{K}$). The only work transferred between the system and the environment is boundary work. The change of the system's pressure P with its volume V during the transition is

$$P = 2 \times 10^5 V^{-1}, \tag{3.158}$$

where P and V are in pascals and cubic meters, respectively. The change of the system entropy S with temperature T during the transition is

$$S = 25 \times 10^{-4} T^2, \tag{3.159}$$

where S and T are in J/K and in kelvins, respectively.

When the system undergoes an irreversible transition between the same equilibrium states, the measured change of the system's pressure with its volume can be approximated by the following polynomial:

$$P = -1.6093279 \times 10^5 + 9.0526735 \times 10^5 V^{-1}$$
$$- 7.7624185 \times 10^5 V^{-2} + 2.3191528 \times 10^5 V^{-3}, \tag{3.160}$$

where, as before, P and V are in pascals and in cubic meters, respectively.

Find the work and heat transfers, and the changes of the system's thermal energy and entropy during: (a) the reversible transition between the equilibrium states 1 and 2, and (b) the irreversible transition between these states.

Solution:

a) the reversible transition:

The boundary work of the reversible transition is (see expression (2.59))

$$W_{\text{rev}} = -\int_{V_1}^{V_2} P(V)dV = -2\times 10^5 \int_{V_1}^{V_2} V^{-1}dV = -2\times 10^5 \ln(V_2/V_1)$$

$$= +138.629\,\text{kJ}. \tag{3.161}$$

The heat of the transition is (see expression (2.58))

$$Q_{\text{rev}} = \int_{S_1}^{S_2} T(S)dS. \tag{3.162}$$

According to expression (3.159), the system entropies at the beginning and at the end of the transition are, respectively,

$$S_1 = 0.225\,\text{kJ/K} \quad \text{and} \quad S_2 = 1.225\,\text{kJ/K}, \tag{3.163}$$

and the dependence of the system's temperature on its total entropy during the transition is

$$T = 20S^{1/2}. \tag{3.164}$$

Subsequently, relationship (3.162) gives

$$Q_{\text{rev}} = 20 \int_{S_1}^{S_2} S^{1/2}dS = (40/3)(S_2^{3/2} - S_1^{3/2}) = +16.655\,\text{kJ}. \tag{3.165}$$

The change of the system internal energy during the reversible transition can be obtained from the First Law of Thermodynamics (see expression (3.58)),

$$\Delta U_{\text{rev}} = W_{\text{rev}} + Q_{\text{rev}} = +138.629\,\text{kJ} + (+16.665\,\text{kJ})$$

$$= +155.284\,\text{kJ}, \tag{3.166}$$

and the change of the system entropy during the transition is

$$\Delta S_{\text{rev}} = S_2 - S_1 = +1\,\text{kJ/K}. \tag{3.167}$$

b) the irreversible transition:

Relationship (3.160) gives the dependence of the system pressure on its volume during the irreversible transition. But the dependence is useless for calculating the transition's boundary work because Classical Thermodynamics has not proposed any expression that allows one to accurately calculate such work even when the initial and final states of the transition are equilibrium states. Similarly, even if we had the dependence of the system temperature on its entropy during the irreversible transition, it would still be impossible to calculate the heat of the transition because Classical Thermodynamics is unable to accurately calculate the heat of such a process. However, the change of the system's thermal energy (a function of state) during the irreversible transition can be obtained easily because it is the same as the change of the internal energy in the corresponding reversible process that begins and ends at the same equilibrium states. In other words,

$$\Delta U_{\text{irr}} = \Delta U_{\text{rev}} = W_{\text{irr}} + Q_{\text{irr}} = W_{\text{rev}} + Q_{\text{rev}} = +155.284\,\text{kJ}. \tag{3.168}$$

Since entropy is a function of state in every (reversible or not) process where the initial and final states are equilibrium states, one has

$$\Delta S_{\text{irr}} = \Delta S_{\text{rev}} = +1\,\text{kJ/K}. \tag{3.169}$$

One should notice that even though $W_{\text{irr}} + Q_{\text{irr}} = W_{\text{rev}} + Q_{\text{rev}}$ in expression (3.168), there is no reason to claim that $W_{\text{irr}} = W_{\text{rev}}$ or $Q_{\text{irr}} = Q_{\text{rev}}$. In fact, these two equalities are never true because of the dissipation effects that are always present in real (that is, irreversible) processes. The difference between Q_{irr} and Q_{rev} is a measure of magnitude of the dissipation effect – see Sections 3.7.2 and 13.1. The fact that $Q_{\text{irr}} \neq Q_{\text{rev}}$ is sometimes hidden in entropic studies. For example, the change of entropy during any (reversible or not) process between given states of equilibrium is the same because entropy is a function of state. However, the contributions of the heat entropy and the natural entropy to the system's entropy change ΔS depend on the degree of irreversibility of the process (see expressions (2.17) and (2.18)).

End of Problem 3.4.

Problem 3.5: $M = 2\,\text{kg}$ of ideal diatomic nitrogen gas in thermodynamic equilibrium at pressure $P_1 = 1\,\text{Atm}$ have been moved isothermally and irreversibly to another equilibrium state where the gas pressure is $P_2 = 2\,\text{Atm}$. What is the change of the system's entropy during the process?

Solution: In principle, isothermal change of a system's entropy can be calculated from one of the following general (and equivalent) relationships:

$$\Delta S = \int_{t_1}^{t_2} \left(\frac{dS}{dt}\right)_T dt = \int_{P_1}^{P_2} \left(\frac{dS}{dP}\right)_T dP = \int_{V_1}^{V_2} \left(\frac{dS}{dV}\right)_T dV. \qquad (3.170)$$

These integrals are difficult to calculate because Classical Thermodynamics does not provide reliable expressions for the dependence of the system entropy S on time t, pressure P, or volume V. However, the problem can be significantly simplified because the transition under consideration takes place between two equilibrium states. Then, ΔS of this irreversible process is the same as it would be in the corresponding reversible process between the states 1 and 2. Therefore, one can use Maxwell's relationship (3.145) and write

$$\Delta S = \int_{P_1}^{P_2} \left(\frac{dS}{dP}\right)_T dP = - \int_{P_1}^{P_2} \left(\frac{dV}{dT}\right)_P dP, \qquad (3.171)$$

where the last integral can easily be calculated if the relationship between V, T and P during the process is known. (Such relationships are called the **Equations of State (EOS)** – see Chapter 5.) Fortunately, we do know the V–T–P relationships for many ideal gases in equilibrium, including ideal nitrogen gas. Subsequently, we can calculate the derivative $(dV/dT)_P$ using the equation of state for the ideal nitrogen. The equation is (see Section 5.1):

$$V = \frac{M R_g T}{P}, \qquad (3.172)$$

where M is the mass of the gas, and R_g (in kJ/kg·K) is the individual gas constant for diatomic nitrogen. Thus,

$$\left(\frac{dV}{dT}\right)_P = \frac{M R_g}{P}, \qquad (3.173)$$

and the change of the system entropy is

$$\Delta S = -M R_g \int_{P_1}^{P_2} \frac{dP}{P} = -M R_g \ln(P_2/P_1) = -412\,\text{J/K}. \qquad (3.174)$$

One can see that the term $-MR_g \ln(P_2/P_1)$ on the right-hand side of expression (3.174) is just the function $f_s(P_1, P_2, T = \text{const})$ – see expression (2.38) and the discussion in Section 2.8.1.

The reader should recall that ΔS for the discussed irreversible transition between equilibrium states could also be calculated in an easier way if the values of S_1 (the system entropy in its initial state) and S_2 (the system entropy in its final state) were already known. In such a case (see, for example, Section 2.8.1 and Problem 3.4)

$$\Delta S = S_2 - S_1. \tag{3.175}$$

The absolute entropies S_1 and S_2 for many single-component ideal gases in equilibrium in a broad range of physical conditions can be found in various databases such as the *NIST-JANAF Thermochemical Tables* (ed. M.W. Chase, Jr.) or *Thermodynamic Properties of Individual Substances* (ed. L.V. Gurvich, I.V. Veyts and C.B. Alcock) – see Chapter 8.

End of Problem 3.5.

Problem 3.6: Consider v mols of a single-phase pure substance in an equilibrium state. The substance equation of state (the relationship between the volume V, pressure P, and temperature T of the substance in a state of equilibrium – see Chapter 5) can be given as

$$\frac{PV^2}{T} = vRb, \tag{3.176}$$

where R and b are some positive constants. Find: (1) what will happen to the pressure of the substance when its temperature is *slightly* decreased in an isochoric way, and (2) what will happen to the volume of the substance when its temperature is *slightly* increased in an isobaric way?

Solution: According to the substance's equation of state (3.176), the change of the substance pressure with temperature during the isochoric ($V = \text{const}$) transition can be given by the following derivative:

$$\left(\frac{\partial P}{\partial T}\right)_V = \frac{vRb}{V^2}. \tag{3.177}$$

Because the system's temperature changes in processes (a) and (b) are very small, one can assume that $\Delta T \simeq dT$. Thus, since the right-hand side of relationship (3.177) is a positive number, a slight isochoric decrease of the system temperature ($dT < 0$) will cause a decrease of the system pressure ($dP < 0$).

The change of the substance volume with temperature during the isobaric ($P = $ const) transition of the system can be expressed by derivative $(\partial V/\partial T)_P$ which, according to expression (3.176), is

$$\left(\frac{\partial V}{\partial T}\right)_P = \frac{1}{2}\left(\frac{vRb}{PT}\right)^{1/2}. \tag{3.178}$$

Since the right-hand side of relationship (3.178) is a positive number, a slight isobaric increase of the system temperature ($dT > 0$) will cause an increase of the system volume ($dV > 0$).

End of Problem 3.6.

Problem 3.7: Derive an expression for the change of the specific internal energy of a single-component non-ideal gas undergoing a real (that is, irreversible) transition between an equilibrium state 1 and an equilibrium state 2 in a closed, nonflow system. Assume that the thermodynamic variables (pressure P, volume V, and absolute temperature T) of the gas in the equilibrium states are related by the so-called **Van der Waals Equation of State** (see expression (5.32)),

$$P = \frac{R_g T}{v - b_g} - \frac{a_g}{v^2}, \tag{3.179}$$

where v is the gas specific volume, R_g is the gas individual constant, and a_g and b_g are some positive constants representing molecular properties of the gas.

Solution: Since we discuss a closed nonflow system, the system's specific internal energy is just its specific thermal energy u. We can write (see expression (1.29)):

$$du = \left(\frac{\partial u}{\partial T}\right)_v dT + \left(\frac{\partial u}{\partial v}\right)_T dv, \tag{3.180}$$

where the first derivative is often called the constant-volume specific heat c_v (see Chapter 6),

$$c_v = (\partial u/\partial T)_v, \tag{3.181}$$

and, according to relationships (4.20) and (3.179),

$$\left(\frac{\partial u}{\partial v}\right)_T = T\left(\frac{\partial P}{\partial T}\right)_v - P = \frac{R_g T}{v - b_g} - P = \frac{a_g}{v^2}. \tag{3.182}$$

Taking the above into account one has

$$du = c_v dT + \frac{a_g}{v^2} dv, \tag{3.183}$$

or

$$u_2 - u_1 = \int_{T_1}^{T_2} c_v dT + a_g \int_{v_1}^{v_2} v^{-2} dv = \int_{T_1}^{T_2} c_v dT - a_g(v_2^{-1} - v_1^{-1}). \tag{3.184}$$

Thus, once the system's specific thermal energy of the system in an equilibrium state 1 is known, the value of the energy for any other equilibrium state 2 can be found if the system's equation of state and the constant-volume specific heat are known.

End of Problem 3.7.

Problem 3.8: Prove that the following relationship is true for any (reversible or not) thermodynamic transition between two equilibrium states of some amount of matter:

$$\left(\frac{\partial P}{\partial V}\right)_S = \frac{C_P}{C_V}\left(\frac{\partial P}{\partial V}\right)_T, \tag{3.185}$$

where P, T, V, and S are the system's pressure, temperature, volume, and entropy, respectively, and C_P and C_V are the system's heat capacities at constant pressure and constant volume, respectively.

Solution: Expressions (4.24) and (4.25) lead to

$$\frac{C_P}{C_V} = \frac{(\partial S/\partial T)_P}{(\partial S/\partial T)_V}. \tag{3.186}$$

In addition, (see Sections 1.4 and 3.7.4),

$$\left(\frac{\partial S}{\partial T}\right)_V \left(\frac{\partial V}{\partial S}\right)_T \left(\frac{\partial T}{\partial V}\right)_S = -1, \tag{3.187}$$

$$\left(\frac{\partial S}{\partial T}\right)_P \left(\frac{\partial P}{\partial S}\right)_T \left(\frac{\partial T}{\partial P}\right)_S = -1, \tag{3.188}$$

$$\left(\frac{\partial V}{\partial S}\right)_T = \left(\frac{\partial V}{\partial P}\right)_T \left(\frac{\partial P}{\partial S}\right)_T, \tag{3.189}$$

$$\left(\frac{\partial T}{\partial P}\right)_S = \left(\frac{\partial T}{\partial V}\right)_S \left(\frac{\partial V}{\partial P}\right)_S, \tag{3.190}$$

$$\left(\frac{\partial S}{\partial T}\right)_V \left(\frac{\partial V}{\partial P}\right)_T \left(\frac{\partial P}{\partial S}\right)_T \left(\frac{\partial T}{\partial V}\right)_S = -1, \tag{3.191}$$

and

$$\left(\frac{\partial S}{\partial T}\right)_P \left(\frac{\partial P}{\partial S}\right)_T \left(\frac{\partial T}{\partial V}\right)_S \left(\frac{\partial V}{\partial P}\right)_S = -1. \tag{3.192}$$

Thus,

$$\frac{(\partial S/\partial T)_P}{(\partial S/\partial T)_V} = \frac{(\partial V/\partial P)_T}{(\partial V/\partial P)_S} = \left(\frac{\partial V}{\partial P}\right)_T \left(\frac{\partial P}{\partial V}\right)_S. \tag{3.193}$$

Comparison of relationships (3.185), (3.186), and (3.193) validates relationship (3.185).

End of Problem 3.8.

Chapter 4

More on Changes of State Properties

4.1 General Remarks

It was shown in the previous chapter that the changes of state properties (internal energy, enthalpy, the Helmholtz free energy, and the Gibbs free energy) in any (reversible or not) thermodynamic process in any (closed or open) system can easily be calculated from general thermodynamic relationships if the initial and final states of the process are equilibrium, quasi-equilibrium or local-equilibrium states. (As mentioned earlier, Classical Thermodynamics treats the *differentially close* (with respect to thermodynamic variables) quasi-equilibrium and local-equilibrium states as equilibrium states.)

Study of path properties (transfers of work, heat, and radiation in thermodynamic systems) using the formalism of Classical Thermodynamics is difficult even in processes between states of equilibrium, because the formalism does not have general expressions for the transfers in real (irreversible) processes. In some cases, the transfers can be obtained from the Energy Balance Equations for the studied systems. In other cases, some of the transfers can be measured or obtained from theories other than Classical Thermodynamics. For example, the non-boundary (shaft) work supplied to a system by an electric motor usually is easy to determine with a good accuracy since the power of such a motor is listed by its manufacturer. Similarly, it is easy to determine the amount of heat released during a combustion process since the amount of heat produced by a unit of mass (or volume) of the fuel used in the process is usually listed by the fuel manufacturer. If the degree of irreversibility of the process of interest is not large, some of its properties of path can be estimated by the theory of reversible processes (see the discussions leading to expressions (2.58), (2.59), (2.71), and (2.72)).

4.2   Thermal Energy, Enthalpy, and Entropy

We derive below expressions for changes of several important thermodynamic properties of state (the thermal energy U, enthalpy H, entropy S, the Helmholtz free energy, the Gibbs free energy) of any system undergoing any transition between any two states of equilibrium, quasi-equilibrium, or local equilibrium. These expressions will be given in terms of measurable thermodynamic variables such as the system pressure P, temperature T, volume V, and the system's material property called the heat capacity.

Some thermodynamic variables describing matter in equilibrium are mathematically more convenient than other variables, and that only *two* such variables (P and T, V and T, or P and V) are needed to describe the thermal properties of a state in a *single-component, single-phase system in equilibrium* (see Section 9.1). The properties of state of such a system can be given as the following functions:

$$U = U(V,T), \tag{4.1}$$

$$H = H(P,T), \tag{4.2}$$

$$S = S(V,T) \qquad \text{or} \qquad S = S(P,T), \tag{4.3}$$

$$A = A(V,T), \tag{4.4}$$

and

$$G = G(P,T). \tag{4.5}$$

We will see below that our choice of the pair of the thermodynamic variables in functions (4.1)–(4.5) leads to significant simplifications of the mathematical formalism of thermodynamic studies.

Since properties (4.1)–(4.5) are functions of state, their differentials are exact differentials. Therefore, one can use general relationship (1.29) and write

$$dU = \left[\frac{\partial U(V,T)}{\partial T}\right]_V dT + \left[\frac{\partial U(V,T)}{\partial V}\right]_T dV, \tag{4.6}$$

$$dH = \left[\frac{\partial H(P,T)}{\partial T}\right]_P dT + \left[\frac{\partial H(P,T)}{\partial P}\right]_T dP, \tag{4.7}$$

$$dS = \left[\frac{\partial S(V,T)}{\partial T}\right]_V dT + \left[\frac{\partial S(V,T)}{\partial V}\right]_T dV \tag{4.8}$$

or

$$dS = \left[\frac{\partial S(P,T)}{\partial T}\right]_P dT + \left[\frac{\partial S(P,T)}{\partial P}\right]_T dP, \tag{4.9}$$

$$dA = \left[\frac{\partial A(V,T)}{\partial T}\right]_V dT + \left[\frac{\partial A(V,T)}{\partial V}\right]_T dV, \tag{4.10}$$

and

$$dG = \left[\frac{\partial G(P,T)}{\partial T}\right]_P dT + \left[\frac{\partial G(P,T)}{\partial P}\right]_T dP, \tag{4.11}$$

where, as before, the subscripts denote the variables to be held constant during the corresponding differentiating.

The volume V of some amount of matter in equilibrium can be replaced by the matter's specific volume $v = V/M$, where M is the mass of the matter. Then the *extensive* variable V can be replaced by the *intensive* variable v, and all three thermodynamic variables present in relationships (4.6)–(4.11) would become *intensive* variables T, P, and v, a situation that is convenient in many thermodynamic calculations.

When the properties U, H, S, A, and G are those of the *entire* matter of the equilibrium system under consideration, the corresponding *specific properties* (the values of the properties per unit of mass (kilogram, mol) of the matter) are usually denoted by lowercase letters,

$$u = U/M, \quad h = H/M, \quad s = S/M, \quad a = A/M, \quad g = G/M. \tag{4.12}$$

Subsequently, relationships (4.6)–(4.11) can be rewritten as

$$du = \left[\frac{\partial u(v,T)}{\partial T}\right]_v dT + \left[\frac{\partial u(v,T)}{\partial v}\right]_T dv, \tag{4.13}$$

$$dh = \left[\frac{\partial h(P,T)}{\partial T}\right]_P dT + \left[\frac{\partial h(P,T)}{\partial P}\right]_T dP, \tag{4.14}$$

$$ds = \left[\frac{\partial s(v,T)}{\partial T}\right]_v dT + \left[\frac{\partial s(v,T)}{\partial v}\right]_T dv \tag{4.15}$$

or

$$ds = \left[\frac{\partial s(P,T)}{\partial T}\right]_P dT + \left[\frac{\partial s(P,T)}{\partial P}\right]_T dP, \tag{4.16}$$

$$da = \left[\frac{\partial a(v,T)}{\partial T}\right]_v dT + \left[\frac{\partial a(v,T)}{\partial v}\right]_T dv, \tag{4.17}$$

and

$$dg = \left[\frac{\partial g(P,T)}{\partial T}\right]_P dT + \left[\frac{\partial g(P,T)}{\partial P}\right]_T dP, \tag{4.18}$$

where the matter's specific volume v can be replaced by the matter's mass density ρ (another intensive variable) because

$$\rho = M/V = 1/v. \tag{4.19}$$

4.2.1 Compressible Substances

Relationships (4.6)–(4.9) have to be transformed into a set of practical expressions giving the changes of the discussed state properties as functions of thermodynamic variables P, T, and V. The resulting functions are f_u, f_h, and f_s (the right-hand sides of relationships (2.34), (2.36), and (2.38), respectively). Once the changes ΔU, ΔH, and ΔS are known, the changes ΔA and ΔG can be obtained from expressions (3.112) and (3.114), respectively.

Expressions (4.6)–(4.9), (3.115)–(3.118), and (3.138)–(3.145) lead to the following very useful general relationships:

$$\left[\frac{\partial U(V,T)}{\partial V}\right]_T = T\left[\frac{\partial P(V,T)}{\partial T}\right]_V - P(V,T), \tag{4.20}$$

$$\left[\frac{\partial H(P,T)}{\partial P}\right]_T = -T\left[\frac{\partial V(P,T)}{\partial T}\right]_P + V(P,T), \tag{4.21}$$

$$\left[\frac{\partial S(V,T)}{\partial V}\right]_T = \left[\frac{\partial P(V,T)}{\partial T}\right]_V \tag{4.22}$$

and

$$\left[\frac{\partial S(P,T)}{\partial P}\right]_T = -\left[\frac{\partial V(P,T)}{\partial T}\right]_P. \tag{4.23}$$

In equations (4.6)–(4.9), two of the derivatives with respect to temperature can be written as the following thermodynamic functions:

$$C_V(V,T) = \left[\frac{\partial U(V,T)}{\partial T}\right]_V, \tag{4.24}$$

and

$$C_P(P,T) = \left[\frac{\partial H(P,T)}{\partial T}\right]_P. \tag{4.25}$$

These two derivatives are very important and widely used in Classical Thermodynamics; C_V is the **heat capacity of the entire matter present in a constant-volume system**, and C_P is the **heat capacity of the entire matter present in a constant-pressure system** (see Chapter 6).

Relationships (4.6)–(4.9) can now be written as

$$dU = C_V(V,T)dT + \left\{T\left[\frac{\partial P(V,T)}{\partial T}\right]_V - P(V,T)\right\}dV, \tag{4.26}$$

$$dH = C_P(P,T)dT + \left\{-T\left[\frac{\partial V(P,T)}{\partial T}\right]_P + V(P,T)\right\}dP, \tag{4.27}$$

$$dS = \frac{C_V(V,T)}{T}dT + \left[\frac{\partial P(V,T)}{\partial T}\right]_V dV \tag{4.28}$$

or

$$dS = \frac{C_P(P,T)}{T}dT - \left[\frac{\partial V(P,T)}{\partial T}\right]_P dP. \tag{4.29}$$

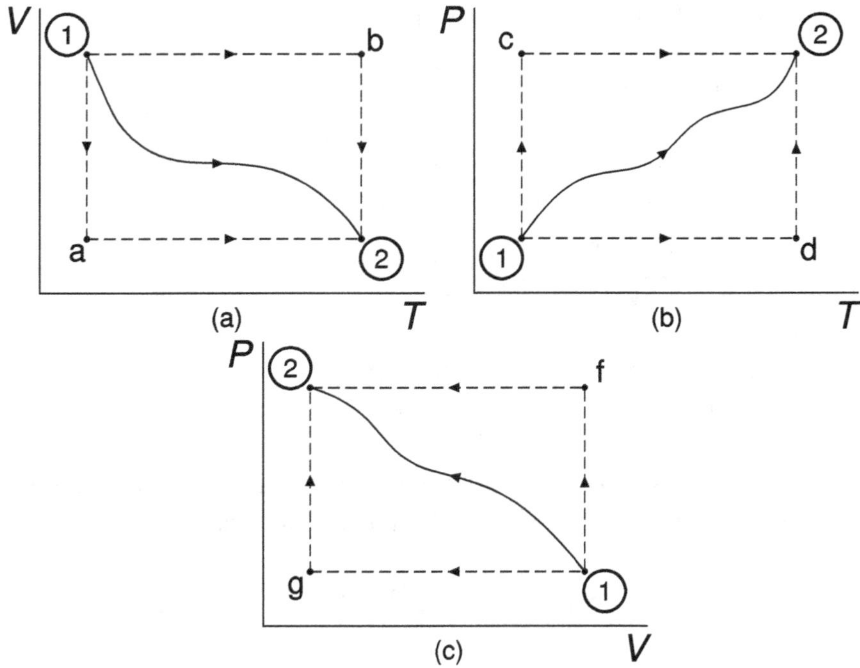

Fig. 4.1: In studies of thermodynamic functions of state, any thermodynamic transition (represented in the figure by the solid line) between an equilibrium state 1 and an equilibrium state 2 can be replaced by a sequence of arbitrary processes (the dashed lines) connecting the states 1 and 2. Examples of such replacements are sequences $1{\rightarrow}a{\rightarrow}2$ (isotherm followed by isochor), $1{\rightarrow}b{\rightarrow}2$ (isochor followed by isotherm), $1{\rightarrow}c{\rightarrow}2$ (isotherm followed by isobar), $1{\rightarrow}d{\rightarrow}2$ (isobar followed by isotherm), $1{\rightarrow}f{\rightarrow}2$ (isochor followed by isobar), and $1{\rightarrow}g{\rightarrow}2$ (isobar followed by isochor).

Let us study a real (irreversible) thermodynamic transition between two equilibrium states of a thermodynamic system. The thermodynamic path of the transition is shown on the V–T, P–T, and P–V diagrams in Fig. 4.1 as the solid line connecting states 1 and 2. Because changes of *thermal properties of state* U, H, and S have no 'memory' of what process has caused the changes, one can replace in the calculations the actual real $1{\rightarrow}2$ transition by a *computationally most convenient* reversible process connecting the states 1 and 2. The most convenient replacements of this kind are the reversible thermodynamic paths shown in the V–T, P–T, and P–V diagrams of Fig. 4.1 as dashed lines. Out of a large number of other possible replacements of the solid-line transition $1{\rightarrow}2$, the dashed-line paths are the most convenient because they proceed with constant values of some thermodynamic variables (V, P, or T).

As can be seen in expression (4.26), the change of the system thermal energy U (property of state) during *any* finite transition of the discussed system from its initial (equilibrium) state 1 to its final (equilibrium) state 2 can be given as the following line integral:

$$\Delta U = \int_{T_1,V_1}^{T_2,V_2} dU = \int_{T_1,V_1}^{T_2,V_2} C_V(V,T)dT$$

$$+ \int_{T_1,V_1}^{T_2,V_2} \left\{ T\left[\frac{\partial P(V,T)}{\partial T}\right]_V - P(V,T)\right\} dV. \tag{4.30}$$

Let us replace the $1{\rightarrow}2$ transition (the solid line in the V–T diagram shown in Fig. 4.1a) by the $1{\rightarrow}a{\rightarrow}2$ path consisting of two consecutive processes represented in the figure by the dashed-line isotherm $1{\rightarrow}a$ followed by the dashed-line isochor $a{\rightarrow}2$. (The path $1{\rightarrow}b{\rightarrow}2$ could have been chosen instead.) Then relationship (4.30) can be rewritten as ($T_a = T_1$, $V_a = V_2$)

$$\Delta U = \int_{T_1,V_1}^{T_a,V_a} dU + \int_{T_a,V_a}^{T_2,V_2} dU$$

$$= \int_{T_1,V_1}^{T_a,V_a} C_V(V,T)dT + \int_{T_1,V_1}^{T_a,V_a} \left\{ T\left[\frac{\partial P(V,T)}{\partial T}\right]_V - P(V,T)\right\} dV$$

$$+ \int_{T_a,V_a}^{T_2,V_2} C_V(V,T)dT + \int_{T_a,V_a}^{T_2,V_2} \left\{ T\left[\frac{\partial P(V,T)}{\partial T}\right]_V - P(V,T)\right\} dV$$

$$= \int_{T_1,V_1}^{T_1,V_2} C_V(V,T)dT + \int_{T_1,V_1}^{T_1,V_2} \left\{ T\left[\frac{\partial P(V,T)}{\partial T}\right]_V - P(V,T)\right\} dV$$

$$+ \int_{T_1,V_2}^{T_2,V_2} C_V(V,T)dT + \int_{T_1,V_2}^{T_2,V_2} \left\{ T\left[\frac{\partial P(V,T)}{\partial T}\right]_V - P(V,T)\right\} dV$$

$$= \int_{V_1}^{V_2} \left\{ T\left[\frac{\partial P(V,T)}{\partial T}\right]_V - P(V,T)\right\}_{T=T_1} dV$$

$$+ \int_{T_1}^{T_2} \{C_V(V,T)\}_{V=V_2}\, dT. \tag{4.31}$$

According to expression (4.27), the change of enthalpy (property of state) during *any* finite transition of the system from an equilibrium state 1 to an equilibrium state 2 can be given as the following line integral:

$$\Delta H = \int_{T_1,P_1}^{T_2,P_2} dH = \int_{T_1,P_1}^{T_2,P_2} C_P(P,T)\,dT$$

$$+ \int_{T_1,P_1}^{T_2,P_2} \left\{ -T \left[\frac{\partial V(P,T)}{\partial T} \right]_P + V(P,T) \right\} dP. \tag{4.32}$$

Let us replace the $1 \to 2$ transition (the solid line in the P–T diagram shown in Fig. 4.1b) by the $1 \to c \to 2$ path consisting of two consecutive processes represented in the figure by the dashed-line isotherm $1 \to c$ followed by the dashed-line isobar $c \to 2$ (the path $1 \to d \to 2$ could have been chosen instead.) Then relationship (4.32) can be written as ($T_c = T_1$, $P_c = P_2$)

$$\Delta H = \int_{T_1,P_1}^{T_c,P_c} dH + \int_{T_c,P_c}^{T_2,P_2} dH$$

$$= \int_{T_1,P_1}^{T_c,P_c} C_P(P,T)\,dT + \int_{T_1,P_1}^{T_c,P_c} \left\{ -T \left[\frac{\partial V(P,T)}{\partial T} \right]_P + V(P,T) \right\} dP$$

$$+ \int_{T_c,P_c}^{T_2,P_2} C_P(P,T)\,dT + \int_{T_c,P_c}^{T_2,P_2} \left\{ -T \left[\frac{\partial V(P,T)}{\partial T} \right]_P + V(P,T) \right\} dP$$

$$= \int_{T_1,P_1}^{T_1,P_2} C_P(P,T)\,dT + \int_{T_1,P_1}^{T_1,P_2} \left\{ -T \left[\frac{\partial V(P,T)}{\partial T} \right]_P + V(P,T) \right\} dP$$

$$+ \int_{T_1,P_2}^{T_2,P_2} C_P(P,T)\,dT + \int_{T_1,P_2}^{T_2,P_2} \left\{ -T \left[\frac{\partial V(P,T)}{\partial T} \right]_P + V(P,T) \right\} dP$$

$$= \int_{P_1}^{P_2} \left\{ -T \left[\frac{\partial V(P,T)}{\partial T} \right]_P + V(P,T) \right\}_{T=T_1} dP$$

$$+ \int_{T_1}^{T_2} \{ C_P(P,T) \}_{P=P_2}\,dT. \tag{4.33}$$

According to expressions (4.28) and (4.29), the change of the system entropy (property of state) during *any* finite transition from an equilibrium state 1 to an equilibrium state 2 can be given as the following line integrals:

$$\Delta S = \int_{T_1,V_1}^{T_2,V_2} dS = \int_{T_1,V_1}^{T_2,V_2} \frac{C_V(V,T)}{T}\,dT + \int_{T_1,V_1}^{T_2,V_2} \left[\frac{\partial P(V,T)}{\partial T} \right]_V dV \tag{4.34}$$

or

$$\Delta S = \int_{T_1,P_1}^{T_2,P_2} dS = \int_{T_1,P_1}^{T_2,P_2} \frac{C_P(P,T)}{T} dT - \int_{T_1,P_1}^{T_2,P_2} \left[\frac{\partial V(P,T)}{\partial T}\right]_P dP. \quad (4.35)$$

First, we derive an expression for the change of the system's entropy using relationship (4.34) where the entropy is given in terms of thermodynamic variables T and V. In order to do so, we replace the $1\rightarrow2$ transition (the solid line in the V–T diagram shown in Fig. 4.1a) by the path $1\rightarrow b\rightarrow2$ consisting of a sequence of two consecutive processes represented in the figure by the dashed-line isochor $1\rightarrow b$ followed by the dashed-line isotherm $b\rightarrow2$ (the path $1\rightarrow a\rightarrow2$ could have been chosen instead.) Then relationship (4.34) can be written as ($T_b = T_2$, $V_b = V_1$)

$$\begin{aligned}
\Delta S &= \int_{T_1,V_1}^{T_b,V_b} dS + \int_{T_b,V_b}^{T_2,V_2} dS \\
&= \int_{T_1,V_1}^{T_b,V_b} \frac{C_P(V,T)}{T} dT + \int_{T_1,V_1}^{T_b,V_b} \left[\frac{\partial P(V,T)}{\partial T}\right]_V dV \\
&\quad + \int_{T_b,V_b}^{T_2,V_2} \frac{C_P(V,T)}{T} dT + \int_{T_b,V_b}^{T_2,V_2} \left[\frac{\partial P(V,T)}{\partial T}\right]_V dV \\
&= \int_{T_1,V_1}^{T_2,V_1} \frac{C_P(V,T)}{T} dT + \int_{T_1,V_1}^{T_2,V_1} \left[\frac{\partial P(V,T)}{\partial T}\right]_V dV \\
&\quad + \int_{T_2,V_1}^{T_2,V_2} \frac{C_P(V,T)}{T} dT + \int_{T_2,V_1}^{T_2,V_2} \left[\frac{\partial P(V,T)}{\partial T}\right]_V dV \\
&= \int_{T_1}^{T_2} \left\{\frac{C_P(V,T)}{T}\right\}_{V=V_1} dT + \int_{V_1}^{V_2} \left\{\left[\frac{\partial P(V,T)}{\partial T}\right]_V\right\}_{T=T_2} dV. \quad (4.36)
\end{aligned}$$

The derivation of an expression for the change of the system's entropy can also be done using relationship (4.35) (where the entropy is given in terms of variables T and P) while replacing the $1\rightarrow2$ transition (the solid line in the P–T diagram shown in Fig. 4.1b) by the $1\rightarrow d\rightarrow2$ path consisting of a sequence of two consecutive processes represented in the figure by the dashed-line isobar $1\rightarrow d$ followed by the dashed-line isotherm $d\rightarrow2$ (the path $1\rightarrow c\rightarrow2$ could have been chosen instead.) Then relationship (4.35) can be written as ($T_d = T_2$,

$P_d = P_1$)

$$\Delta S = \int_{T_1,P_1}^{T_d,P_d} dS + \int_{T_d,P_d}^{T_2,P_2} dS$$

$$= \int_{T_1,P_1}^{T_d,P_d} \frac{C_P(P,T)}{T} dT - \int_{T_1,P_1}^{T_d,P_d} \left[\frac{\partial V(P,T)}{\partial T}\right]_P dP$$

$$+ \int_{T_d,P_d}^{T_2,P_2} \frac{C_P(P,T)}{T} dT - \int_{T_d,P_d}^{T_2,P_2} \left[\frac{\partial V(P,T)}{\partial T}\right]_P dP$$

$$= \int_{T_1,P_1}^{T_2,P_1} \frac{C_P(P,T)}{T} dT - \int_{T_1,P_1}^{T_2,P_1} \left[\frac{\partial V(P,T)}{\partial T}\right]_P dP$$

$$+ \int_{T_2,P_1}^{T_2,P_2} \frac{C_P(P,T)}{T} dT - \int_{T_2,P_1}^{T_2,P_2} \left[\frac{\partial V(P,T)}{\partial T}\right]_P dP$$

$$= \int_{T_1}^{T_2} \left\{\frac{C_P(P,T)}{T}\right\}_{P=P_1} dT - \int_{P_1}^{P_2} \left\{\left[\frac{\partial V(P,T)}{\partial T}\right]_P\right\}_{T=T_2} dP. \tag{4.37}$$

Relationships (4.31), (4.33), (4.36), and (4.37) tell us that **changes of the thermal properties of state during any transition between two equilibrium states of any system can be calculated in a straightforward way if the equation of state and heat capacities of the 'working' substance of the system are known.** One should notice that the integrands of the final integrals in the relationships must be calculated only for the initial and final states of the process under consideration, and that these states are in equilibrium. Therefore, use of the relationships requires knowledge of the equations of state and heat capacities of the system matter in equilibrium at P_1, T_1, V_1 and in equilibrium at P_2, T_2, V_2. Equations of states and heat capacities of substances in equilibrium are discussed in Chapters 5 and 6, respectively.

The right-hand sides of relationships (4.31), (4.33), and (4.37) are just the functions f_u, f_h, and f_s given in expressions (2.34), (2.36), and (2.38), respectively. If, however, the values of the system's thermal energy U_1 (of the initial state of the process taking place in the system) and the thermal energy U_2 (of the final state of the process) are already known (say, from Quantum Statistical Mechanics, or from some measurements), then the change of the system's thermal energy during the $1\rightarrow2$ process can be calculated more easily from expression (2.35) than from expression (4.31). Similarly, if the enthalpies H_1 and H_2 and entropies S_1 and S_2 are already known, the changes

ΔH and ΔS can be obtained much easier from expressions (2.37) and (2.39) than from expressions (4.33) and (4.36), respectively. Equilibrium values of the molar thermal energies, enthalpies, and entropies of many single-phase, single-component substances in a broad range of thermodynamic variables are available in literature (see, for example, the databases discussed in Chapter 8).

4.2.2 Incompressible Substances

In some thermodynamic processes, mass densities change insignificantly and, therefore, can be assumed constant. (The constancy of the mass density ρ of some fixed amount of matter guarantees the constancy of the matter's specific volume v – see expression (4.19).) Such substances (most liquids and solids) are called the **incompressible substances**. The assumption that $\rho = $ const and $v = 1/\rho = $ const significantly simplifies the expressions describing thermodynamic processes in incompressible matter. For example, the differential change of the specific thermal energy during an elementary thermodynamic transition between two equilibrium states of an incompressible matter can be given as (see expression (4.26) and Chapter 6)

$$du = c_v dT, \tag{4.38}$$

where $c_v = C_V/M$; C_V is the matter's **constant-volume heat capacity**, c_v is the corresponding **constant-volume specific heat capacity** (often called, somewhat confusingly, the **constant-volume specific heat**), and M is the mass of the matter. (As mentioned earlier, upper-case letters are usually used to denote thermodynamic properties of the entire matter under consideration while lower-case letters are used to denote the matter's *specific* properties, that is, the values of the properties per unit of mass.)

Since the specific enthalpy is defined as $h = u + Pv$ (see expression (3.119)), the constancy of v leads to

$$dh = du + v dP = c_v dT + v dP. \tag{4.39}$$

The differential change of the specific enthalpy also can be given as (see expression (4.27) and Chapter 6)

$$dh = \left(\frac{\partial h}{\partial T}\right)_P dT + \left(\frac{\partial h}{\partial P}\right)_T dP = c_p dT + \left(\frac{\partial h}{\partial P}\right)_T dP, \tag{4.40}$$

where $c_p = C_P/M$; C_P is the system's **constant-pressure heat capacity**, and c_p is the corresponding **constant-pressure specific heat capacity** (often called the **constant-pressure specific heat**).

Comparison of the terms containing dT in relationships (4.39) and (4.40) yields

$$c_p = c_v = c_{\text{inc}} \qquad \text{incompressible matter.} \qquad (4.41)$$

In other words, the constant-volume specific heat c_v of incompressible matter in a state of equilibrium is the same as the matter's constant-pressure specific heat c_p.

According to expressions (4.38) and (4.39), the changes of the specific thermal energy and the specific enthalpy in any (reversible or not) thermodynamic transition between an equilibrium state 1 and an equilibrium state 2 of incompressible matter are, respectively,

$$\Delta u = u_2 - u_1 = \int_{T_1}^{T_2} c_{\text{inc}} dT, \qquad (4.42)$$

and

$$\Delta h = h_2 - h_1 = \int_{T_1}^{T_2} c_{\text{inc}} dT + v(P_2 - P_1). \qquad (4.43)$$

Expression (4.28) can be written in terms of specific properties ($dV \rightarrow dv$). Remembering that the specific volume of an incompressible substance is constant, the change of the specific entropy during a thermodynamic transition from an equilibrium state 1 to an equilibrium state 2 can be given as

$$\Delta s = s_2 - s_1 = \int_{T_1}^{T_2} \frac{c_{\text{inc}}}{T} dT. \qquad (4.44)$$

In many applications of incompressible systems, the changes of the specific heat capacity c_{inc} with temperature are small enough to assume that

$$c_{\text{inc}} = \text{const.} \qquad (4.45)$$

Then, relationships (4.42), (4.43), and (4.44) can be written, respectively, as

$$\Delta u = c_{\text{inc}}(T_2 - T_1), \qquad (4.46)$$

$$\Delta h = c_{\text{inc}}(T_2 - T_1) + v(P_2 - P_1), \qquad (4.47)$$

and

$$\Delta s = c_{\text{inc}} \ln(T_2/T_1). \qquad (4.48)$$

4.3 Changes of the Free Energies

The expressions for the differentials of the Helmholtz free energy and the Gibbs free energy (see expressions (4.10) and (4.11), respectively) cannot be simplified much. However, the second derivatives of the free energies with respect to the thermodynamic variables can be calculated from the following general relationships valid in any system undergoing a thermodynamic transition between equilibrium states:

$$\frac{\partial^2 A(V,T)}{\partial V \partial T} = -\left[\frac{\partial P(V,T)}{\partial T}\right]_V, \tag{4.49}$$

and

$$\frac{\partial^2 G(P,T)}{\partial P \partial T} = \left[\frac{\partial V(P,T)}{\partial T}\right]_P. \tag{4.50}$$

The changes of the Helmholtz free energy and the Gibbs free energy during a finite thermodynamic transition can be obtained by integrating expressions (4.49)–(4.50). In some transitions, the differentials dA and dG can be given in convenient forms. For example, in isothermal processes one has (see expressions (3.140))

$$\left[\frac{\partial A(V,T)}{\partial V}\right]_T = -P \rightarrow dA = -PdV, \tag{4.51}$$

and

$$\left[\frac{\partial G(P,T)}{\partial P}\right]_T = V \rightarrow dG = VdP. \tag{4.52}$$

Similarly, one can write for isochoric and isobaric processes, respectively (see expression (3.141)),

$$\left[\frac{\partial A(V,T)}{\partial T}\right]_V = -S \rightarrow dA = -SdT, \tag{4.53}$$

and

$$\left[\frac{\partial G(P,T)}{\partial T}\right]_P = -S \rightarrow dG = -SdT. \tag{4.54}$$

As discussed in Section 3.7.2, the Helmholtz free energy of any substance in equilibrium is

$$A = U - TS, \tag{4.55}$$

where, as before, U and S are the thermal energy and entropy of the system, respectively. Thus, the change of the system's Helmholtz free energy (function of state) during any thermodynamic transition between equilibrium states 1 and 2 is

$$\Delta A = A_2 - A_1 = (U_2 - T_2 S_2) - (U_1 - T_1 S_1) = \Delta U - \Delta(TS), \qquad (4.56)$$

where A_1 and A_2 are the Helmholtz free energies of the system in the states, respectively,

$$\Delta U = U_2 - U_1, \qquad (4.57)$$

and

$$\Delta(TS) = T_2 S_2 - T_1 S_1. \qquad (4.58)$$

In isothermal processes, relationship (4.56) becomes

$$\Delta A = \Delta U - T\Delta S, \qquad (4.59)$$

where the change of the system entropy during the process can be given as

$$\Delta S = S_2 - S_1. \qquad (4.60)$$

Expressing the Helmholtz free energy as a function of volume and temperature, one can write (see Section 1.4)

$$d(\Delta A) = \left[\frac{\partial(\Delta A)}{\partial T}\right]_V dT + \left[\frac{\partial(\Delta A)}{\partial V}\right]_T dV. \qquad (4.61)$$

According to expression (4.59), the first derivative on the right-hand side of relationship (4.61) is

$$\left[\frac{\partial(\Delta A)}{\partial T}\right]_V = \left[\frac{\partial(\Delta U)}{\partial T}\right]_V - \Delta S - T\left[\frac{\partial(\Delta S)}{\partial T}\right]_V, \qquad (4.62)$$

where

$$\left[\frac{\partial(\Delta U)}{\partial T}\right]_V = \left[\frac{\partial U_2}{\partial T}\right]_V - \left[\frac{\partial U_1}{\partial T}\right]_V = (C_V)_2 - (C_V)_1 = \Delta C_V, \qquad (4.63)$$

and $(C_V)_1$ and $(C_V)_2$ are the constant-volume heat capacities of the considered system in its initial and final states, respectively (see expressions (4.24) and (4.25)).

We know from the discussion in Section 4.2.1 that

$$\left[\frac{\partial(\Delta S)}{\partial T}\right]_V = \left[\frac{\partial S_2}{\partial T}\right]_V - \left[\frac{\partial S_1}{\partial T}\right]_V = \frac{(C_V)_2}{T} - \frac{(C_V)_1}{T} = \frac{\Delta C_V}{T}, \tag{4.64}$$

so that relationship (4.62) can also be written as

$$\left[\frac{\partial(\Delta A)}{\partial T}\right]_V = \Delta C_V - \Delta S - T\frac{\Delta C_V}{T} = -\Delta S. \tag{4.65}$$

The second derivative on the right-hand side of relationship (4.61) can be given as (see expression (4.51))

$$\left[\frac{\partial(\Delta A)}{\partial V}\right]_T = \left[\frac{\partial A_2}{\partial V}\right]_T - \left[\frac{\partial A_1}{\partial V}\right]_T = -P_2 - (-P_1) = -(P_2 - P_1), \tag{4.66}$$

where P_1 and P_2 are pressures of the initial and final states of the transition, respectively.

Relationships (4.59) and (4.65) lead to the following expression for the change of the Helmholtz free energy during an isothermal transition between two states of equilibrium:

$$\Delta A = \Delta U + T\left[\frac{\partial(\Delta A)}{\partial T}\right]_V. \tag{4.67}$$

This is the **Gibbs-Helmholtz Equation for the Helmholtz free energy**.

The Gibbs free energy of a substance in an equilibrium state is (see Section 3.7.2)

$$G = H - TS, \tag{4.68}$$

where, as before, H and S are the enthalpy and entropy of the system in this state, respectively.

The change of the Gibbs free energy (a function of state) during a finite thermodynamic transition between equilibrium states 1 and 2 is

$$\Delta G = G_2 - G_1 = (H_2 - T_2 S_2) - (H_1 - T_1 S_1) = \Delta H - \Delta(TS), \tag{4.69}$$

where G_1 and G_2 are the Gibbs free energies of the system in the states, respectively, and

$$\Delta H = H_2 - H_1 \tag{4.70}$$

is the change of the system's enthalpy during the process.

Expressing the Gibbs free energy as a function of pressure P and temperature T, one can write (see Section 1.4)

$$d(\Delta G) = \left[\frac{\partial(\Delta G)}{\partial T}\right]_P dT + \left[\frac{\partial(\Delta G)}{\partial P}\right]_T dP. \qquad (4.71)$$

If the process under consideration is an isothermal transition, then, according to relationship (4.69), the first derivative on the right-hand side of expression (4.71) is

$$\left[\frac{\partial(\Delta G)}{\partial T}\right]_P = \left[\frac{\partial(\Delta H)}{\partial T}\right]_P - \Delta S - T\left[\frac{\partial(\Delta S)}{\partial T}\right]_P, \qquad (4.72)$$

while (see Section 4.2.1)

$$\left[\frac{\partial(\Delta H)}{\partial T}\right]_P = \left[\frac{\partial H_2}{\partial T}\right]_P - \left[\frac{\partial H_1}{\partial T}\right]_P = (C_P)_2 - (C_P)_1 = \Delta C_P, \qquad (4.73)$$

where $(C_P)_1$ and $(C_P)_2$ are the constant-pressure heat capacities of the system in its initial and final states, respectively.

Since

$$\left[\frac{\partial(\Delta S)}{\partial T}\right]_P = \left[\frac{\partial S_2}{\partial T}\right]_P - \left[\frac{\partial S_1}{\partial T}\right]_P = \frac{(C_P)_2}{T} - \frac{(C_P)_1}{T} = \frac{\Delta C_P}{T}, \qquad (4.74)$$

relationship (4.72) becomes

$$\left[\frac{\partial(\Delta G)}{\partial T}\right]_P = \Delta C_P - \Delta S - T\frac{\Delta C_P}{T} = -\Delta S. \qquad (4.75)$$

The second derivative on the right-hand side of expression (4.71) can be written as (see expression (4.52))

$$\left[\frac{\partial(\Delta G)}{\partial P}\right]_T = \left[\frac{\partial G_2}{\partial P}\right]_T - \left[\frac{\partial G_1}{\partial P}\right]_T = V_2 - V_1 = \Delta V, \qquad (4.76)$$

where V_1 and V_2 are volumes of the system in its initial and final states of the transition, respectively.

Relationships (4.71) and (4.75) lead to

$$\Delta G = \Delta H + T\left[\frac{\partial(\Delta G)}{\partial T}\right]_P. \qquad (4.77)$$

This is the **Gibbs-Helmholtz Equation for the Gibbs free energy.**

4.4 Summary

In this chapter we have derived a set of easy-to-use expressions for changes of thermal properties of state ΔF ($F \equiv U,H,S,A,G$) – see expressions (4.31), (4.33), (4.36)–(4.37), (4.67), and (4.77). They are valid in any (reversible or not) thermodynamic process between two equilibrium (quasi-equilibrium, local-equilibrium) states of any system. The origins of the expressions are in the Gibbs Equations (see Section 3.7.2).

The change ΔF in a process between two equilibrium (quasi-equilibrium, local-equilibrium) states can also be calculated from expressions (2.37), (2.39), (2.41), or (2.43) if the *absolute* values of F_1 (for the initial state of the process) and F_2 (for the final state of the process) are known. Absolute values of molar properties $\bar{F}$ for a large number of single-phase, single-component substances in equilibrium in a wide range of pressure and temperature are available in thermodynamic and thermochemical literature – see Chapter 8.

Knowledge of the changes ΔF is important in thermodynamic studies because the changes are parts of the Balance Equations for Matter, Energy, and Entropy which are studied in Section 3.5. The other parts of the equations are thermodynamic properties of path (heat Q, boundary work W, and non-boundary work K). While expressions (4.31), (4.33), (4.36)–(4.37), (4.67), and (4.77) accurately predict the changes of the properties of state in any process between two equilibrium (quasi-equilibrium, local-equilibrium) states of any system, heat and work transfers can be calculated only for processes for which thermodynamic paths are known. Since determination of these transfers is often difficult, the Balance Equations are usually formed in a way that includes some of the transfers as unknowns that can be found from solutions of the equations.

Chapter 5

Equations of State

We discuss in this chapter equations of state (relationships between pressure P, temperature T, and volume V) of single-component gases in an equilibrium, quasi-equilibrium, or local equilibrium. (Since the formalism of Classical Thermodynamics does not distinguish between these three states, the term 'equilibrium' is often used in literature to denote any of them – see Section 1.2.2) Equations of state are necessary for successful integration of expressions (4.31), (4.33), (4.36), and (4.37) since the expressions contain the derivatives $\partial P(V,T)/\partial T$ and $\partial V(P,T)/\partial T$.

5.1   Ideal Gases

An **ideal gas** is a gas where: 1) the mean energies of the intermolecular interactions are much smaller than the particles' mean translational energies; 2) the particles volumes are negligibly smaller than the space between them (that is why the gases are often called 'rarefied gases'); 3) the particles have no internal structures and, therefore, can be treated as structureless objects ('material points' characterized only by their masses) that interact with other particles only through elastic collisions, that is, collisions with no loss of kinetic energy. Most gases at pressures lower than several bars and temperatures not very much lower than room temperature can be treated as ideal gases.

Classical Thermodynamics gives the following relationship between mass M, volume V, pressure P, and absolute temperature T of any ideal gas in a state of equilibrium:

$$PV = MR_gT, \tag{5.1}$$

where R_g is the **individual gas constant** (in kJ/kg·K) which depends on the kind of gas under consideration. Relationship (5.1), called the **Equation of State (EOS) for ideal gas,** has been validated by numerous experiments in many different gases. **Equations of states are almost always formulated for single-phase, single-component substances in states of equilibrium.** Applications of the equations to mixtures of gases are discussed in Section 9.3.

Since the mass density and specific volume of a substance in equilibrium are $\rho = M/V$ and $v = 1/\rho$, respectively, relationship (5.1) can be rewritten as

$$P = \rho R_g T \qquad \text{or} \qquad Pv = R_g T. \tag{5.2}$$

Mass (in kg) of an amount of a single-component gas can be given as

$$M = v\mu, \tag{5.3}$$

where v is the number of mols of the gas, and μ is the gas' **molar mass** (in kg/kmol). According to the definition of 'mol' (see Section 2.1), the unit of the molar mass should be g/mol. However, it is a common (but informal) practice in thermodynamic literature to have the unit as kg/mol or kg/kmol which causes a great deal of confusion in thermodynamic calculations.

Equation of state for a single-component ideal gas in equilibrium can be written in the following form:

$$PV = v\mu R_g T = vRT = \frac{M}{\mu}RT, \tag{5.4}$$

where the constant

$$R = \mu R_g \tag{5.5}$$

is called the **universal gas constant** (or **molar gas constant**) because it has the same value for all gases. The constant is

$$R = 8.314\,\frac{\text{kJ}}{\text{kmol·K}} = 8.314\,\frac{\text{kPa·m}^3}{\text{kmol·K}} = 1.987\,\frac{\text{kcal}}{\text{kmol·K}}. \tag{5.6}$$

Since the gas **molar volume** (that is, the volume of one mol of the gas) is $\bar{V} = V/v$ (in m³/kmol), the equation of state for one mol of any ideal gas in equilibrium can be given as

$$P\bar{V} = RT. \tag{5.7}$$

Equation of state (5.1) for a fixed amount of an ideal gas in equilibrium can be written as

$$\frac{PV}{T} = \text{const} \qquad \text{or} \qquad \frac{P_1V_1}{T_1} = \frac{P_2V_2}{T_2} = \frac{P_3V_3}{T_3} = \cdots , \qquad (5.8)$$

where subscripts 1, 2, 3,... denote the gas's different equilibrium states, each being defined by a different set of thermodynamic variables P, V, and T. Relationship (5.8) also describes thermodynamic properties of every *reversible* transition in any ideal gas because each such transition is a sequence of quasi-equilibrium states.

When an ideal gas moves isothermally between various equilibrium states ($T_1 = T_2 = T_3 = ...$), expression (5.8) can be written as

$$PV = \text{const} \qquad \text{or} \qquad P_1V_1 = P_2V_2 = P_3V_3 = \cdots . \qquad (5.9)$$

Relationship (5.9) tells us that an isothermal transition of an ideal gas from one equilibrium state to another equilibrium state does not change the value of the product of the gas pressure and volume. The relationship is called the **Boyle-Mariotte Law**.

When an ideal gas moves thermodynamically between various equilibrium states in an isochoric way ($V_1 = V_2 = V_3 = ...$), expression (5.8) becomes

$$\frac{P}{T} = \text{const} \qquad \text{or} \qquad \frac{P_1}{T_1} = \frac{P_2}{T_2} = \frac{P_3}{T_3} = \cdots . \qquad (5.10)$$

Thus, an isochoric transition of an ideal gas from one equilibrium state to another equilibrium state does not change the ratio of the gas pressure to its temperature. The relationship is called the **Charles Law**.

When an ideal gas moves thermodynamically between various equilibrium states in an isobaric way ($P_1 = P_2 = P_3 = ...$), expression (5.8) gives

$$\frac{V}{T} = \text{const} \qquad \text{or} \qquad \frac{V_1}{T_1} = \frac{V_2}{T_2} = \frac{V_3}{T_3} = \cdots . \qquad (5.11)$$

Relationship (5.11) tells us that an isobaric transition of an ideal gas between two equilibrium (local-equilibrium) states does not change the ratio of the gas volume to its temperature. The relationship is called the **Gay-Lussac Law**.

Problem 5.1: Find the individual gas constant for ideal diatomic oxygen (O_2) in a state of equilibrium.

Solution: According to the *Periodic Chart of Elements*, the atomic mass of monatomic oxygen gas (O) is $\mu_O = 15.999 \, kg/kmol$. Thus, the molar mass of diatomic oxygen is

$$\mu_{O_2} = 2\mu_O = 2 \times 15.999 \, \frac{kg}{kmol} = 31.998 \, \frac{kg}{kmol}, \tag{5.12}$$

and the individual gas constant for ideal diatomic oxygen is (see expression (5.5))

$$(R_g)_{O_2} = \frac{R}{\mu_{O_2}} = \frac{8.314 \, kJ/kmol \cdot K}{31.998 \, kg/kmol} = 0.259 \, \frac{kJ}{kg \cdot K}. \tag{5.13}$$

End of Problem 5.1.

Problem 5.2: Find the values of the individual gas constants per one molecule of diatomic oxygen gas (O_2) and per one molecule of carbon dioxide gas (CO_2).

Solution: The individual gas constant per one molecule of any gas can be obtained by dividing R (the molar (universal) gas constant) by N_A (the number of particles in one mol of any gas),

$$k = \frac{R}{N_A} = \frac{8.314 \, J/mol \cdot K}{6.022 \times 10^{23} \, particles/mol} = \frac{1.3807 \times 10^{-23} \, J}{K \cdot particle}. \tag{5.14}$$

The number $N_A = 6.022 \times 10^{23}$ particles/mol is called the **Avogadro's number** – see expression (2.2).

Ratio (5.14), called the **Boltzmann constant**, is usually written as $k = 1.3807 \times 10^{-23} \, J/K$. It plays an important role in Molecular Thermodynamics and other thermal sciences. For example, the mean kinetic (translational) energy of particles in any ideal gas in equilibrium at temperature T is

$$\langle \epsilon \rangle = \frac{3}{2} kT, \tag{5.15}$$

regardless of the gas pressure.

End of Problem 5.2.

5.2 'Molecular' Equation of State of Ideal Gas

Assuming that an ideal gas under consideration consists of only one kind of particle (a single-component system), the mass density of the gas of mass M and volume V is

$$\rho = \frac{M}{V} = \frac{Nm}{V} = nm, \tag{5.16}$$

where N is the total number of particles in the gas, m is the mass of a single particle, and $n = N/V$ is the number of particles in a unit volume of the gas. The latter number is called the **particle number density** or just the **particle density**; note that the mass density and particle density are constant across the gas volume when the gas is in a state of equilibrium.

Equation (5.1) and relationships (5.15) and (5.16) lead to

$$PV = NkT \qquad \text{or} \qquad P = nkT = \frac{2}{3}n\langle\epsilon\rangle = \rho\frac{kT}{m}. \tag{5.17}$$

Applying expressions (5.7), (5.14), and (5.17) to one mol of ideal gas in equilibrium gives

$$P\bar{V} = N_{A}kT, \tag{5.18}$$

where, as before, k is the Boltzmann constant, N_{A} is Avogadro's number (see expression (2.2)), and $\bar{V}$ is the molar volume of the gas. (Overbars are used in the book to mark molar values of various properties of substances.)

Since the equation of state for ideal gas is widely used in science and technology, it is useful to give it in the following practical form of accuracy better than a few percent:

$$n \simeq 7 \times 10^{27}\,\frac{P}{T}, \tag{5.19}$$

where n is the particle density in $1/m^3$, P is the gas pressure in bars, and T is the gas temperature in kelvins.

Problem 5.3: Calculate the number of atoms in $1\,m^3$ of ideal helium gas in equilibrium at temperature $t = 0\,°C$ and pressure $P = 1$ Atm.

Solution: The Boltzmann constant $k = 1.38 \times 10^{-23}$ J/K·particle, $0\,°C = 273.15\,K$, and 1 Atm $= 1.013 \times 10^5\,N/m^2 \simeq 1$ bar. Thus, relationship (5.17) gives

$$n = \frac{P}{kT} = \frac{1.013 \times 10^5 \, \text{N/m}^2}{(1.38 \times 10^{-23} \, \text{J/K·particle})(273.15 \, \text{K})} = 2.68 \times 10^{25} \, \frac{\text{particles}}{\text{m}^3},$$

$$(5.20)$$

where we could use one of the following commonly accepted equivalent notations for units of the substance's particle density n:

$$1 \, \text{particle/m}^3 \equiv 1/\text{m}^3 \equiv \text{m}^{-3}. \tag{5.21}$$

As expected, the number of atoms in $1 \, \text{m}^3$ of ideal helium gas in equilibrium under the considered physical conditions is very large.

End of Problem 5.3.

5.3   Avogadro's Law

Consider two ideal gases (each being a single-component system) of different mass densities (ρ_1 and ρ_2) but in equilibrium states that have the same pressures and temperatures ($P_1 = P_2$ and $T_1 = T_2$). The ratio of the two mass densities is (see expressions (5.2) and (5.5))

$$\frac{\rho_1}{\rho_2} = \frac{v_2}{v_1} = \frac{\mu_1}{\mu_2}, \tag{5.22}$$

where v_1 and v_2 are the specific volumes (in m^3/kg), and μ_1 and μ_2 are the molar masses (in kg/kmol) of the gases. Subsequently,

$$\mu_1 v_1 = \mu_2 v_2 = \text{const} = \bar{V}_1 = \bar{V}_2, \tag{5.23}$$

where $\bar{V}_1$ and $\bar{V}_2$ are the **molar volumes** (in m^3/kmol) of the gases.

Relationship (5.23) tells us that at given pressure and temperature, the product μv has the same value in every ideal gas in equilibrium. Since the product is just the volume of one mol (kmol) of (any) ideal gas, relationship (5.23) can be stated as **Avogadro's Law: the molar volumes of all ideal gases in equilibrium are the same as long as the pressures and temperatures of the gases are the same.**

Problem 5.4: Find the molar volume of an ideal gas in equilibrium at pressure $P_r = 1$ Atm and temperature $T_r = 298.15$ K. Such pressure and temperature are often called the **room pressure** and **room temperature**, respectively.

Solution: The equation of state for 1 mol of the gas under the room conditions is (see expression (5.7))

$$P_r \bar{V} = RT_r, \tag{5.24}$$

so

$$\bar{V} = R\frac{T_r}{P_r} = 8.314 \, \frac{\text{J}}{\text{mol·K}} \times \frac{298.15 \, \text{K}}{1.01 \times 10^5 \, \text{N/m}^2} = 24.53 \, \text{m}^3/\text{kmol}. \tag{5.25}$$

In other words, **one kilomol of every ideal gas in equilibrium at room pressure and room temperature has volume of 24.53 m³**; the molar volume $\bar{V}$ does not depend on the kind of the gas under consideration (the universal gas constant R is the same for all gases) but only on the gas physical conditions such as pressure and temperature.

End of Problem 5.4.

Problem 5.5: Calculate: 1) the sum of translational (kinetic) energies of all particles in one cubic foot of an ideal gas in equilibrium at temperature $t = 20\,°C$ and pressure $P = 1$ bar; 2) the kinetic energy of a student of mass $M = 80\,$kg running with speed $\omega = 10\,$m/s; 3) the energy (in form of heat) produced by a complete combustion of one cubic foot of gasoline at temperature $t = 20\,°C$ and pressure $P = 1$ bar when the gasoline mass density is $750\,$kg/m³.

Solution:

1) The total kinetic (translational) energy of all particles in one cubic foot of the ideal gas in equilibrium at absolute temperature T is (see expression (5.15))

$$E_{\text{particles}} = N\langle\epsilon\rangle = N\frac{3}{2}kT, \tag{5.26}$$

where N is the number of particles in one cubic foot of the gas, $\langle\epsilon\rangle$ is the mean kinetic energy of the particles, and k is the Boltzmann constant.

The equation of state for the gas can be written as (see expression (5.17))

$$P = \frac{N}{V}kT, \tag{5.27}$$

and, according to expression (5.26), the total translational energy of all particles in one cubic foot of the gas is (1 foot = 0.305 m)

$$E_{\text{particles}} = \frac{3}{2}PV = \frac{3}{2}10^5\,\frac{\text{N}}{\text{m}^2}(0.305\,\text{m})^3 = 4.25\,\text{kJ}. \tag{5.28}$$

2) The kinetic energy of the running student is:

$$E_{\text{student}} = \frac{M\omega^2}{2} = \frac{(80\,\text{kg})(10\,\text{m/s})^2}{2} = 4\,\text{kJ}; \tag{5.29}$$

3) Since the mass density of the gasoline at pressure of 1 bar and temperature of 20 °C is 750 kg/m^3, the mass of one cubic foot of the gasoline at this pressure is 21.3 kg. Combustion of 1 kg of the gasoline produces about 4.5×10^7 J of heat. Thus, the heat produced by complete combustion of one cubic foot of the gasoline at pressure of 1 bar is

$$E_{\text{gasoline}} = (21.3\,\text{kg})(4.5 \times 10^7\,\text{J/kg}) = 956\,\text{MJ}. \tag{5.30}$$

End of Problem 5.5.

5.4   Non-Ideal Gases

In some applications, gases in equilibrium do not 'follow' relationship (5.1). Such gases are called **non-ideal gases** or **real gases**.

Classical Thermodynamics has proposed several kinds of equations of state for single-component non-ideal gases in equilibrium. The first of them, sometimes called the **Clausius Equation of State**, was proposed by German scholar Rudolf Clausius (1822–1888) in the following form:

$$P = \frac{R_g T}{v - b_o}, \tag{5.31}$$

where, as before, P, v, T, and R_g are pressure, specific volume, temperature, and the individual gas constant, respectively, and b_o is a constant.

Dutch scientist Johannes Van der Waals (1837–1923) has improved equation (5.31) by introducing an additional correction. His equation, called the **Van der Waals Equation of State**, is

$$P = \frac{R_g T}{v - b_g} - \frac{a_g}{v^2}, \tag{5.32}$$

where the constant a_g accounts for the presence of the intermolecular forces, and the constant b_g accounts for the size of the gas particles.

Equations (5.31)–(5.32) have been used extensively to study a variety of non-ideal gases as well as some liquids (liquids can sometimes be treated as 'very dense gases') with a rather modest degree of success.

Another class of equations of state for non-ideal gases are the so-called **Virial Equations of State** which have the following general form of series expansion:

$$Pv = R_g T \left(1 + \frac{b}{v} + \frac{c}{v^2} + \frac{d}{v^3} + \cdots \right), \tag{5.33}$$

where the coefficients $b, c, d, \ldots$ are called the first, the second, the third, $\ldots$, **virial coefficients**.

A convenient and quite accurate form of equation of state for non-ideal gases in equilibrium was obtained by introducing the so-called **compressibility factor** $Z(P, T)$ defined as

$$PV = Z(P, T) M R_g T. \tag{5.34}$$

Note that the compressibility factor Z is a function of P and T, and that relationship (5.34) becomes the equation of state for ideal gas when $Z(P, T) = 1$. Many experimental studies have shown that formulating the general equation (5.34) in terms of the so-called **reduced pressure** P_R and **reduced temperature** T_R gives the compressibility factor's dependence on these two variables that is practically identical in almost all single-component non-ideal gases. This surprising dependence of the compressibility factor of real gases on their reduced pressures and reduced temperatures is called the **Principle of Corresponding States**. The dependence, summarizing the results of a large number of experiments, is shown in Fig. 5.1.

The reduced pressure and reduced temperature are defined as

$$P_R = \frac{P}{P_C} \qquad \text{and} \qquad T_R = \frac{T}{T_C}, \tag{5.35}$$

where P_C and T_C are the substance **critical pressure** and **critical temperature**, respectively (see Section 9.2). Each pure substance has unique values of its critical pressure and critical temperature.

Experiments found that three gases (hydrogen, helium, and neon) do not follow well the Principle of Corresponding States. However, this complication

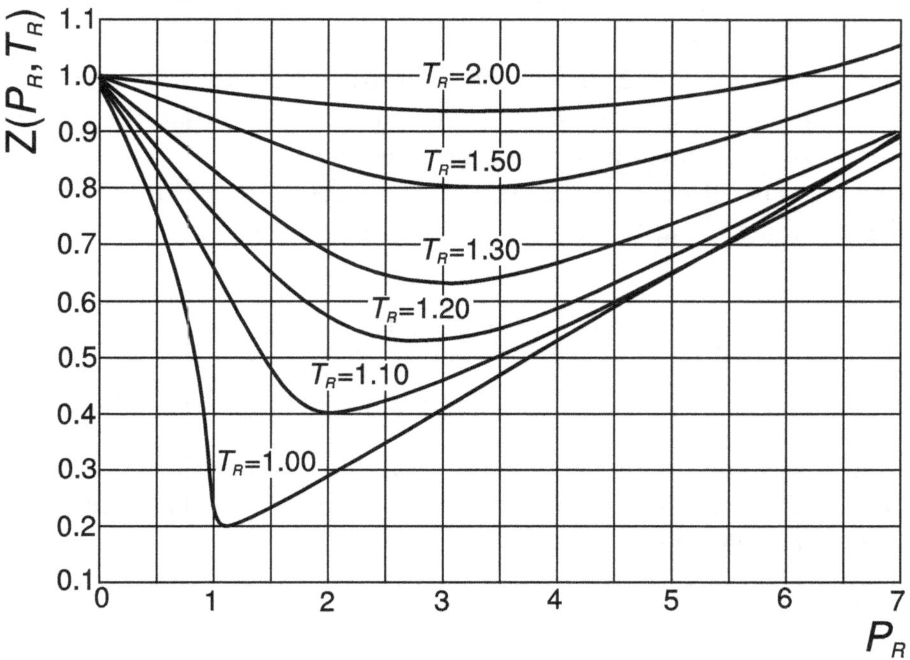

Fig. 5.1: The measured compressibility factor $Z(P,T)$ (see relationship (5.34)) of single-component real gases in equilibrium at pressure P and temperature T. The reduced pressure $P_R = P/P_C$, and reduced temperature $T_R = T/T_C$, where P_C and T_C are the gas's critical pressure and critical temperature, respectively.

can be removed by replacing expressions (5.35) by their modified form,

$$P_R = \frac{P}{P_C + a'} \qquad \text{and} \qquad T_R = \frac{T}{T_C + a'}, \qquad (5.36)$$

where $a' = 8$ when P and P_C are in bars, and T and T_C are in kelvins.

A substantial amount of work was done on development of the so-called *Generalized Cubic Equations of State* for real gases. Some of them are frequently used in thermodynamic studies because of their moderate mathematical complexity and good accuracy (usually close to one percent). A typical Generalized Cubic Equation of State for one mol of a single-component non-ideal gas in equilibrium can be written as

$$P = \frac{RT}{\bar{V} - t_c} - \frac{f(T)}{(\bar{V} + g_c - t_c)(\bar{V} + h_c - t_c)}, \tag{5.37}$$

where, as before, $\bar{V}$ is the gas molar volume, R is the molar gas constant (8.314 J/mol·K), $f(T)$ is some function of temperature, and g_c, h_c, and t_c are coefficients representing the effects of the gas particles' sizes and particle-particle interactions. The two most frequently used Cubic Equations of State are those proposed by Redlich and Kwong, and by Peng and Robinson.

The **Peng-Robinson Equation of State** for one mol of a non-ideal, single-component gas in equilibrium is

$$P = \frac{RT}{\bar{V} - b} - \frac{a(T)}{\bar{V}(\bar{V} + b) + b(\bar{V} - b)}, \tag{5.38}$$

where function $a(T)$ and constant b, obtainable from the stability criteria of the substance critical state, are

$$a(T) = 0.45724 \frac{R^2 T_C^{\,2}}{P_C} \left[1 + \kappa_a (1 - \sqrt{T/T_C}) \right]^2, \tag{5.39}$$

$$b = 0.0778 \frac{R T_C}{P_C}, \tag{5.40}$$

and

$$\kappa_a = 0.37464 + 1.54226 \omega_a - 0.26992 \omega_a^2, \tag{5.41}$$

where ω_a is a constant called the 'acentric factor'.

Most non-ideal gases have the acentric factors between 0 and 0.5 which correspond to values of κ_a between 0.4 and 1.1; the mean values of these parameters ($\langle \omega_a \rangle = 0.25$ and $\langle \kappa_a \rangle = 0.75$) are often used in practical calculations. Acentric factors for different gases can be found in thermodynamic literature.

Defining the compressibility factor $Z(P,T)$ as a measure of departure of non-ideal gas properties from its properties obtained from the ideal-gas model, one can write

$$Z(P,T) = P\bar{V}/RT, \tag{5.42}$$

and the equation of state (5.38) can be written as the following cubic equation:

$$Z^3 + \alpha Z^2 + \beta Z + \gamma = 0, \tag{5.43}$$

where

$$\alpha = B - 1, \tag{5.44}$$

$$\beta = A - 3B^2 - 2B, \tag{5.45}$$

$$\gamma = B^3 + B^2 - 2AB, \tag{5.46}$$

with

$$A = \frac{a(T)P}{R^2 T^2} = \frac{0.45724 g_a P_R}{T_R^2}, \tag{5.47}$$

$$B = \frac{bP}{RT} = \frac{0.07780 P_R}{T_R}, \tag{5.48}$$

and

$$g_a = \left[1 + \kappa_a \left(1 - \sqrt{T_R} \right) \right]^2. \tag{5.49}$$

The cubic equation (5.43) has between one and three real solutions, depending on the value of

$$D = D_1{}^3 + D_2{}^2, \tag{5.50}$$

where

$$D_1 = \frac{3\beta - \alpha^2}{9}, \tag{5.51}$$

and

$$D_2 = \frac{9\alpha\beta - 27\gamma - 2\alpha^3}{54}. \tag{5.52}$$

When $D < 0$, equation (5.43) has three real and different solutions:

$$Z_1 = 2\sqrt{-D_1} \cos(\theta/3) - \alpha/3, \tag{5.53}$$

$$Z_2 = 2\sqrt{-D_1} \cos[(\theta + 2\pi)/3] - \alpha/3, \tag{5.54}$$

and

$$Z_3 = 2\sqrt{-D_1} \cos[(\theta + 4\pi)/3] - \alpha/3, \tag{5.55}$$

where

$$\theta = \cos^{-1}[D_2/(-D_1)^{3/2}]. \tag{5.56}$$

When $D = 0$, equation (5.43) has three (two of them identical) real solutions,

$$Z_1 = (D_2 + \sqrt{D})^{1/3} + (D_2 - \sqrt{D})^{1/3} - \alpha/3, \qquad (5.57)$$

and

$$Z_2 = Z_3 = \frac{1}{2}[(D_2 + \sqrt{D})^{1/3} + (D_2 - \sqrt{D})^{1/3}] - \alpha/3. \qquad (5.58)$$

When $D > 0$, there is only one real solution,

$$Z_1 = (D_2 + \sqrt{D})^{1/3} + (D_2 - \sqrt{D})^{1/3} - \alpha/3. \qquad (5.59)$$

When three real solutions of equation (5.43) exist, the largest solution gives the compressibility factor for the gaseous phase of the substance. When two real solutions of equation (5.43) exist, then the larger one gives Z for the gaseous phase.

Problem 5.6: Find the volume of 2 pounds of carbon dioxide gas in equilibrium at pressure of 100 Atm and temperature of 60 °C.

Solution: Since the pressure of the gas is rather high, a non-ideal gas model should give a more realistic thermodynamic picture of the gas than the ideal-gas model. Therefore, we will find the volume of the gas using the Principle of Corresponding States.

The critical pressure and critical temperature of CO_2 are, respectively,

$$P_C = 72.9 \, \text{bar} \qquad \text{and} \qquad T_C = 304 \, \text{K}. \qquad (5.60)$$

Assuming that 1 Atm = 1 bar (the error of such approximation is about one percent) one obtains (see expression (5.35))

$$P_R = \frac{100 \, \text{bar}}{72.9 \, \text{bar}} = 1.37 \qquad \text{and} \qquad T_R = \frac{(60 + 273.15) \, \text{K}}{304 \, \text{K}} = 1.1. \qquad (5.61)$$

The molecular mass of carbon dioxide can be obtained from the *Periodic Chart of Elements*,

$$\mu_{CO_2} = \mu_C + 2\mu_O = 12.011 \, \frac{\text{kg}}{\text{kmol}} + 2 \times 15.999 \, \frac{\text{kg}}{\text{kmol}} = 44.01 \, \frac{\text{kg}}{\text{kmol}}, \qquad (5.62)$$

so that the individual gas constant for the gas is

$$(R_g)_{CO_2} = \frac{R}{\mu_{CO_2}} = \frac{8.314 \, \text{kJ/kmol·K}}{44.01 \, \text{kg/kmol}} = 0.189 \, \frac{\text{kJ}}{\text{kg·K}}. \qquad (5.63)$$

The value of the compressibility factor Z for the reduced parameters (5.61) can be found from the general (experimental) plot of the factor shown in Fig. 5.1. This value is $Z \simeq 0.5$. Subsequently, use of the equation of state (5.34) gives the following volume of the CO_2 gas discussed here:

$$V = \frac{ZMR_gT}{P} = \frac{0.5(2 \times 0.454 \, \text{kg})(0.189 \, \text{kJ/kg·K})[(60 + 273.15) \, \text{K}]}{10^7 \, \text{N/m}^2}$$

$$= 0.003 \, \text{m}^3, \tag{5.64}$$

while the volume of the gas obtained from the ideal-gas model ($Z = 1$) is $0.006 \, \text{m}^3$.

End of Problem 5.6.

Chapter 6

Heat Capacities

6.1 General Relationships

The constant-volume heat capacity C_V and the constant-pressure heat capacity C_P of a substance in an equilibrium, quasi-equilibrium, or local-equilibrium state are of great importance in many thermodynamic studies – see relationships (4.31), (4.33), (4.36), and (4.37). The constant-volume heat capacity of some amount of such a substance is defined as (U is the thermal energy of the substance – see expression (4.24))

$$C_V = \left(\frac{\partial U}{\partial T} \right)_V , \tag{6.1}$$

and the constant-pressure heat capacity of the substance is defined as (H is the enthalpy of the substance – see expression (4.25))

$$C_P = \left(\frac{\partial H}{\partial T} \right)_P . \tag{6.2}$$

It is often convenient to use the **specific heat capacities** (that is, the heat capacities per a unit of mass of the system's matter) instead of the **heat capacities** C_V and C_P which characterize the *entire* matter of the system. The constant-volume specific heat capacity (it is usually called, somewhat confusingly, the **constant-volume specific heat**) is

$$c_V = C_V/M = \left(\frac{\partial u}{\partial T} \right)_V , \tag{6.3}$$

131

where M is the mass of the substance and u (in kJ/kg) is its specific thermal energy. The constant-pressure specific heat capacity (it is usually called the **constant-pressure specific heat**) is

$$c_P = C_P/M = \left(\frac{\partial h}{\partial T}\right)_P, \tag{6.4}$$

where h (in kJ/kg) is the specific enthalpy of the substance.

Other commonly used thermodynamic properties are the specific heat capacities of substances given on a molar basis. They are called the constant-volume molar heat capacity $\bar{c}_V$ and the constant-pressure molar heat capacity $\bar{c}_P$, or, frequently, just the **constant-volume molar heat** and the **constant-pressure molar heat**, respectively. The constant-volume molar heat is

$$\bar{c}_V = C_V/v = \left(\frac{\partial \bar{u}}{\partial T}\right)_V, \tag{6.5}$$

and the constant-pressure molar heat is

$$\bar{c}_P = C_P/v = \left(\frac{\partial \bar{h}}{\partial T}\right)_P. \tag{6.6}$$

where v is the number of kilomols of the substance, and $\bar{u}$ (in kJ/kmol) and $\bar{h}$ (in kJ/kmol) are the substance's molar thermal energy and molar enthalpy, respectively.

Since U and H are functions of state, their changes during *any* transition between an equilibrium state 1 and an equilibrium state 2 of *any* thermodynamic system are (see expressions (4.31) and (4.33), respectively):

$$\Delta U = \int_{T_1}^{T_2} \{C_V(V,T)\}_{V=V_2}\, dT$$
$$+ \int_{V_1}^{V_2} \left\{ T\left[\frac{\partial P(V,T)}{\partial T}\right]_V - P(V,T) \right\}_{T=T_1} dV, \tag{6.7}$$

and

$$\Delta H = \int_{T_1}^{T_2} \{C_P(P,T)\}_{P=P_2}\, dT$$
$$+ \int_{P_1}^{P_2} \left\{ -T\left[\frac{\partial V(P,T)}{\partial T}\right]_P + V(P,T) \right\}_{T=T_1} dP, \tag{6.8}$$

where P_1, T_1, and V_1 are pressure, temperature, and volume of the initial state, respectively, and P_2, T_2, and V_2 are pressure, temperature and volume of the final state.

It should be emphasized that **relationships (6.7) and (6.8) are valid in all thermodynamic transitions in all systems where the transitions' initial and final states are equilibrium, quasi-equilibrium, or local-equilibrium states**. One should notice that parts of the integrals on the right-hand sides of the relationships are evaluated under some specific (isochoric, isobaric, or isothermal) conditions. However, the sums of the parts make the relationships valid in all processes between equilibrium, quasi-equilibrium, or local-equilibrium. Also, the **changes (6.7) and (6.8) are independent of the physical structure of the systems where the processes take place**. Therefore, use of the relationships does not require any information about the systems.

6.2 Heat Capacities and Heat Transfers

We show in this section some examples of usefulness of the concept of heat capacity for studying energy flows in thermodynamic systems. Applying the First Law of Thermodynamics (the Energy Balance Equation) to *isochoric* ($\Delta V = 0$) transition in a closed, nonflow system with transfer of heat Q but with no non-boundary work transfer K (boundary work W is always zero in isochoric transitions) gives (see expression (3.58))

$$Q = \Delta U. \tag{6.9}$$

Note that the system's 'motion properties' (the matter's kinetic and potential energies – see expressions (2.6) and (2.7), respectively) are absent in the Energy Balance Equation (6.9) because the discussed system is a closed, nonflow system.

Relationships (6.7) and (6.9) lead to the following very convenient expression for the heat of any isochoric transition between equilibrium state 1 and equilibrium state 2 of a closed, nonflow system with no work transfer:

$$Q = \Delta U = \int_{T_1}^{T_2} C_V(T)\,dT, \tag{6.10}$$

or, when C_V does not change much within the considered temperature range $T_1 \leq T \leq T_2$ (then C_V can be taken as a constant equal to the mean value of the heat capacity within the range), as

$$Q = \Delta U = C_V(T_2 - T_1), \tag{6.11}$$

or

$$Q = \Delta U = U_2 - U_1, \tag{6.12}$$

if U_1 and U_2, the thermal energies of the initial and final states of the discussed process, are known. These energies can be found in thermodynamic databases discussed in Chapter 8.

Let us consider an *isobaric* transition ($\Delta P = 0$) between two equilibrium states of a closed, nonflow system with no non-boundary work. Definition of enthalpy allows one to express the isobaric change of the system enthalpy as (see expression (3.32)):

$$\Delta H = \Delta(U + PV) = \Delta U + P\Delta V + V\Delta P = \Delta U + P\Delta V. \tag{6.13}$$

Thus, the First Law of Thermodynamics (the Principle of Conservation of Energy) for the isobaric transition in the closed, nonflow system without non-boundary work transfer can be written as (see expression (3.58))

$$Q + W = \Delta U, \tag{6.14}$$

or

$$Q + W = \Delta H - P\Delta V. \tag{6.15}$$

If the boundary work transfer in the transition can be treated as a reversible process, then $W = -P\Delta V$ (see expression (2.59)) and the heat of the transition can be given as (see expression (6.8))

$$Q = \Delta H = \int_{T_1}^{T_2} C_P(T)dT, \tag{6.16}$$

or (when C_P is independent on temperature)

$$Q = \Delta H = C_P(T_2 - T_1), \tag{6.17}$$

or

$$Q = \Delta H = H_2 - H_1 \tag{6.18}$$

when the enthalpies of the transition's initial and final (equilibrium) states are known (see expression (2.37) and Chapter 8).

The calculations become more complicated when the boundary work of the discussed isobaric transition cannot be treated as a reversible process. But, when the first and second gradients of temperature during the transition are not significant (the gradients of pressure are zero in isobaric processes), the boundary work is close to that of the corresponding reversible transition ($W = -P\Delta V$ – see the discussions associated with expressions (2.71), (2.72), and (3.62)). Then, relationship (6.15) can be written as

$$Q \simeq \Delta H - (-P\Delta V) - P\Delta V \simeq \Delta H, \tag{6.19}$$

and relationships (6.8) and (6.19) yield the heat of the isobaric transition as

$$Q \simeq \Delta H = \int_{T_1}^{T_2} C_P(T)dT, \tag{6.20}$$

or, when C_P does not change much within the considered temperature range $T_1 \leq T \leq T_2$,

$$Q \simeq \Delta H = C_P(T_2 - T_1). \tag{6.21}$$

Of course,

$$Q \simeq \Delta H = H_2 - H_1, \tag{6.22}$$

when H_1 and H_2 of the isobaric process are known.

6.3 Non-Ideal Substances

Expressions for the heat capacities C_V and C_P of a substance in equilibrium can be found with the help of general relationships (4.28) and (4.29). Use of these two relationships and expressions (1.23) and (1.25)–(1.26) leads to

$$\left(\frac{\partial C_V}{\partial V}\right)_T = T\left(\frac{\partial^2 P}{\partial T^2}\right)_V, \tag{6.23}$$

$$\left(\frac{\partial C_P}{\partial P}\right)_T = -T\left(\frac{\partial^2 V}{\partial T^2}\right)_P, \tag{6.24}$$

$$C_P - C_V = T\left(\frac{\partial V}{\partial T}\right)_P\left(\frac{\partial P}{\partial T}\right)_V \tag{6.25}$$

and

$$C_P - C_V = -T \left(\frac{\partial V}{\partial T}\right)_P^2 \left(\frac{\partial P}{\partial V}\right)_T. \tag{6.26}$$

Relationships (6.25)–(6.26) are called the **Mayer Equations**.

One can see that the difference $C_P - C_V$ is always positive or zero. The latter happens when either $T = 0\,\mathrm{K}$ while the other terms in equation (6.26) are finite, or when derivative $(\partial V/\partial T)_P$ is zero, which occurs when the substance under consideration is in the state of its maximum density.

The difference $C_P - C_V$ is small in liquids and solids because their derivatives $(\partial V/\partial T)_P$ are always small. Therefore, the single term 'heat capacity' is usually used for C_V as well as for C_P in thermodynamic studies of solids and liquids (see Section 4.2.2).

Integrating relationship (6.24) from zero pressure to the actual pressure of the system gives the following heat capacity of any substance in equilibrium at pressure P and temperature T:

$$C_P(P,T) = C_P(P=0,T) - T \int_0^P \left(\frac{\partial^2 V}{\partial T^2}\right)_P dP, \tag{6.27}$$

where $C_P(P=0,T)$ is the constant-pressure heat capacity of a zero-pressure substance (such a substance is sometimes called 'calorically perfect ideal gas'). In practical calculations, $C_P(P=0,T)$ can be replaced by heat capacity $C_P(P=P_{\mathrm{low}},T)$ where pressure P_{low} is low enough to have the heat capacity value close to that of the perfect ideal gas. In other words, in practical calculations relationship (6.27) must be replaced by

$$C_P(P,T) = C_P(P=P_{\mathrm{low}},T) - T \int_{P_{\mathrm{low}}}^P \left(\frac{\partial^2 V}{\partial T^2}\right)_P dP. \tag{6.28}$$

The heat capacities of substances at constant pressure show singularities in their temperature-dependences. These singularities occur during changes of phase of the substances at temperatures that are unique for each substance. This can be seen in Fig. 6.1 which shows the temperature-dependence of the molar constant-pressure heat capacity of a substance called diatomic nitrogen at constant pressure of 1 bar. At the pressure, the singularities (the phase changes) occur at $T_1 = 35.62\,\mathrm{K}$ (solid→solid allotropic transition), $T_2 = 63.15\,\mathrm{K}$ (solid→liquid phase transition), and $T_3 = 77.33\,\mathrm{K}$ (liquid→gas phase transition). Phase transitions are further discussed in Chapter 9.

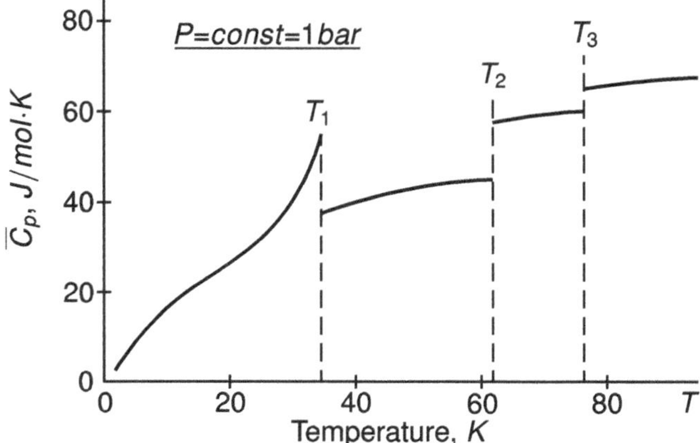

Fig. 6.1: The change of the molar heat capacity of diatomic nitrogen when heated isobarically at pressure $P = 1$ bar to temperature T; solid→solid allotropic transition occurs at $T_1 = 35.62$ K, solid→liquid transition occurs at $T_2 = 63.15$ K, and liquid→gas transition occurs at $T_3 = 77.33$ K.

6.4 Ideal Gases

When the substance under consideration is an ideal gas, the substance's thermal energy U and enthalpy H ($H = U + PV$, where $PV = \nu RT$) depend only on the gas temperature. Subsequently, the **heat capacities C_V and C_P of every ideal gas depend only on temperature** – see relationships (6.1) and (6.2). Such a gas is called a **thermally perfect ideal gas**. The heat capacities that are weak functions of temperature within some temperature range are often assumed constant (temperature-independent) within the range. An gas having a constant heat capacity is called a **calorically perfect gas**. An ideal gas having a constant heat capacity is called a **calorically perfect ideal gas**.

In every ideal gas one has (see relationship (5.4))

$$\left(\frac{\partial^2 V}{\partial T^2}\right)_P = 0. \tag{6.29}$$

This and the pressure-independence of the heat capacity of the ideal gas in equilibrium (local-equilibrium) allows one to write expression (6.28) as

$$C_P(P,T) = C_P(P = \text{any pressure}, T) = C_P(P = P^\circ, T), \tag{6.30}$$

where $P° = 1$ bar is the standard pressure, the usual choice of pressure when thermochemical properties are tabularized. In common thermochemical notation, relationship (6.30) can be written as (see Section 8.2.2)

$$C_P(P = \text{any pressure}, T) = C_P(P=P°, T) \equiv C_P^{\circ}(T), \qquad (6.31)$$

where values of the standard-state heat capacity $C_P^{\circ}(T)$ can be found in the databases discussed in Chapter 8.

If an ideal gas in equilibrium is a calorically perfect monoatomic gas then, according to Quantum Statistical Mechanics, the heat capacities of M kilograms (or v mols) of the gas can be given as

$$C_V = \frac{3}{2}MR_g = \frac{3}{2}vR \qquad \text{gas of atoms,} \qquad (6.32)$$

and

$$C_P = \frac{5}{2}MR_g = \frac{5}{2}vR \qquad \text{gas of atoms,} \qquad (6.33)$$

where, as before, R_g and R are the individual and molar (universal) gas constants, respectively.

The heat capacities of a calorically perfect ideal polyatomic gas in equilibrium are:

$$C_V = \frac{5}{2}MR_g = \frac{5}{2}vR \qquad \text{gas of diatomic molecules,} \qquad (6.34)$$

$$C_P = \frac{7}{2}MR_g = \frac{7}{2}vR \qquad \text{gas of diatomic molecules,} \qquad (6.35)$$

$$C_V \simeq \frac{6}{2}MR_g \simeq \frac{6}{2}vR \qquad \text{gas of triatomic molecules,} \qquad (6.36)$$

and

$$C_P \simeq \frac{8}{2}MR_g \simeq \frac{8}{2}vR \qquad \text{gas of triatomic molecules.} \qquad (6.37)$$

Quantum Statistical Mechanics does not provide simple expressions for C_V and C_P of calorically perfect ideal gases of molecules made of more than three atoms. Sometimes, crude estimates of the heat capacities of such gases can be obtained from expressions (6.36) and (6.37).

One should notice that in calorically perfect ideal gases of monoatomic, diatomic, and triatomic molecules one has

$$C_P - C_V = MR_g = M\frac{R}{\mu} = vR, \qquad (6.38)$$

where μ is the molar mass of the gas. The relationship confirms the fact that in a calorically perfect ideal gas, one always has $C_P > C_V$ (which is also true in thermally ideal gases – see discussion after expression (6.26)).

Since the **isentropic exponent** is defined as $\kappa = C_P/C_V$ (see Section 7.5), one can write for calorically perfect ideal gases:

$$\kappa = \frac{5}{3} = 1.67 \qquad\qquad \text{gas of atoms,} \qquad\qquad (6.39)$$

$$\kappa = \frac{7}{5} = 1.40 \qquad\qquad \text{gas of diatomic molecules,} \qquad\qquad (6.40)$$

$$\kappa \simeq \frac{4}{3} \simeq 1.33 \qquad\qquad \text{gas of triatomic molecules.} \qquad\qquad (6.41)$$

Problem 6.1: A car is parked outside for part of a hot, sunny day. Should the car's window be slightly open while parked?

Solution: The heat entering the interior of the car while parked is mainly the radiative heat Q. No non-boundary work transfer takes place during the heating.

When all windows of the parked car are closed, the air in the car is a closed, non-flow system heated in an *isochoric* manner. Then, according to expression (6.11), the temperature increase of the air inside the car while parked is

$$T_V - T_\circ = Q/C_V, \qquad\qquad (6.42)$$

where T_V is the final temperature of the air in the car after the isochoric heating stops, $T_\circ$ is the initial temperature of the air when the heating starts, and the air's constant-volume heat capacity C_V is assumed temperature-independent.

When one of the car's windows is slightly open, the pressure of the air inside the car is the same as the (constant) pressure of the air outside. In addition, the mass of the air in the car changes very little during the heating. Therefore, it is justified to assume that the behavior of the air inside the car during the heating is an *isobaric* process where (see expression (6.17))

$$T_P - T_\circ = Q/C_P, \qquad\qquad (6.43)$$

where T_P is the final temperature of the air in the car after the isobaric heating stops, and the air's constant-pressure heat capacity C_P is assumed temperature-independent.

Since properties of the air are dominated by diatomic molecules (N_2, O_2), its heat capacities can be given as (see expressions (6.34) and (6.35))

$$C_V = \frac{5}{2}\nu R \qquad \text{and} \qquad C_P = \frac{7}{2}\nu R, \qquad (6.44)$$

where ν is the number of mols of the air inside the car, and R is the molar (universal) gas constant.

Expressions (6.42)–(6.43) lead to the following relationship between the temperatures T_V and T_P if the heating process begins when the temperature of the air in the car is T_o,

$$\frac{T_V}{T_o} = 1 + \left(\frac{T_P}{T_o} - 1\right)\kappa, \qquad (6.45)$$

where $\kappa = C_P/C_V$. Since T_P/T_o is greater than one, the derivative of T_V/T_o with respect to T_P/T_o is equal to κ which is always greater than one (see discussion following relationship (6.26)); for air, $\kappa = 1.4$. Thus, parking the car with all the windows closed will increase the temperature of the air in the car significantly more than it would have if the car was parked with a window slightly open.

End of Problem 6.1.

Chapter 7

Common Thermodynamic Processes

7.1 Models of Thermodynamic Systems

A thermodynamic process in a physical system changes the system's thermodynamic variables (pressure, temperature, volume), changes its properties of state (the internal energy, enthalpy, etc.), and exchanges heat, work and/or radiation with the environment. The magnitudes of the changes can be found from the Balance Equations for matter, energy, and entropy. The equations are mathematical representations of the Principles of Thermodynamics (Principle of Conservation of Matter, Principle of Conservation of Energy, and Principle of Production of Natural Entropy). Some terms in the Balance Equations represent the properties of state, others represent properties of path (see Section 3.5).

The internal energy $\mathcal{E}$, thermal energy U, enthalpy H, entropy S, the Helmholtz free energy A, and the Gibbs free energy G of *any* (closed or open) physical system undergoing *any* thermodynamic process that begins and ends in an equilibrium (quasi-equilibrium, local-equilibrium) state are functions of state – see Section 2.8.1. This means that the *changes* of the functions during any such process depend neither on the kind of process nor on the structure of the system in which the process takes place – the changes depend only on the properties of the process's initial and final states (see Section 4.1).

Thermodynamic functions of path (work, heat) present in the system's Balance Equations depend not only on the characteristics of initial and final states of the considered process but also on the character of the process, that is, on the process's thermodynamic path.

The isochoric transition shown in Fig. 7.1a takes place in a closed system that receives some heat Q and some non-boundary work K done by a paddlewheel which is driven by an outside motor. (Note that the non-boundary work is easy to estimate since powers of motors driving various technological devices are usually specified by the motors' manufacturers.) Fig. 7.1 shows three kinds of thermodynamic processes common in applications (in thermodynamic studies, the possible radiation transfers R are usually ignored or included in the heat transfers). The transition starts at a time t_1 when the system's matter M_1 is in an equilibrium (local-equilibrium) state at P_1, T_1, V_1, and it ends at a time t_2 when the matter is in another state of equilibrium at P_2, T_2, V_2. The internal energy, enthalpy and entropy of the system in its initial state are $\mathcal{E}_1$, H_1, and S_1, respectively, and they are $\mathcal{E}_2$, H_2, S_2 when the system is in its final state. The internal energies $\mathcal{E}_1$ and $\mathcal{E}_2$ are the total (kinetic+potential+thermal) energies of the system's matter; the two symbols should be replaced by U_1 and U_2, respectively, if the matter locked up in the system is at rest and, therefore, does not change its kinetic and potential energies. (U_1 and U_2 are the thermal energies of the system.) In other words, the internal energy of every closed, nonflow system is the thermal energy of its matter, $\mathcal{E}_i = U_i$ – see Section 2.2.

The thermodynamic transition shown in Fig. 7.1b takes place in a closed system that releases some heat Q, receives some boundary work W (which causes motion of the piston to the right), and receives some non-boundary work K done by a turbine driven by an outside motor. The process starts at a time t_3 when the moving matter of mass M_3 locked up in the system is in an equilibrium state at P_3, T_3, V_3. The process ends at a time t_4 when the matter is in another state of equilibrium at P_4, T_4, V_4. The internal energy, enthalpy, and entropy of the system in its initial state are $\mathcal{E}_3$, H_3, and S_3, respectively, and they are $\mathcal{E}_4$, H_4, S_4 when the system is in its final state.

The thermodynamic process shown in Fig. 7.1c takes place in an open system that absorbs some heat Q, produces some boundary work W, and some non-boundary work K which is transferred by a rotating shaft from the moving paddlewheel to an external device (a gear). (Note that the non-boundary work of this process is difficult to calculate because the motion of the paddlewheel depends in a complicated way on changes of the thermal and kinetic properties of the flowing matter.) The entire observed process starts at a time t_5 when the matter (in a state of local equilibrium at P_5, T_5, ρ_5, ε_5, h_5, s_5) enters the system,

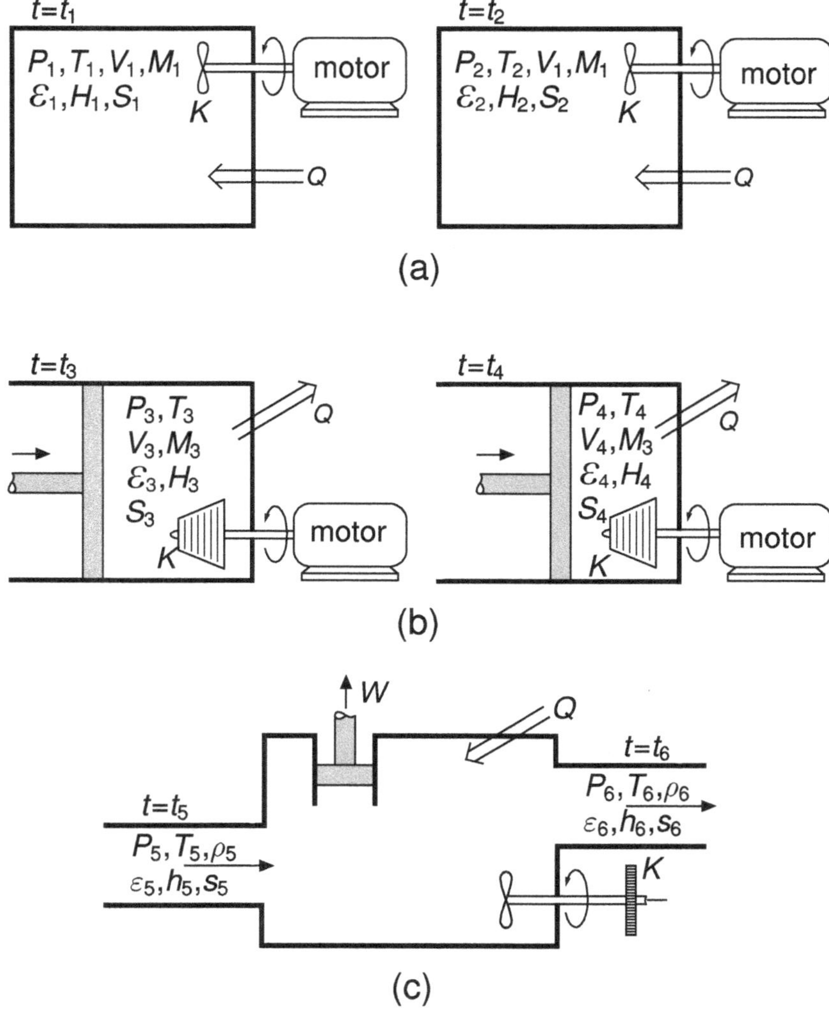

Fig. 7.1: Examples of thermodynamic processes with various work and heat transfers in closed systems (a) and (b), and in open system (c).

and it ends at a time t_6 when the matter (now in a state of local equilibrium at P_6, T_6, ρ_6, ε_6, h_6, s_6) leaves the system; ρ_i, ε_i, h_i, and s_i are the mass density, the specific internal energy, specific enthalpy, and specific entropy of the moving matter. Note that it is usually more convenient in studying flows to deal with flowing matter's *specific* properties (ρ,ε,h,s) and with per-unit-of-mass transfers w (boundary work), k (non-boundary work), and q (heat) than to deal with the *entire* system's properties such as M (mass), $\mathcal{E}$ (the internal energy), H (enthalpy), S (entropy), W (boundary work), K (non-boundary work), and Q (heat) – see Section 3.5.2 and Problem 3.1.

7.2   Changes of Properties of State

In general, the internal energy of the matter 'working' in a thermodynamic system can be given as (see Section 2.2, and expressions (2.4) and (3.53)):

$$\mathcal{E} = U + E_{kin} + E_{pot}, \qquad (7.1)$$

where U is the matter's *thermal energy*, and ($E_{kin}+E_{pot}$) is the matter's *motion energy*, a sum of its kinetic and potential (gravitational) energies. Thus, the change of the system internal energy during the process is (see expression (3.54))

$$\Delta\mathcal{E} = \Delta U + \Delta E_{kin} + \Delta E_{pot}. \qquad (7.2)$$

One should recall that not only U but also $E_{kin} = M\omega^2/2$ (M is the mass of the matter and ω is the speed of the center of the mass) and $E_{pot} = Mgz$ (g is the gravitational acceleration, and z is the elevation of the center of the mass of the matter) are functions of state. Therefore, the changes of the three energies during a transition between two states of equilibrium (or local-equilibrium) depend only on the characteristics of the states.

If the absolute values of $\mathcal{E}_1$, H_1, and S_1 (the initial state) and of $\mathcal{E}_2$, H_2, and S_2 (the final state) of a process of interest are already known (from some other calculations or measurements), then the changes of these properties during the process can easily be calculated as (see relationships (2.35), (2.37), and (2.39))

$$\Delta\mathcal{E} = \mathcal{E}_2 - \mathcal{E}_1, \qquad (7.3)$$
$$\Delta H = H_2 - H_1, \qquad (7.4)$$

and

$$\Delta S = S_2 - S_1. \qquad (7.5)$$

One should emphasize again that **relationships (7.3)–(7.5) are valid in any (reversible or not) process in any (closed or not) system when the process**

begins and ends in some states of equilibrium (quasi-equilibrium, local equilibrium). This is because the change of every function of state in such a process depends neither on the character of the process nor on the system's structure.

In the next few chapters of this book we will focus mainly on processes in closed thermodynamic systems. Processes in open systems (flows) will be studied in Chapters 11 and 13.

If a process under consideration takes place in a closed, nonflow system (that is, a system where the changes ΔE_{kin} and ΔE_{pot} are negligible in comparison with the change of the thermal energy ΔU), then expression (7.3) reduces to

$$\Delta \mathcal{E} = \Delta U = U_2 - U_1. \tag{7.6}$$

When the values of the thermodynamic properties of state in the initial and final states of a transition are not known, then expressions (7.4)–(7.6) cannot be used. In such a case, the changes of the thermal energy U, enthalpy H, and entropy S during a process between two equilibrium states can be obtained from relationships (4.31), (4.33), (4.36), and (4.37) if the values of the states' thermodynamic values (P_1, T_1, V_1, P_2, T_2, V_2) are known,

$$\Delta U = \int_{T_1}^{T_2} \{C_V(V,T)\}_{V=V_2} \, dT$$

$$+ \int_{V_1}^{V_2} \left\{ T \left[\frac{\partial P(V,T)}{\partial T} \right]_V - P(V,T) \right\}_{T=T_1} dV, \tag{7.7}$$

$$\Delta H = \int_{T_1}^{T_2} \{C_P(P,T)\}_{P=P_2} \, dT$$

$$+ \int_{P_1}^{P_2} \left\{ -T \left[\frac{\partial V(P,T)}{\partial T} \right]_P + V(P,T) \right\}_{T=T_1} dP, \tag{7.8}$$

$$\Delta S = \int_{T_1}^{T_2} \left\{ \frac{C_V(V,T)}{T} \right\}_{V=V_1} dT$$

$$+ \int_{V_1}^{V_2} \left\{ \left[\frac{\partial P(V,T)}{\partial T} \right]_V \right\}_{T=T_2} dV, \tag{7.9}$$

or

$$\Delta S = \int_{T_1}^{T_2} \left\{ \frac{C_P(P,T)}{T} \right\}_{P=P_1} dT$$

$$- \int_{P_1}^{P_2} \left\{ \left[\frac{\partial V(P,T)}{\partial T} \right]_P \right\}_{T=T_2} dP, \qquad (7.10)$$

where C_V and C_P are the system's constant-volume and constant-pressure heat capacities, respectively.

The fact that relationships (7.7)–(7.10) include constant-volume heat capacity C_V and constant-pressure heat capacity C_P is sometimes erroneously interpreted as the limitation of validity of the relationships to isochoric and isobaric processes only. In reality, the relationships are valid in *all* thermodynamic transitions between states of equilibrium (quasi-equilibrium, local-equilibrium). The presence of C_V and C_P in the relationships is just a result of the replacement of the general thermodynamic paths studied in Section 4.2 by mathematically convenient sequences of processes involving isochors and isobars (see Fig. 4.1).

The Helmholtz free energy A and the Gibbs free energy G also are functions of state in every process that begins and ends in equilibrium (quasi-equilibrium, local-equilibrium) states. Therefore, the changes of the free energies in any transition between two such states can be obtained from relationships (4.49) and (4.50), respectively. However, if the absolute values of the free energies for the initial and final states of the transition are known, the changes of the free energies during the transition can be found as (also see expressions (3.112) and (3.114))

$$\Delta A = A_2 - A_1, \qquad (7.11)$$

and

$$\Delta G = G_2 - G_1. \qquad (7.12)$$

Databases of the absolute values of U, H, S, A, and G for substances in states of equilibrium are discussed in Chapter 8.

7.3   Transfers of Properties of Path

Boundary work W, non-boundary work K, and heat Q of all processes in all thermodynamic systems are functions of paths even when the initial and final states of the processes are equilibrium states. Therefore, these path properties cannot be calculated from analytical expressions as simple as relationships (7.7)–(7.10) that describe the changes of thermodynamic functions of state. (However, one should recall that some work and heat of *reversible* transitions can be calculated from simple analytical expressions – see relationships (2.58)–(2.59) and discussions in Section 2.8.2 and Chapter 13.) Fortunately, work and heat of real processes can often be measured or calculated by methods other than those proposed by Classical Thermodynamics. For example, non-boundary (shaft) works of real processes in some systems can easily be determined when the work is supplied to the system by a motor (the powers of motors are usually listed by their manufacturers). Similarly, the heat transfers in combustion processes can often be determined with good accuracy because the amounts of heat produced by combustion of specific amounts of the fuels used in the processes are listed by the fuels' manufacturers.

In some special cases, the transfers of path properties in real thermodynamic processes can be easily calculated. For example, the boundary work of any isochoric process in any system is

$$W = 0 \qquad (7.13)$$

because the change of the system volume during every isochoric process is zero.

The degree of irreversibility in real processes is often proportional to the first and second gradients of pressure and temperature during the processes (see Section 2.8.2). Therefore, the processes where the gradients are insignificant can be treated as reversible transitions. Subsequently, the boundary work transfer in most real *isobaric* transitions in closed systems where the temperature gradients are small can be given as (see the discussion leading to expression (2.71))

$$W \simeq -P(V_2 - V_1), \qquad (7.14)$$

where V_1 and V_2 are the initial and final volumes of the system during the process, respectively, and P is the process's constant pressure.

When pressure gradients in a closed system undergoing a real *isothermal* transition are small, one can assume that the transfer of heat during the transition takes place in a reversible manner. Then, the heat transfer during the isotherm can be given as (see expression (2.72)),

$$Q \simeq T(S_2 - S_1), \qquad (7.15)$$

where S_1 and S_2 are entropies of the initial and final states of the process, respectively.

The Principle of Conservation of Matter (see Section 3.5.1) and the Principle of Conservation of Energy (the First Law of Thermodynamics – see Section 3.5.2) are valid in all non-relativistic processes in all systems. The Balance Equations for Matter and Energy in open systems will be further studied in Chapters 11 and 13. In this section we discuss the balances of thermal energy U, enthalpy H, entropy S, boundary work W, non-boundary work K, and heat Q in real (that is, irreversible) transitions between equilibrium states of closed, nonflow systems that do not exchange radiative energy with the environment. Such systems are common in thermodynamic applications.

The Balance Equation for Matter in any closed system undergoing $1{\rightarrow}2$ transition is

$$\Delta M = 0 \quad \rightarrow \quad M_2 = M_1, \tag{7.16}$$

where M is the total mass of the system's matter, and M_1 and M_2 are masses of the matter in the transition initial and final states, respectively.

According to the Principle of Conservation of Energy, the internal energy change $\Delta\mathcal{E}$, the total work L, and heat Q of any closed, nonflow system with no radiation transfer can be balanced as follows:

$$\Delta\mathcal{E} = \Delta U = L + Q, \tag{7.17}$$

where, as before, U is the system thermal energy, and L is the sum of boundary work W and non-boundary work K done on or done by the system,

$$L = W + K. \tag{7.18}$$

When the initial and final states of the considered transition are equilibrium states, ΔU in expression (7.17) can be obtained either from expression (7.7) or from expression (7.6) if the values of U_1 and U_2 are known. However, calculation of work and heat transfers is usually more difficult because both transfers are functions of path, that is, the knowledge of the characteristics of the initial and final states of the transition is not sufficient to do the calculation even when these states are equilibrium states.

7.4 Processes in Non-Ideal Matter

Applying relationships (7.7)–(7.10), (7.13) and (7.17)–(7.18) to isothermal, isobaric, isochoric, and isentropic transitions between an equilibrium state 1 and an equilibrium state 2 of any closed, nonflow thermodynamic system leads to:

1) isothermal transitions ($T = \text{const}$, $dT = 0$):

$$\Delta U = \int_{V_1}^{V_2} \left\{ T \left[\frac{\partial P(V,T)}{\partial T} \right]_V - P(V,T) \right\} dV, \tag{7.19}$$

$$\Delta H = \int_{P_1}^{P_2} \left\{ -T \left[\frac{\partial V(P,T)}{\partial T} \right]_P + V(P,T) \right\} dP, \tag{7.20}$$

$$\Delta S = \int_{V_1}^{V_2} \left[\frac{\partial P(V,T)}{\partial T} \right]_V dV \tag{7.21}$$

or

$$\Delta S = - \int_{P_1}^{P_2} \left[\frac{\partial V(P,T)}{\partial T} \right]_P dP. \tag{7.22}$$

The transition's total (net) heat transfer Q or the total (net) work transfer $L = W + K$ can be calculated from the Energy Balance Equation (7.17) (the First Law of Thermodynamics),

$$Q = \Delta U - L \qquad \text{or} \qquad L = \Delta U - Q, \tag{7.23}$$

or, when the concept of enthalpy is used,

$$Q = \Delta H - (P_2 V_2 - P_1 V_1) - L \quad \text{or} \quad L = \Delta H - (P_2 V_2 - P_1 V_1) - Q. \tag{7.24}$$

2) isochoric transitions ($V = \text{const}$, $dV = 0$):

$$\Delta U = \int_{T_1}^{T_2} C_V(V,T) dT, \tag{7.25}$$

$$\Delta H = \Delta U + V(P_2 - P_1), \tag{7.26}$$

and

$$\Delta S = \int_{T_1}^{T_2} \frac{C_V(V,T)}{T} dT. \tag{7.27}$$

The boundary work of the transition is (see expression (7.13))

$$W = 0, \tag{7.28}$$

and the transition's heat Q or non-boundary work K can be obtained from the First Law of Thermodynamics (also see relationship (2.52)),

$$Q = \Delta U - K \qquad \text{or} \qquad K = \Delta U - Q, \tag{7.29}$$

or

$$Q = \Delta H - V(P_2 - P_1) - K \qquad \text{or} \qquad K = \Delta H - V(P_2 - P_1) - Q. \tag{7.30}$$

3) isobaric transitions ($P = \text{const}, dP = 0$):

$$\Delta H = \int_{T_1}^{T_2} C_P(P,T)\,dT, \tag{7.31}$$

$$\Delta U = \Delta H - P(V_2 - V_1), \tag{7.32}$$

and

$$\Delta S = \int_{T_1}^{T_2} \frac{C_P(P,T)}{T}\,dT. \tag{7.33}$$

The transition's heat Q and its total work L can be obtained from the First Law of Thermodynamics (also see relationship (2.52)),

$$Q = \Delta U - L \qquad \text{or} \qquad L = \Delta U - Q, \tag{7.34}$$

which also can be written as,

$$Q = \Delta H - P(V_2 - V_1) - L \qquad \text{or} \qquad L = \Delta H - P(V_2 - V_1) - Q. \tag{7.35}$$

4) isentropic transitions ($S = \text{const}, dS = 0$):

There are two kinds of isentropic transitions, that is, the transitions during which the system's entropy does not change:

a) irreversible isentropic transition (often called *irreversible isentrop*) where, in an elementary process, $dS = 0$, $\delta S_{\text{nat}} > 0$, $\delta Q \neq 0$ (as before, δS_{nat} is the (differential) amount of the natural entropy added to the system during the process, and δQ is the heat of the process). If the transition is a finite process then

$$\Delta S = 0, \tag{7.36}$$

$$Q = \Delta U - L = \Delta H - (P_2 V_2 - P_1 V_1) - L, \tag{7.37}$$

or

$$L = \Delta U - Q = \Delta H - (P_2 V_2 - P_1 V_1) - Q, \tag{7.38}$$

where ΔU and ΔH can be obtained from general expressions (7.7) and (7.8), respectively.

b) reversible isentropic transition (often called *adiabat-isentrop* or *reversible isentrop*) where $dS = 0$, $\delta S_{\text{nat}} = 0$, and $\delta Q = 0$. (This model of isentropic processes is very often used in thermodynamic studies.) If the transition is a finite process, then

$$\Delta S = 0, \tag{7.39}$$

$$Q = 0, \tag{7.40}$$

and

$$L = \Delta U = \Delta H - (P_2 V_2 - P_1 V_1), \tag{7.41}$$

where ΔU and ΔH are given in expressions (7.7) and (7.8), respectively.

Since no natural entropy is produced during the reversible isentrop, there is no dissipation of energy during the process (see Sections 3.5.3 and 13.1). Also, expressions (4.28) and (4.29) lead to the following useful relationship valid in the process:

$$dP = -\frac{C_P}{C_V} \frac{(\partial P/\partial T)_V}{(\partial V/\partial T)_P} dV. \tag{7.42}$$

7.5 Processes in Ideal Gases

We discuss in this section the balancing of the internal energy, enthalpy, entropy, work, and heat of real thermodynamic transitions between equilibrium states of ideal gas in closed, nonflow systems with no radiation transfer. We do this using general relationships (7.7)–(7.10).

The equation of state for M kilograms (or ν kmols) of any single-component ideal gas in equilibrium is (see Section 5.1)

$$P = \frac{MR_gT}{V} \qquad \text{or} \qquad P = \frac{\nu RT}{V}, \tag{7.43}$$

where, as before, P, V and T are pressure, volume, and absolute temperature of the gas, respectively, R_g (in kJ/kg·K is the individual gas constant, and $R = 8.314\,$kJ/kmol·K is the universal (molar) gas constant. Thus,

$$T\left[\frac{\partial P(V,T)}{\partial T}\right]_V = \frac{MR_gT}{V} = P. \tag{7.44}$$

According to relationship (7.7), the change of the thermal energy of any *thermally perfect ideal gas* (that is, an ideal gas where heat capacities are functions of temperature only – see Section 6.1) undergoing *any* thermodynamic transition between an equilibrium state 1 and an equilibrium state 2 is

$$\Delta U = \int_{T_1}^{T_2} C_V(T)\,dT + \int_{V_1}^{V_2} [P - P]\,dV = \int_{T_1}^{T_2} C_V(T)\,dT, \tag{7.45}$$

where, as before, $C_V(T)$ is the temperature-dependence of the constant-volume heat capacity of the gas.

Equation of state (7.43) gives

$$-T\left[\frac{\partial V(P,T)}{\partial T}\right]_P = -\frac{MR_gT}{P} = -V, \tag{7.46}$$

and the change of enthalpy of a thermally ideal gas during *any* thermodynamic transition between an equilibrium state 1 and an equilibrium state 2 is (see expression (7.8)),

$$\Delta H = \int_{T_1}^{T_2} C_P(T)\,dT + \int_{P_1}^{P_2} [-V + V]\,dP = \int_{T_1}^{T_2} C_P(T)\,dT, \tag{7.47}$$

where $C_P(T)$ is the temperature-dependence of the constant-pressure heat capacity of the gas.

Relationships (7.45) and (7.47) tell us a lot about thermodynamic properties of ideal gases. One can see there that the **internal energy U and enthalpy H of every ideal gas in equilibrium are functions of temperature only** (this was confirmed experimentally by James Joule in 1843). One should notice that when U depends only on temperature, then H also depends only on temperature. This is because $H = U + PV$ and, according to the equation of state for ideal gas, $PV = vRT$. In addition, the **heat capacities C_V and C_P of every ideal gas in equilibrium are functions of temperature only** – see the Mayer Equation (6.25) and expression (6.31).

The heat capacities of ideal gases in equilibrium (matter in such systems is uniformly distributed in space) can be expressed by the specific heat capacities (often called 'specific heats') $c_V(T)$ and $c_P(T)$ (both in kJ/kg·K – Section 6.1),

$$C_V(T) = Mc_V(T) \qquad \text{and} \qquad C_P(T) = Mc_P(T), \qquad (7.48)$$

or by the molar heat capacities (often called 'molar heats') $\bar{c}_V(T)$ and $\bar{c}_P(T)$ (both in kJ/kmol·K) ,

$$C_V(T) = v\bar{c}_V(T) \qquad \text{and} \qquad C_P(T) = v\bar{c}_P(T), \qquad (7.49)$$

where, as before, M is the mass of the system's matter, v is the number of kilomols of the matter, and overbars mark molar values of the heat capacities. (Note that v can be taken as the number of mols but then the molar gas constant must be taken as $R = 8.314\,\text{J/mol·K}$.)

Expressions (7.9) and (7.10) in conjunction with expressions (7.44) and (7.46) give the following two ways of calculating the change of the system's entropy during a thermodynamic transition of M kilograms of a thermally ideal gas between equilibrium states 1 and 2:

a) when the gas properties are described by variables V and T,

$$\Delta S = \int_{T_1}^{T_2} \frac{C_V(T)}{T} dT + MR_g \ln(V_2/V_1), \qquad (7.50)$$

or

b) when the gas properties are described by variables P and T,

$$\Delta S = \int_{T_1}^{T_2} \frac{C_P(T)}{T} dT - MR_g \ln(P_2/P_1). \qquad (7.51)$$

Summarizing the above, one can write for an ideal gas in equilibrium:

$$U = U(T) \qquad , \qquad H = H(T), \qquad (7.52)$$

and

$$C_V = C_V(T) \quad , \quad C_P = C_P(T). \tag{7.53}$$

Expressions (7.50) and (7.51) tell us that entropy of any ideal gas in equilibrium depends on two thermodynamic variables (V and T or P and T),

$$S = S(V,T) \quad \text{or} \quad S = S(P,T). \tag{7.54}$$

Relationships (7.45), (7.47), (7.50)–(7.51) are very important in thermodynamic studies of processes in ideal gases. Therefore, it is useful to list them in one place;

$$\Delta U = \int_{T_1}^{T_2} C_V(T)\,dT, \tag{7.55}$$

$$\Delta H = \int_{T_1}^{T_2} C_P(T)\,dT, \tag{7.56}$$

$$\Delta S = \int_{T_1}^{T_2} \frac{C_V(T)}{T}\,dT + MR_g \ln(V_2/V_1), \tag{7.57}$$

or

$$\Delta S = \int_{T_1}^{T_2} \frac{C_P(T)}{T}\,dT - MR_g \ln(P_2/P_1), \tag{7.58}$$

together with the Energy Balance Equation (the First Law of Thermodynamics) for heat Q and work L in a closed, nonflow system undergoing a thermodynamic transition with no radiation transfer (see expression (3.58)),

$$Q = \Delta U - L, \tag{7.59}$$

or, using the concept of enthalpy,

$$Q = \Delta H - (P_2V_2 - P_1V_1) - L. \tag{7.60}$$

Expressions (7.55)–(7.60) lead to the following relationships describing real isothermal, isochoric, isobaric, and isentropic transitions between an equilibrium state 1 and an equilibrium state 2 of any ideal gas in any closed, nonflow system with no radiation transfer (note that the following expressions can be applied, in conjunction with relationships (3.50)–(3.54), to processes in closed systems with moving matter):

1) isothermal transitions ($T = $ const, $dT = 0$):

$$\Delta U = 0, \tag{7.61}$$

$$\Delta H = 0, \tag{7.62}$$

$$\Delta S = M R_g \ln(V_2/V_1), \tag{7.63}$$

or

$$\Delta S = -M R_g \ln(P_2/P_1), \tag{7.64}$$

$$L = W + K = -Q. \tag{7.65}$$

2) isochoric transitions ($V = $ const, $dV = 0$):

$$\Delta U = \int_{T_1}^{T_2} C_V(T)dT, \tag{7.66}$$

$$\Delta H = \Delta U + V(P_2 - P_1) = \Delta U + M R_g(T_2 - T_1), \tag{7.67}$$

$$\Delta S = \int_{T_1}^{T_2} \frac{C_V(T)dT}{T}, \tag{7.68}$$

$$W = 0, \tag{7.69}$$

$$Q + K = \Delta U. \tag{7.70}$$

3) isobaric transitions ($P = $ const, $dP = 0$):

$$\Delta H = \int_{T_1}^{T_2} C_P(T)dT, \tag{7.71}$$

$$\Delta U = \Delta H - P(V_2 - V_1) = \Delta H - M R_g(T_2 - T_1), \tag{7.72}$$

$$\Delta S = \int_{T_1}^{T_2} \frac{C_P(T)dT}{T}, \tag{7.73}$$

$$Q + W + K = \Delta H - P(V_2 - V_1). \tag{7.74}$$

4) isentropic transitions ($S = $ const, $dS = 0$):

a) irreversible isentropic transitions ($dS = 0$, $\delta S_{\text{nat}} > 0$, $\delta Q \neq 0$):

$$\Delta S = 0, \tag{7.75}$$

$$\Delta U = \int_{T_1}^{T_2} C_V(T)dT, \tag{7.76}$$

$$\Delta H = \Delta U + V(P_2 - P_1) = \Delta U + M R_g(T_2 - T_1), \tag{7.77}$$

and

$$L = \Delta U - Q = \Delta H - (P_2 V_2 - P_1 V_1) - Q. \tag{7.78}$$

b) adiabatic-isentropic transitions ($dS = \delta S_{\text{nat}} = \delta Q = 0$):

$$\Delta S = 0, \tag{7.79}$$

$$\Delta U = \int_{T_1}^{T_2} C_V(T)\,dT, \tag{7.80}$$

$$\Delta H = \Delta U + V(P_2 - P_1) = \Delta U + MR_g(T_2 - T_1), \tag{7.81}$$

$$Q = 0, \tag{7.82}$$

$$L = \Delta U = \Delta H - (P_2 V_2 - P_1 V_1). \tag{7.83}$$

As mentioned earlier, heat capacities C_V and C_P often do not change much with temperature when its range is not significant. Then, one can assume that they are constant within the range and equal to their range-averaged values. (One should recall that any gas of constant heat capacities is called the 'calorically perfect gas' – see Section 6.1.) Assuming constancy of heat capacities in relationships (7.57) and (7.58) (also see expression (7.42)) leads to the following expression, valid in adiabatic-isentropic transitions between equilibrium states 1 and 2 of a calorically perfect ideal gas:

$$\kappa = \frac{C_P}{C_V} = -\frac{\ln(P_2/P_1)}{\ln(V_2/V_1)}, \tag{7.84}$$

where the temperature-independent ratio κ is called the **isentropic exponent**. Subsequently,

$$\ln(P_2/P_1) = -\ln[(V_2/V_1)^\kappa]. \tag{7.85}$$

This allows one to write the so-called **isentropic relations** for calorically perfect ideal gases undergoing adiabatic-isentropic processes between equilibrium states,

$$\frac{P_2}{P_1} = \left(\frac{V_1}{V_2}\right)^\kappa, \tag{7.86}$$

$$\frac{T_2}{T_1} = \left(\frac{V_1}{V_2}\right)^{\kappa-1}, \tag{7.87}$$

and

$$\frac{T_2}{T_1} = \left(\frac{P_2}{P_1}\right)^{(\kappa-1)/\kappa}. \tag{7.88}$$

The above discussion suggests a generalization of expressions (7.86)–(7.88) on processes other than adiabatic isentrops. The generalization leads to the following expressions (called the **polytropic relations**) that are often used (sometimes without a rational justification) to study various irreversible processes where heat capacities can be assumed constant:

$$\frac{P_2}{P_1} = \left(\frac{V_1}{V_2}\right)^{m}, \tag{7.89}$$

$$\frac{T_2}{T_1} = \left(\frac{V_1}{V_2}\right)^{m-1}, \tag{7.90}$$

and

$$\frac{T_2}{T_1} = \left(\frac{P_2}{P_1}\right)^{(m-1)/m}, \tag{7.91}$$

where m is a constant called the **polytropic exponent** that can have values between $-\infty$ and $+\infty$. The values of the exponent for some common thermodynamic transitions are:

process	m
isobar	0
isotherm	1
adiabat-isentrop	κ
isochor	∞

where κ is the isentropic exponent defined in relationship (7.84).

One should recall that the changes of state properties in processes between equilibrium or local-equilibrium states of thermodynamic systems depend neither on the processes thermodynamic paths nor on the physical structures of the systems – see discussion in Section 2.8.1. Thus, the expressions used above for calculating the changes of state properties apply to any (reversible or not) process between any equilibrium (local-equilibrium) states of any system. The system's Energy Balance Equation (such as equation (7.74) or (7.78)) includes the changes of state properties but also the transfers of the path properties (heat

Q, works W and K). Therefore, a hard-to-calculate path property transfer can be obtained from a solution of the Energy Balance Equation that includes an easy-to-calculate change of a state property.

Problem 7.1: One mol of a single-phase pure substance (a closed, nonflow system) was moved thermodynamically from an equilibrium state 1 (of temperature $T_1 = 800\,\text{K}$ and volume $V_1 = 0.1\,\text{m}^3$) to an equilibrium state 2 (of temperature $T_2 = 100\,\text{K}$ and volume $V_2 = 0.3\,\text{m}^3$). The substance's equation of state is

$$PV^2 = vRTb, \tag{7.92}$$

where $R = 8.314\,\text{J/mol·K}$ is the molar (universal) gas constant, P, V, and T are the system's pressure, volume, and absolute temperature, respectively; v is the number of mols of the substance, and b is a constant equal to $0.5\,\text{m}^3$. The measured heat released during the transition is $4\,\text{kJ}$. (One should notice that any heat transfer that was *measured* must have come from a *real* (that is, *irreversible*) transition, and that expression (7.92) tells us that the discussed substance is a non-ideal gas.) The only kind of work that can be done on or done by the system is boundary work. The molar constant-pressure heat capacity and molar constant-volume heat capacity in the considered range of temperature are weakly dependent on thermodynamic variables other than temperature. Therefore, the molar heat capacities $\bar{c}_V$ and $\bar{c}_P$ (see expressions (6.3)–(6.6)) can be approximated by the following polynomials where the molar heat capacities are in J/mol·K and the absolute temperature T is in kelvins):

$$\bar{c}_P(P, T_1 \leq T \leq T_2) \simeq a_1 + b_1 T + c_1 T^2 + d_1 T^3, \tag{7.93}$$

and

$$\bar{c}_V(V, T_1 \leq T \leq T_2) \simeq a_2 + b_2 T + c_2 T^2 + d_2 T^3, \tag{7.94}$$

where $a_1 = 19.875$, $b_1 = 5.021 \times 10^{-2}$, $c_1 = 1.268 \times 10^{-5}$, $d_1 = -1.110 \times 10^{-8}$, $a_2 = 11.561$, $b_2 = 5.021 \times 10^{-2}$, $c_2 = 1.268 \times 10^{-5}$, and $d_2 = -1.1 \times 10^{-8}$.

Find the changes of the system's internal energy, enthalpy, entropy, and the boundary work transfer during the process.

Solution: First, we calculate the initial and final pressures of the transition. Using equation of state (7.92) one has

$$P_1 = \frac{vRbT_1}{V_1^2} = \frac{1 \times 8.314 \times 0.5 \times 800}{(0.1)^2} = 3.32\,\text{bar}, \tag{7.95}$$

and

$$P_2 = \frac{\nu R b T_2}{V_2^2} = \frac{1 \times 8.314 \times 0.5 \times 100}{(0.3)^2} = 0.05 \,\text{bar}. \qquad (7.96)$$

The changes of the system's internal energy (in this case, it is the thermal energy), enthalpy, and entropy can be obtained from general relationships (7.7)–(7.9), respectively,

$$\Delta U = \int_{T_1}^{T_2} \nu \bar{c}_V(T) dT$$
$$+ \int_{V_1}^{V_2} \left\{ T \left[\frac{\partial P(V,T)}{\partial T} \right]_V - P(V,T) \right\}_{T=T_1} dV, \qquad (7.97)$$

$$\Delta H = \int_{T_1}^{T_2} \nu \bar{c}_P(T) dT$$
$$+ \int_{P_1}^{P_2} \left\{ -T \left[\frac{\partial V(P,T)}{\partial T} \right]_P + V(P,T) \right\}_{T=T_1} dP, \qquad (7.98)$$

and

$$\Delta S = \int_{T_1}^{T_2} \frac{\nu \bar{c}_V(T)}{T} dT + \int_{V_1}^{V_2} \left\{ \left[\frac{\partial P(V,T)}{\partial T} \right]_V \right\}_{T=T_2} dV, \qquad (7.99)$$

where we will use, according to the problem formulation, $\nu = 1$ (also see expressions (7.48) and (7.49)).

The right-hand sides of relationships (7.97)–(7.99) require evaluation of several derivatives. This can be done using equation of state (7.92),

$$P = \frac{\nu R T b}{V^2}, \qquad (7.100)$$

which leads to

$$V = \pm \left(\frac{\nu R T b}{P} \right)^{1/2}, \qquad (7.101)$$

where, in agreement with reality, the positive value of V is chosen for the discussion that follows.

Differentiating of expressions (7.100) and (7.101) with respect to temperature gives

$$\left[\frac{\partial P(V,T)}{\partial T} \right]_V = \frac{\nu R b}{V^2}, \qquad (7.102)$$

and

$$\left[\frac{\partial V(P,T)}{\partial T}\right]_P = \frac{1}{2}\left(\frac{vRb}{PT}\right)^{1/2}.$$ (7.103)

Thus, expressions (7.97)–(7.99) lead to

$$\Delta U = \int_{T_1}^{T_2} v(a_2 + b_2T + c_2T^2 + d_2T^3)\,dT$$
$$+ \int_{V_1}^{V_2}\left\{T\frac{vRb}{V^2} - \frac{vRTb}{V^2}\right\}_{T=T_1} dV = a_2(T_2 - T_1)$$
$$+ \frac{6b_2(T_2^2 - T_1^2) + 4c_2(T_2^3 - T_1^3) + 3d_2(T_2^4 - T_1^4)}{12}$$
$$=-24.94\,\text{kJ},$$ (7.104)

$$\Delta H = \int_{T_1}^{T_2} v(a_1 + b_1T + c_1T^2 + d_1T^3)\,dT$$
$$+ \int_{P_1}^{P_2}\left\{-T\frac{1}{2}\left(\frac{vRb}{PT}\right)^{1/2} + \left(\frac{vRTb}{P}\right)^{1/2}\right\}_{T=T_1} dP = a_1(T_2 - T_1)$$
$$+ \frac{6b_1(T_2^2 - T_1^2) + 4c_1(T_2^3 - T_1^3) + 3d_1(T_2^4 - T_1^4)}{12}$$
$$+ [RbT_1(P_2 - P_1)]^{1/2} = -60.10\,\text{kJ},$$ (7.105)

and

$$\Delta S = \int_{T_1}^{T_2}\left\{\frac{v(a_2 + b_2T + c_2T^2 + d_2T^3)}{T}\right\}_{V=V_1} dT$$
$$+ \int_{V_1}^{V_2}\left\{\frac{vRb}{V^2}\right\}_{T=T_2} dV = a_2\ln(T_2/T_1) + b_2(T_2 - T_1)$$
$$+ (c_2/2)(T_2^2 - T_1^2) + (d_2/3)(T_2^3 - T_1^3)$$
$$+ Rb(V_1^{-1} - V_2^{-1}) = -0.033\,\text{kJ/K}.$$ (7.106)

Work is a function of path and, therefore, it depends on the kind of transition taking place between states 1 and 2. Classical Thermodynamics does not have any general expressions able to calculate boundary work (and the other kinds of work) from the values of thermodynamic variables and material properties of thermodynamic systems during their irreversible transitions. However, using the First Law of Thermodynamics (which is valid in all processes in all systems), one can write for the considered process that

$$W = \Delta U - Q = -24.94\,\text{kJ} - (-4\,\text{kJ}) = -20.94\,\text{kJ}. \tag{7.107}$$

In other words, according to the International Convention of the Sign of Energy, the system has done 20.94 kJ of boundary work during the $1 \rightarrow 2$ transition.

End of Problem 7.1.

Problem 7.2: In a closed, nonflow thermodynamic system, 2 kJ of heat Q is added slowly and isothermally to two mols of an ideal gas in order to move it thermodynamically from an equilibrium state 1 (of temperature $T_1 = 300$ K and pressure $P_1 = 2$ bar) to an equilibrium state 2 (of pressure $P_2 = 1.5$ bar). What is the total (boundary plus non-boundary) work L done on or done by the system during the process? What fraction of the total work is the non-boundary work K?

Solution: Since the considered gas is an ideal gas, its equation of state is (see expression (7.43))

$$PV = vRT, \tag{7.108}$$

where v is the number of mols of the gas, and $R = 8.314\,\text{J/mol·K}$ is the molar (universal) gas constant. Thus, one can write for the initial and final states of the process that

$$V_1 = \frac{vRT_1}{P_1} = \frac{2\times8.314\times300}{2\times10^5} = 0.025\,\text{m}^3, \tag{7.109}$$

and

$$V_2 = \frac{vRT_2}{P_2} = \frac{2\times8.314\times300}{1.5\times10^5} = 0.033\,\text{m}^3. \tag{7.110}$$

According to the First Law of Thermodynamics for closed, nonflow systems, the total work transferred between the system and the environment is

$$L = \Delta U - Q = 0\,\text{kJ} - (+2\,\text{kJ}) = -2\,\text{kJ} \tag{7.111}$$

where ΔU is the change of the system's internal (thermal) energy during the process; $\Delta U = 0$ because the process under consideration is an *isothermal* transition while the thermal energy of any ideal gas in equilibrium is a function of temperature only.

Since the $1 \rightarrow 2$ transition is isothermal, slow, and with no significant change of pressure, one can assume that the transition is close to a reversible process (see Section 2.8.2). Then, the transition's boundary work transfer can be calculated from expression (2.59) as

$$
\begin{aligned}
W \simeq W_{\text{rev}} &= -\int_{V_1}^{V_2} P dV = -\nu R T_1 \int_{V_1}^{V_2} V^{-1} dV \\
&= -\nu R T_1 \ln[V_2/V_1] = -2 \times 8.314 \times 300 \times \ln[0.033/0.025] \\
&= -1.435 \,\text{kJ}.
\end{aligned}
\tag{7.112}
$$

Thus, the system is able to produce $2\,\text{kJ}$ of work; $1.435\,\text{kJ}$ of it is the boundary work W, and the rest is the non-boundary work K,

$$
K = L - W = -2\,\text{kJ} - (-1.435)\,\text{kJ} = -0.565\,\text{kJ}. \tag{7.113}
$$

End of Problem 7.2.

Chapter 8

Absolute Values of State Properties

In Chapter 7 we derived expressions for *changes* of properties of state in thermodynamic processes between equilibrium (local-equilibrium) states. In this chapter we discuss the methods of finding the *absolute* values of the properties in such states. The knowledge of the absolute values is very useful because the change ΔF of a function of state F ($F \equiv U, H, S, A$ or G) during any transition between an equilibrium (local-equilibrium) state 1 and an equilibrium (local-equilibrium) state 2 can be quickly calculated as (see expressions (2.35), (2.37), (2.39), (2.41), and (2.43))

$$\Delta F = F_2 - F_1, \tag{8.1}$$

where F_1 and F_2 are the absolute values of the property F in the initial and final states, respectively.

We will focus our attention now on the *absolute* enthalpy H and the *absolute* entropy S of substances in equilibrium because these two properties play important roles in many applications. The other state properties can be calculated using values of H and S (see Section 3.7.2),

$$U = H - PV, \tag{8.2}$$

$$A = U - TS, \tag{8.3}$$

$$G = H - TS = U + PV - TS, \tag{8.4}$$

where U is the absolute thermal energy, A is the absolute Helmholtz free energy, and G is the absolute Gibbs free energy of the substance in a given state of equilibrium. Since (see Section 3.7.4)

$$S = -\left(\frac{\partial G}{\partial T}\right)_P \qquad \text{and} \qquad V = \left(\frac{\partial G}{\partial P}\right)_T, \tag{8.5}$$

one has

$$U = G - PV + TS = G - P\left(\frac{\partial G}{\partial P}\right)_T - T\left(\frac{\partial G}{\partial T}\right)_P, \tag{8.6}$$

$$H = G + TS = G - T\left(\frac{\partial G}{\partial T}\right)_P, \tag{8.7}$$

and

$$A = G - PV = G - P\left(\frac{\partial G}{\partial P}\right)_T. \tag{8.8}$$

Expressions (8.5)–(8.8) show the importance and general character of the Gibbs free energy in equilibrium systems described by thermodynamic variables P and T.

8.1 General Expressions for Functions of State

The differential change of enthalpy of a substance with respect to temperature during an *isobaric* transition between two states of equilibrium can be obtained from relationship (6.2),

$$\left(\frac{\partial H}{\partial T}\right)_P = C_P(P,T), \tag{8.9}$$

where, as before, C_P is the constant-pressure heat capacity of the substance. Integrating this expression between an equilibrium state (of pressure P and temperature $T = 0\,\mathrm{K}$) and an equilibrium state of pressure P and temperature T gives

$$H(P,T) = H(P,0) + \int_0^T C_P(P,T)\,dT, \tag{8.10}$$

where $H(P,T)$ is the *absolute* enthalpy of the substance in equilibrium at pressure P and temperature T, and $H(P,0)$ is the *absolute* enthalpy of the substance in equilibrium at pressure P and temperature $T = 0$.

Relationship (8.10) should be used with caution because the substance at pressure P might have already passed through some phase transitions (solid→liquid,

liquid→gas, or solid→gas) while isobarically changing its temperature from absolute zero to T – see Chapter 9. (In Classical Thermodynamics, we usually consider only solid⇌liquid, liquid⇌gas, and solid⇌gas phase transitions. Some solids can change their electrical and/or mechanical properties (for example, elasticity) while remaining in solid phase during an isobaric change of the solid's temperature. Such transitions are called the solid⇌solid *allotropic* phase transitions.)

The constant-pressure heat capacity $C_P(P,T)$ has one or more discontinuities associated with the phase transitions – see Fig. 6.1. Therefore, one has to replace general relationship (8.10) by its more detailed version (a shorter version of the relationship can be written for the substance where some phase transitions are absent during the isobaric change of the substance temperature from zero to T),

$$H(P,T) = H(P,0) + \int_0^{T_1} C_P(P,T)dT + \Delta H_{s\to s}(P,T_1)$$

$$+ \int_{T_1}^{T_2} C_P(P,T)dT + \Delta H_{s\to l}(P,T_2) + \int_{T_2}^{T_3} C_P(P,T)dT$$

$$+ \Delta H_{l\to g}(P,T_3) + \int_{T_3}^{T} C_P(P,T)dT, \tag{8.11}$$

where $\Delta H_{s\to s}$, $\Delta H_{s\to l}$ and $\Delta H_{l\to g}$ are called the **latent enthalpies** or **latent heats**. They represent isobaric-isothermal changes of the substance's enthalpy during the complete solid→solid phase transition at temperature T_1, the complete solid→liquid phase transition at temperature T_2, and the complete liquid→gas phase transition at temperature T_3, respectively (see Fig. 6.1). The values of temperatures T_1, T_2, and T_3 are unique for every pure substance at a given (constant) pressure. (Thermal properties of multi-phase substances are studied in Chapter 9.)

A pure substance undergoing an isobaric process in a wide temperature range can be in solid phase (when $T < T_2$), liquid phase (when $T_2 < T < T_3$), or gas phase (when $T > T_3$). Therefore, some terms in expression (8.11) are not needed in studies of isobaric processes in some ranges of temperature. Also, particles of a single-component gas at temperatures well above T_3 can start to dissociate and produce a mixture, that is, become a multi-component system.

Summarizing the above, one can say that the qualitative dependences of heat capacities of pure substances at constant pressure are similar to the corresponding dependence of diatomic nitrogen shown in Fig. 6.1. The heat capacities have discontinuities (at temperatures T_1, T_2, and T_3) during *isobaric* heating of the substance. Each of the discontinuities represents an isobaric-isothermal **phase transition**.

The *absolute* entropy S of a pure substance in equilibrium at pressure P and temperature T can be obtained from expression (4.29),

$$S(P,T) = S(P,0) + \int_0^{T_1} \frac{C_P(P,T)}{T} dT + \frac{\Delta S_{s \to s}(P,T_1)}{T}$$
$$+ \int_{T_1}^{T_2} \frac{C_P(P,T)}{T} dT + \frac{\Delta S_{s \to l}(P,T_2)}{T} + \int_{T_2}^{T_3} \frac{C_P(P,T)}{T} dT$$
$$+ \frac{\Delta S_{l \to g}(P,T_3)}{T} + \int_{T_3}^{T} \frac{C_P(P,T)}{T} dT, \tag{8.12}$$

where $\Delta S_{s \to l}$ and $\Delta S_{l \to g}$ are isobaric-isothermal changes of the entropy during the complete solid$\to$solid phase transition, the complete solid$\to$liquid phase transition, and the complete liquid$\to$gas phase transition, respectively. The meaning of the subscripts is the same as in relationship (8.11).

The absolute values of thermodynamic functions of state of many pure substances in equilibrium at pressure P and temperature T have been tabulated. One of the best databases of this kind is the *NIST-JANAF Thermochemical Tables* listing the values of several properties of state for hundreds of ideal gases in equilibrium in a broad range of temperature T (up to several thousand kelvins) at pressure $P = P^\circ = 1$ bar. The pressure P° is called the **standard pressure**, and any substance at this pressure is said to be in its **standard state** regardless of the temperature of the substance. (The standard pressure of early tabulations of thermodynamic properties was usually taken as 1 Atm).

One should recall (see Section 7.5) that the **thermal energy and enthalpy of any ideal gas in equilibrium are functions of temperature only. Therefore, the values of these two properties at given temperature and pressure are the same as the values of the properties at the same temperature and any other pressure including the standard pressure P°.**

As mentioned earlier, calculation of $H(P,T)$ and $S(P,T)$ requires use of methods of Quantum Statistical Mechanics. Since Classical Thermodynamics has no tools to deal with quantum-mechanical properties of matter, it has developed its own thermochemical concepts and procedures, including the very powerful concept of the so-called **reaction of formation**, that allow one to calculate the absolute values of the state properties without use of Quantum Statistical Mechanics. The thermochemical approach is discussed in the next section.

8.2 The Thermochemical Approach

8.2.1 The Reference Levels

The thermochemical approach to formation of molecules is based on the concept of the **reaction of formation**, and on the related concepts of the **energy reference level**, the **enthalpy reference level**, and the **entropy reference level**. In the approach, the *change* of the thermal energy of any substance during any thermodynamic process between an equilibrium state 1 and an equilibrium state 2 can be given as

$$\Delta U_{1\to2} = U_2 - U_1 = [U_b + U_{b\to2}] - [U_b + U_{b\to1}], \qquad (8.13)$$

where U_1 and U_2 are the *absolute* thermal energies of the initial and final states of the process, respectively, U_b is the energy reference level, and $U_{b\to1}$ and $U_{b\to2}$ are the states' *relative* thermal energies measured from the reference level U_b. If the reference level is fixed during the entire process under consideration, then

$$\Delta U_{1\to2} = U_2 - U_1 = U_{b\to2} - U_{b\to1}. \qquad (8.14)$$

It is a common, but informal, practice (also used in this book) to treat the *relative* energies $U_{b\to1}$ and $U_{b\to2}$ as the *absolute* thermal energies of states 1 and 2, respectively, and denote them just U_1 and U_2, respectively. (The relative energies $U_{b\to1}$ and $U_{b\to2}$ are called the *thermochemical energies* of the states.) Such practice is allowed in thermodynamic studies since it has no impact on the solutions of the Energy Balance Equations which deal only with the *changes* (not the *absolute* values) of the state properties, and the changes do not depend on the value of the energy reference level U_b – see Section 3.5.

In most thermodynamic and thermochemical studies, dealing with enthalpy is more convenient than dealing with internal energy. Therefore, the enthalpy reference level is usually of greater interest than the energy reference level. One can write, analogously to relationships (8.13) and (8.14), that the *change* of the enthalpy of any system undergoing any thermodynamic transition between two states of equilibrium is

$$\Delta H_{1\to2} = H_2 - H_1 = [H_b + H_{b\to2}] - [H_b + H_{b\to1}] = H_{b\to2} - H_{b\to1}, \quad (8.15)$$

where H_1 and H_2 are the *absolute* enthalpies of the initial and final states of the process, respectively, H_b is the reference level from which enthalpy is measured, and the *relative* enthalpies $H_{b\to1}$ and $H_{b\to2}$ are (informally) treated in thermodynamic calculations as the absolute enthalpies of states 1 and 2. These relative enthalpies are then denoted H_1 and H_2 and used in thermodynamic studies (also in this text) as if they were the absolute enthalpies of states 1 and 2, respectively. The relative entalpies $H_{b \to 1}$ and $H_{b \to 2}$ are sometimes called the *thermochemical enthalpies* to emphasize the fact that they are not the absolute enthalpies of the substance.

The entropy of a pure substance in equilibrium can also be measured from some entropy reference level. Then the change of the entropy of the substance during any thermodynamic transition between an equilibrium state 1 and an equilibrium state 2 can be given as

$$\Delta S_{1 \to 2} = S_2 - S_1 = [S_b + S_{b \to 2}] - [S_b + S_{b \to 1}] = S_{b \to 2} - S_{b \to 1}, \qquad (8.16)$$

where S_1 and S_2 are the *absolute* entropies of the transition's initial and final states, respectively, S_b is the entropy reference level, and the *relative* entropies $S_{b \to 1}$ and $S_{b \to 2}$ are often treated in thermodynamic and thermochemical studies as if they were *absolute* entropies of the states 1 and 2. These relative entropies are sometimes called the *thermochemical entropies*.

8.2.2 Terminology of Thermochemistry

Tables of absolute values of thermodynamic properties of state for substances in equilibrium usually list the values of the enthalpies at some specific pressure (the so-called **standard pressure** $P = P^\circ = 1$ bar) and within a temperature range up to a few thousand kelvins. Subsequently, the enthalpy, entropy, the Helmholtz free energy, and the Gibbs free energy of the substances in the standard states (that is, states at the standard pressure $P^\circ = 1$ bar) are marked by superscript $^\circ$ and often written as

$$H^\circ(T) \equiv H(P = P^\circ, T), \quad S^\circ(T) \equiv S(P = P^\circ, T), \qquad (8.17)$$

and

$$A^\circ(T) \equiv A(P = P^\circ, T), \quad G^\circ(T) \equiv G(P = P^\circ, T). \qquad (8.18)$$

The term 'compound' is often used in thermochemical literature for any pure substance consisting of one kind of molecules (complex structures), each made of 'elements' (simpler structures) which are parts of the molecules. (This terminology is somewhat confusing because term 'element' is often reserved for atoms – see Section 2.1 and *Periodic Chart of Elements*.) Thus, in thermodynamic and thermochemical literature reaction $C + O_2 \rightarrow CO_2$ is often seen as production of compound CO_2 from elements C and O_2.

8.2.3 Reaction of Formation

Let us consider the following chemical reaction at a (constant) standard pressure $P^\circ = 1$ bar and a (constant) temperature T:

$$S(s) + H_2SO_4(l) + \frac{1}{2}O_2(g) \rightarrow 2SO_2(g) + H_2O(l). \tag{8.19}$$

where the letter 'g' in the parenthesis following the chemical symbol of a component indicates that the component is in its gas phase; letters 'l' and 's' are reserved to indicate the liquid and solid phases of the component, respectively.

To study the effects of reaction (8.19), one must know the state property of each of the reaction's components (reactants $S(s)$, $H_2SO_4(l)$, and $O_2(g)$, and products $SO_2(g)$ and $H_2O(l)$). Such a property can be found from the so-called **reaction of formation** at standard pressure P° and temperature T. As an example, we use the thermochemical approach to find the absolute enthalpy of only one component participating in reaction (8.19) – the product $H_2O(l)$. The approach can be applied to any standard-state component at any (practically meaningful) temperature T.

There are several different chemical reactions that can produce 1 mol of $H_2O(l)$ at constant pressure $P = P^\circ = 1$ bar and constant temperature T. However, the most convenient of them is the one that has only two reactants, 1 mol of $H_2(g)$ and half a mol of $O_2(g)$, and only one product – 1 mol of $H_2O(l)$. Such a reaction can be written as

$$H_2(g) + \frac{1}{2}O_2(g) \rightarrow H_2O(l). \tag{8.20}$$

One of useful features of reaction (8.20) is the fact that it produces only 1 mol of $H_2O(l)$, an amount that is already suitable for tabularization. Another useful fact is that the elements $H_2(g)$ and $O_2(g)$ are the most chemically stable (that is, the least chemically active) substances able to produce $H_2O(l)$ at pressure P° and temperature T. This guarantees completion of the reaction (8.20) according to its stoichiometric equation. In addition, since pressure and temperature are constant during the reaction, the initial total enthalpy of the elements $H_2(g)$ and

$O_2(g)$ can be taken as a fixed *enthalpy reference level* from which all enthalpies of all particles participating in the reaction are measured. Of course, the most convenient choice of the value of the enthalpy reference level is zero. Then, the total change of enthalpy during the process of formation of *one* mol of compound $H_2O(l)$ at the constant pressure P° and at a constant temperature T is equal to the entire enthalpy of the compound. This enthalpy is called the compound's **enthalpy of formation** $\Delta_f H^\circ_{H_2O(l)}(T)$. Thus,

$$\Delta_f H^\circ_{H_2O(l)}(T) = H^\circ_{\text{product}}(T) - H^\circ_{\text{reactants}}(T)$$

$$= \bar{H}^\circ_{H_2O(l)}(T) - \left[\bar{H}^\circ_{H_2(g)}(T) + \frac{1}{2}\bar{H}^\circ_{O_2(g)}(T)\right]$$

$$= \bar{H}^\circ_{H_2O(l)}(T) - 0 = \bar{H}^\circ_{H_2O(l)}(T), \tag{8.21}$$

where, as before, the overbar indicates the molar value of enthalpy, and the following common notation is used:

$$\Delta_f H^\circ_{H_2O(l)}(T) \equiv \Delta_f H_{H_2O(l)}(P = P^\circ = 1\,\text{bar}, T), \tag{8.22}$$

where the subscript f emphasizes the fact that the reaction under consideration is a reaction of formation; no overbar is associated with term $\Delta_f H^\circ(T)$ because the subscript f indicates that the change of the enthalpy is that of a reaction of formation which ends up, by definition, as only *one* mol of the desired product.

Since there is no change of pressure and temperature during the reaction of formation, the change $\Delta_f H^\circ_{H_2O(l)}(T)$ is due only to the chemical composition of the single compound produced by the reaction. Also, $\Delta_f H^\circ_{H_2O(l)}(T)$ (the molar enthalpy of $H_2O(l)$), as well as the enthalpies of formation of other substances, can be accurately measured because they are equal to the heat transfers of the corresponding reactions of formation in closed, nonflow thermodynamic systems where there are neither radiation nor non-boundary work transfers. This is because the isobaric-isothermal character of every reaction of formation allows one to treat the reaction as a reversible process. Then, the absence of non-boundary work transfer gives $L = W = -P(V_2 - V_1)$ and $\Delta H = Q$ – see expressions (2.71) and (7.35).

Result (8.21) leads to the statement that the molar enthalpy of any pure substance in equilibrium at the standard-state pressure P° and temperature T is equal to the enthalpy of formation of the substance at the pressure and temperature.

One should notice the simplicity of reactions of formation: only 'elements' are present on the reactants side and only a single 'compound' (1 mol of it) is present on the products side. Some chemical reactions that may seem to qualify as 'reactions of formation' cannot proceed in the way described by their stoichiometric equations. An example of such a reaction, impossible in practice, is the one that would produce liquid-phase hydrogen peroxide from the very stable elements $H_2(g)$ and $O_2(g)$,

$$H_2(g) + O_2(g) \rightarrow H_2O_2(l). \tag{8.23}$$

Liquid hydrogen peroxide can be made by some processes but not by the reaction (8.23).

According to the International Convention for the Sign of Energy, the enthalpy of formation of a substance can be a positive or a negative number depending on whether the reaction is *endothermic* (absorbing heat) or *exothermic* (releasing heat), respectively. Also, the enthalpies of formation of chemically stable elements are zero (see Section 8.2.2 and Problem 8.3),

$$\begin{aligned} \Delta_f H^\circ(T) &= 0 \quad &\text{chemically stable elements,} \\ \Delta_f H^\circ(T) &\neq 0 \quad &\text{compounds.} \end{aligned} \tag{8.24}$$

Thus, chemically stable elements of reactions of formation of compounds at standard pressure P° and temperature T can easily be identified in thermochemical databases because the values of the elements' enthalpies of formation listed there are equal to zero – see Table 8.1 in Appendix A.

8.2.4 The Molar Enthalpies of Substances

The molar absolute enthalpy of a single-phase, single-component substance in equilibrium at pressure P and temperature T can be given as a sum of $\Delta_f H^\circ(T_o)$ (the enthalpy of formation of the substance at the standard pressure P° and reference temperature T_o), and the so-called **sensible enthalpy** $[\bar{H}(P,T) - \bar{H}^\circ(T_o)]$ which is the change of the substance enthalpy when it is moved thermodynamically from the state (P°, T_o) to the actual state (P, T). In other words, the molar entropy is

$$\bar{H}(P,T) = \Delta_f H^\circ(T_o) + [\bar{H}(P,T) - \bar{H}^\circ(T_o)], \tag{8.25}$$

where the sensible enthalpy can be obtained from expression (7.8) as

$$[\bar{H}(P,T) - \bar{H}^\circ(T_\circ)] = \int_{T_\circ,P^\circ}^{T,P} d\bar{H} = \int_{T_\circ}^{T} \bar{c}_P(P,T)dT$$

$$+ \int_{P_\circ}^{P} \left\{ -T \left[\frac{\partial \bar{V}(P,T)}{\partial T} \right]_P + \bar{V}(P,T) \right\}_{T=T_\circ} dP. \quad (8.26)$$

One should notice that sensible enthalpy (8.26) can easily be calculated if the equation of state (relationship between P, T, and V) and the constant-pressure molar heat of the substance under consideration are known.

The most common choice of the **reference temperature** in thermodynamic studies of standard-state systems is either $T_\circ$ = 298.15 K (25 °C) or $T_\circ$ = 273.15 K (0 °C).

When the considered substance is a single-component ideal gas in equilibrium at pressure P and temperature T, the last integral in relationship (8.26) is zero because the equation of state for one mol of any ideal gas is $\bar{V}(P,T) = RT/P$. Also, since the constant-pressure heat capacity and enthalpy of any ideal gas are pressure-independent functions of temperature only, one has in such a gas (see Section 7.5 and expressions (7.52) and (7.53))

$$\bar{c}_P(P,T) = \bar{c}_P(\text{any value of } P,T) = \bar{c}_P(P^\circ,T) \equiv \bar{c}_P^\circ(T), \quad (8.27)$$

and

$$\bar{H}(P,T) = \bar{H}(\text{any value of } P,T) = \bar{H}(P^\circ,T) \equiv \bar{H}^\circ(T). \quad (8.28)$$

Subsequently, the sensible enthalpy in any transition between two $((P^\circ, T_\circ)$ and $(P,T))$ equilibrium states of any ideal gas can be given as

$$[\bar{H}(P,T) - \bar{H}^\circ(T_\circ)] = [\bar{H}^\circ(T) - \bar{H}^\circ(T_\circ)] = \int_{T_\circ}^{T} \bar{c}_P^\circ(T)dT. \quad (8.29)$$

The enthalpy of formation $\Delta_f H(T_\circ)$ and standard-state sensible enthalpy $[\bar{H}^\circ(T) - \bar{H}^\circ(T_\circ)]$ of single-phase, single-component substances in equilibrium are tabulated in various thermochemical databases. Tables 8.1 and 8.2 (see Appendix A) show examples of such data for ideal diatomic oxygen gas ($O_2(g)$) and ideal carbon monoxide gas ($CO(g)$), respectively (the data are taken from the *NIST-JANAF Thermochemical Tables*). The Tables also list the standard-state values of $\bar{c}_P^\circ$ (the constant-pressure molar heat capacity), $\bar{S}^\circ$ (the standard-state molar entropy – see the next Section), $\bar{G}^\circ$ (the standard-state molar Gibbs free energy), $\Delta_f H^\circ$ (the enthalpy of formation), and $\Delta_f G^\circ$ (the Gibbs free energy of formation) of a large number of standard-state pure substances in equilibrium at temperature T.

One can see in Table 8.1 of Appendix A that molecules of ideal diatomic oxygen gas are chemically stable because their formation enthalpies $\Delta_f H^\circ$ are zero in the entire range of temperatures considered in the table (see expressions (8.24)). Also, one can see there that the sensible enthalpy of one mol of ideal carbon monoxide gas in equilibrium at the standard (and any other) pressure and temperature T is:

$$
\begin{aligned}
&\text{at } T = 100\,\text{K} &&\rightarrow && [\bar{H}^\circ(T) - \bar{H}^\circ(T_o)] = -5.77\,\text{kJ/mol},\\
&\text{at } T = 200\,\text{K} &&\rightarrow && [\bar{H}^\circ(T) - \bar{H}^\circ(T_o)] = -2.86\,\text{kJ/mol},\\
&\text{at } T = 298.15\,\text{K} &&\rightarrow && [\bar{H}^\circ(T) - \bar{H}^\circ(T_o)] = 0,\\
&\text{at } T = 500\,\text{K} &&\rightarrow && [\bar{H}^\circ(T) - \bar{H}^\circ(T_o)] = 6.08\,\text{kJ/mol}.
\end{aligned}
$$

8.2.5 The Molar Entropies of Substances

The absolute entropy of one mol of a pure substance in equilibrium at pressure P and temperature T can be given as the substance's entropy at the same temperature T and standard pressure P° plus the change of the entropy of the substance when it is moved thermodynamically from the (equilibrium) state (P°, T) to the actual (equilibrium) state (P, T). In other words,

$$\bar{S}(P,T) = \bar{S}^\circ(T) + [\bar{S}(P,T) - \bar{S}^\circ(T)], \tag{8.30}$$

where the change $[\bar{S}(P,T) - \bar{S}^\circ(T)]$ is often called the **sensible entropy** of the substance.

The sensible entropy can be calculated as the change of the system entropy in isothermal transition from equilibrium state (P°, T) to equilibrium state (P, T) (see relationship (7.10)),

$$[\bar{S}(P,T) - \bar{S}^\circ(T)] = \int_{T,P^\circ}^{T,P} d\bar{S} = \int_{P^\circ}^{P} d\bar{S} = -\int_{P^\circ}^{P} \left[\frac{\partial \bar{V}(P,T)}{\partial T}\right]_P dP, \tag{8.31}$$

where knowledge of the substance's equation of state is sufficient to calculate the integral.

When the substance under consideration is an ideal gas in equilibrium, then $\bar{V}(P,T) = RT/P$ and expression (8.31) becomes

$$[\bar{S}(P,T) - \bar{S}^\circ(T)] = -R \ln(P/P^\circ), \tag{8.32}$$

and relationship (8.30) gives

$$\bar{S}(P,T) = \bar{S}^\circ(T) - R\ln(P/P^\circ). \tag{8.33}$$

Thus, the **entropy of an ideal gas in equilibrium is a function of both pressure and temperature (but the gas's enthalpy and thermal energy are functions of temperature only).**

The standard-state molar entropies $\bar{S}^\circ(T)$ of many single-component ideal gases in equilibrium are listed in thermodynamic databases such as the *NIST-JANAF Thermochemical Tables*. Some of the entropies of ideal carbon monoxide in equilibrium are reproduced in Table 8.2 in Appendix A. One can see there that

$$
\begin{aligned}
\text{at } T = 100\,\text{K} \quad &\rightarrow \quad \bar{S}^\circ(T) = 165.85\,\text{J/mol·K},\\
\text{at } T = 200\,\text{K} \quad &\rightarrow \quad \bar{S}^\circ(T) = 186.02\,\text{J/mol·K},\\
\text{at } T = 298.15\,\text{K} \quad &\rightarrow \quad \bar{S}^\circ(T) = 197.65\,\text{J/mol·K},\\
\text{at } T = 500\,\text{K} \quad &\rightarrow \quad \bar{S}^\circ(T) = 212.83\,\text{J/mol·K}.
\end{aligned}
$$

The reader should notice that the value of $\bar{S}^\circ(T = T_\circ = 298.15\,\text{K})$ is not zero.

Problem 8.1: Find the enthalpy, entropy, the Gibbs free energy, and the constant-pressure heat capacity of one mol of ideal carbon monoxide gas in equilibrium at temperature $T = 1000\,\text{K}$ and pressure $P = 3\,\text{bar}$.

Solution: The value of the enthalpy of one mol of a single-component ideal gas in equilibrium at pressure P (and at any other pressure, including P°) and temperature T is given as (see relationship (8.25))

$$\bar{H}(P,T) = \Delta_f H^\circ(T_\circ) + [\bar{H}^\circ(T) - \bar{H}^\circ(T_\circ)], \tag{8.34}$$

where the enthalpy of formation of the gas at standard-state pressure $P^\circ = 1\,\text{bar}$ and reference temperature $T_\circ = 298.15\,\text{K}$ is (see column 6 in Table 8.2 in Appendix A)

$$\Delta_f H^\circ(T_\circ) = -110.52\,\text{kJ/mol}, \tag{8.35}$$

and the molar sensible enthalpy is (see column 5 in Table 8.2 in Appendix A)

$$[\bar{H}^\circ(T = 1000\,\text{K}) - \bar{H}^\circ(T_\circ = 298.15\,\text{K})] = 21.69\,\text{kJ/mol}. \tag{8.36}$$

Thus, according to expression (8.34), the enthalpy of one mol of ideal-gas carbon monoxide in equilibrium at temperature $T = 1000$ K and pressure $P = 3$ bar (or at any other pressure) is

$$\bar{H}(P = 3\,\text{bar}, T = 1000\,\text{K}) = -110.52\,\text{kJ/mol} + 21.69\,\text{kJ/mol}$$
$$= -88.83\,\text{kJ/mol}. \tag{8.37}$$

The entropy of one mol of the gas in equilibrium at temperature T and pressure P is (see expression (8.32))

$$\bar{S}(P, T) = \bar{S}^{\circ}(T) - R\ln(P/P^{\circ}), \tag{8.38}$$

where $R = 8.314$ J/mol·K is the molar gas constant, and, as before, $P^{\circ} = 1$ bar.

The standard-state molar entropy of the gas in equilibrium at $T = 1000$ K and $P^{\circ} = 1$ bar is (see column 3 in Table 8.2 in Appendix A)

$$\bar{S}^{\circ}(T = 1000\,\text{K}) = 234.57\,\text{J/mol·K}. \tag{8.39}$$

Thus (see expression (8.38)),

$$\bar{S}(P = 3\,\text{bar}, T = 1000\,\text{K})$$
$$= 234.57\,\text{J/mol·K} - (8.314\,\text{J/mol·K})\ln(3\,\text{bar}/1\,\text{bar})$$
$$= 225.41\,\text{J/mol·K}. \tag{8.40}$$

The Gibbs free energy of one mol of the gas can be found from definition (3.114) which, using results (8.37) and (8.40), leads to

$$\bar{G}(P = 3\,\text{bar}, T = 1000\,\text{K}) = \bar{H}(P = 3\,\text{bar}, T = 1000\,\text{K}) +$$
$$- T\bar{S}(P = 3\,\text{bar}, T = 1000\,\text{K}) =$$
$$= -88.83\,\text{kJ/mol} - (1000\,\text{K})(0.22\,\text{kJ/mol·K}) =$$
$$= -313.83\,\text{kJ/mol}. \tag{8.41}$$

The constant-pressure molar heat capacity of the ideal gas is (see expressions (6.3)–(6.6), (6.30), and column 2 in Table 8.2 in Appendix A)

$$\bar{c}_P(P, T) = \bar{c}_P^{\circ}(T) = 33.18\,\text{J·K}^{-1}\text{mol}^{-1}. \tag{8.42}$$

The above values of the molar enthalpy, entropy, and the Gibbs free energy of the ideal-gas carbon monoxide can be used to calculate, while considering the **departure functions** discussed in Section 8.3, the corresponding thermodynamic functions of real (non-ideal) carbon monoxide in a state of equilibrium at pressure P and temperature T.

End of Problem 8.1.

Problem 8.2: Write the reaction of formation of ideal-gas carbon monoxide (CO) at temperature $T = 1000\,\text{K}$, and find the molar enthalpy of formation of the gas at this temperature. Is the reaction exothermic or endothermic?

Solution: According to the *NIST-JANAF Thermochemical Tables*, the most chemically stable chemical elements able to produce a CO(g) compound at pressure $P° = 1$ bar and temperature $T = 1000\,\text{K}$ are solid-phase carbon in form of graphite (C(s)) and gas-phase diatomic oxygen (O_2(g)); according to the *Tables*, the standard-state enthalpies of formation of these two elements-reactants at 1000 K are equal to zero. The choice of the chemically stable reactants for reactions of formation is sometimes not an easy task. For example, the fact that the most stable carbon element at the discussed conditions is solid-phase graphite is contrary to the common perception created by the jewelry industry and Hollywood productions ('diamonds are forever') which suggests that the most stable form of carbon is diamond.

The reaction of formation of carbon monoxide in gaseous phase is

$$C(s) + \frac{1}{2}O_2(g) \rightarrow CO(g). \tag{8.43}$$

The *NIST-JANAF Thermochemical Tables* list the value of the enthalpy of formation of ideal-gas carbon monoxide in equilibrium at $P = P° = 1$ bar and $T = 1000\,\text{K}$ as $\Delta_f H°(T) = -111.98\,\text{kJ/mol}$ (the value is reproduced in Table 8.2 of Appendix A). The pressure and temperature during reaction (8.43) are constant, the reaction components do not change their kinetic and gravitational energies, and there is no transfer of non-boundary work in the system. Therefore, the heat of the process is equal to the change of the reaction enthalpy. The change is negative which means that, according to the International Convention for the Sign of Energy, the heat (111.98 kJ of it) has been released during the reaction. Thus, the reaction is exothermic.

End of Problem 8.2.

Problem 8.3: Find the enthalpy of 5 kg of ideal atomic oxygen gas (O(g)) in equilibrium at pressure $P = P° = 1$ bar and temperature $T = T_o = 298.15$ K. Write the reaction of formation of the gas under such conditions.

Solution: The chemically stable element able to produce gas-phase atomic oxygen at $P = P° = 1$ bar and $T = 298.15$ K is diatomic oxygen gas (O_2(g)) because, according to the *NIST-JANAF Thermochemical Tables*, the enthalpy of formation of diatomic oxygen gas under these conditions is equal to zero (see expression (8.24) and Table 8.1 in Appendix A). Therefore, the reaction of formation of the monatomic oxygen (which under the discussed physical conditions can be considered an ideal gas) is

$$\frac{1}{2}O_2(g) \rightarrow O(g). \tag{8.44}$$

According to the *NIST-JANAF Thermochemical Tables*, the enthalpy of formation of monatomic oxygen is $\Delta_f H_O°(T_o) = 249.170$ kJ/mol while the number of mols of atomic oxygen considered here is

$$\nu_O = \frac{M}{\mu} = \frac{5\,\text{kg}}{16\,\text{kg/kmol}} = 312\,\text{mol}, \tag{8.45}$$

where M is the mass of the gas, and μ is the molecular (molar) mass of the monatomic oxygen. Thus, the enthalpy of 5 kg of the ideal gas at pressure of 1 bar and temperature of 298.15 K is (see relationship (8.21))

$$H_O°(T_o) = \nu_O \Delta_f H_O°(T_o) = (312\,\text{mol})(249.170\,\text{kJ/mol}) = 77.74\,\text{MJ}. \tag{8.46}$$

End of Problem 8.3.

8.3 Departure Functions of Real Gases

The absolute value of a state property of a *non-ideal* substance in a state of equilibrium can be calculated in the following way. First, one calculates the absolute value of the property within the frame of the ideal-gas model discussed above. Then, the result is 'corrected' by the so-called **thermodynamic departure functions** or **thermodynamic departure factors**.

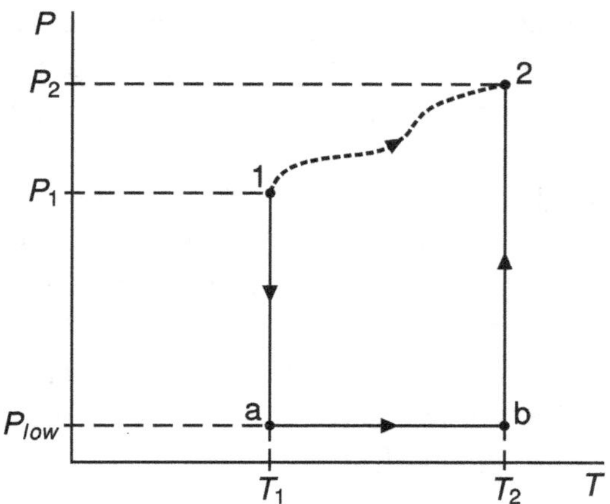

Fig. 8.1: In studies of thermodynamic functions of state, any transition between an equilibrium state 1 and an equilibrium state 2 (the transition is represented by the dotted curve) can be replaced by a sequence of three subprocesses ($1 \rightarrow a$ (isotherm), $a \rightarrow b$ (isobar) and $b \rightarrow 2$ (isotherm)). The isobaric subprocess's pressure $P = P_{low}$ must be low enough to assure that the substance under consideration can be treated during subprocess $a \rightarrow b$ as an ideal gas.

We learned earlier that *changes* of thermodynamic functions of state in any (reversible or not) process in any system moving thermodynamically between an equilibrium state 1 and an equilibrium state 2 can be calculated using any thermodynamic path between these two states. One of the best choices is to replace the actual $1 \rightarrow 2$ transition (represented in Fig. 8.1 by the dotted curve) by a computationally convenient set of three successive subprocesses (shown in Fig. 8.1 as solid lines): the isotherm $1 \rightarrow a$, the isobar $a \rightarrow b$, and the isotherm $b \rightarrow 2$. The pressure of the isobaric subprocess $a \rightarrow b$ can be taken as $P_a = P_b = P_{low}$, a pressure that must be low enough for the substance to be considered an ideal gas during the $a \rightarrow b$ transition. Of course, $P_{low} = 0$ is the most convenient choice to use in mathematical transformations.

The three-step replacement $1 \rightarrow a \rightarrow b \rightarrow 2$ shown in Fig. 8.1 leads to the following change of the substance's entropy during the dotted-curve transition (entropy is an additive property):

$$\Delta S_{1 \rightarrow 2} = \Delta S_{1 \rightarrow a} + \Delta S_{a \rightarrow b} + \Delta S_{b \rightarrow 2}, \tag{8.47}$$

where (see Section 7.4)

$$\Delta S_{1 \to a} = \int_{P_1, T_1}^{P_a = P_{\text{low}}, T_a = T_1} dS = \int_{P_1}^{P_{\text{low}}} \left(\frac{\partial S}{\partial P} \right)_{T = T_1} dP$$

$$= \int_{P_1}^{P_{\text{low}}} \left[-\left(\frac{\partial V}{\partial T} \right)_P \right]_{T = T_1} dP, \tag{8.48}$$

$$\Delta S_{a \to b} = \int_{P_a = P_{\text{low}}, T_a = T_1}^{P_b = P_{\text{low}}, T_b = T_2} dS = \int_{T_1}^{T_2} \left(\frac{\partial S_{\text{ideal}}}{\partial T} \right)_P dT$$

$$= \int_{T_1}^{T_2} \frac{C_{P,\text{ideal}}(T)}{T} dT, \tag{8.49}$$

and

$$\Delta S_{b \to 2} = \int_{P_b = P_{\text{low}}, T_b = T_2}^{P_2, T_2} dS = \int_{P_{\text{low}}}^{P_2} \left(\frac{\partial S}{\partial P} \right)_{T = T_2} dP$$

$$= \int_{P_{\text{low}}}^{P_2} \left[-\left(\frac{\partial V}{\partial T} \right)_P \right]_{T = T_2} dP, \tag{8.50}$$

where the subscript *ideal* denotes properties of ideal gas, and the derivative $(\partial V / \partial T)_P$ can be obtained from the equation of state for v mols of a non-ideal substance (see Section 5.4),

$$V = vZ(P, T)RT/P, \tag{8.51}$$

where R is the universal (molar) gas constant, and $Z(P, T)$ is the compressibility factor that represents the degree of departure of the substance's actual properties from its ideal-gas properties ($Z(P, T) = 1$ for all ideal gases).

The entropy change (8.47) can be calculated without the 'excursion' to the hypothetical ideal-gas transition $a \to b$. It can be done, for example, in a two-step process consisting of an isobaric (at $P = P_1$) transition from temperature T_1 to temperature T_2 followed by an isothermal (at $T = T_2$) transition from pressure P_1 to pressure P_2. However, the 'excursion' allows one to relate the properties of a non-ideal substance to properties of ideal gas. This is a great convenience in many thermodynamic studies.

Using relationships (8.48)–(8.50), adding to and subtracting from the expressions, respectively,

$$vR \int_{P_{\text{low}}}^{P_1} \frac{dP}{P} \quad \text{and} \quad vR \int_{P_{\text{low}}}^{P_2} \frac{dP}{P},$$

and remembering that values of P_1, T_1, P_2, T_2 are arbitrary, leads to

$$S(P,T) - S_{\text{ideal}}(P,T) = -\nu RT \int_{P_{\text{low}}}^{P} \left[\frac{\partial Z(P,T)}{\partial T} \right]_P \frac{dP}{P}$$
$$- \nu R \int_{P_{\text{low}}}^{P} [Z(P,T) - 1] \frac{dP}{P}. \tag{8.53}$$

By a similar procedure, one obtains for enthalpy

$$H(P,T) - H_{\text{ideal}}(P,T) = -\nu RT^2 \int_{P_{\text{low}}}^{P} \left[\frac{\partial Z(P,T)}{\partial T} \right]_P \frac{dP}{P}. \tag{8.54}$$

When the compressibility factor $Z(P,T)$ is the one in the Peng-Robinson equation of state (see Section 5.4), expressions (8.53) and (8.54) become, respectively,

$$S(P,T) - S_{\text{ideal}}(P,T) = \nu \frac{da/dT}{2\sqrt{2}b} \ln \left[\frac{Z(P,T) + (1 + \sqrt{2})B}{Z(P,T) + (1 - \sqrt{2})B} \right]$$
$$+ \nu R \ln(Z(P,T) - B), \tag{8.55}$$

and

$$H(P,T) - H_{\text{ideal}}(P,T) = \nu \frac{T(da/dT) - a}{2\sqrt{2}b} \ln \left[\frac{Z(P,T) + (1 + \sqrt{2})B}{Z(P,T) + (1 - \sqrt{2})B} \right]$$
$$+ \nu RT(Z(P,T) - 1). \tag{8.56}$$

ne should recall that the Gibbs free energy $G(P,T)$ of a substance in
'ynamic equilibrium can be obtained from the free energy definition (8.4),

$$G(P,T) = H(P,T) - TS(P,T). \tag{8.57}$$

(8.53)–(8.56) are called the **thermodynamic departure func-
namic departure factors**. They allow one to calculate the
single-component non-ideal substance in equilibrium
ssure P and temperature T.

Chapter 9

Multi-Component, Multi-Phase Systems

9.1 Phase Equilibrium and the Gibbs Phase Rule

Consider a multi-phase, multi-component, and non-reactive system in a state of equilibrium. Thermodynamic studies of such a system often require exploration of the following question: what is the number of *intensive* thermodynamic variables (temperature, pressure, specific volume) that can be changed while moving the system to another state of equilibrium without changing the numbers of the phases and components? To answer this question requires some information on properties of solid, liquid, and gas phases of substances.

As discussed earlier, a substance is called a 'pure substance' ('component') if all particles of the substance have the same chemical structure during the *entire* thermodynamic transition under consideration even when the substance changes its phase during the process. For example, the substance identifiable by the chemical symbol H_2O can be considered a pure substance in a broad range of not-too-high temperatures. This is because the structure of the H_2O molecules forming the substance is the same in solid, liquid, and vapor phases when its temperature is not high. The substance H_2O cannot be called a 'pure substance' at high temperatures because then many of the H_2O molecules will dissociate and the products of the dissociation will participate in chemical reactions. As a result, the high-temperature substance will become a mixture that consists of not only H_2O molecules but also other species such as H, O, OH, O_2, O_3, H_2, and H_2O_2.

A well-mixed mixture of gaseous components (such as air) can often be considered a 'pure substance' if reliable *average* values of the mixture's properties can be calculated or measured. Then the mixture is treated as a single-component substance consisting of fictitious particles representing the overall properties of the mixture (see Problem 9.5). Miscible liquids also can be well mixed to form a 'pure substance'. A good example is vodka, a solution of a liquid alcohol in water. Non-miscible liquids (for example, oil and water) do not form solutions that can be treated as pure substances.

Matter can exist in solid, liquid, and gaseous phases. (It also can exist as 'plasma' – a partially or fully ionized gas which is not common in most thermodynamic applications.) A stable **phase** of a pure substance is any region of the substance where thermodynamic properties are uniform (homogeneous) throughout the entire region. Another characteristic feature of every phase is that it is separated from other phases by a clearly identifiable boundary at which the density of the substance changes discontinuously. Several single-phase regions of the same component which are separated in space are treated in thermodynamic studies as belonging to *the same* phase. For example, four ice cubes floating in water are considered one (solid) phase of the substance that has the chemical symbol H_2O.

Determining the numbers of components and phases that can coexist in stable equilibrium in complex thermodynamic systems can be a difficult task. A helpful criterion for existence of such equilibrium is the **Gibbs Phase Rule**. Consider a substance that consists of α components and β phases and where some chemical reactions can take place. The substance is in the so-called **phase equilibrium** if the temperatures and pressures of all phases of all components present in the substance are the same. (Usually, studies of phase equilibria do not impose special constraints, such as a requirement that some chemical reactions can occur only in some phases.) Thermodynamic properties of the multi-phase, multi-component system in phase equilibrium can be described by a set of *intensive* thermodynamic variables such as pressure P, temperature T, and the specific volume v. Josiah Willard Gibbs (1839–1903) showed that the number of intensive variables w that can be changed independently without destroying phase equilibrium co-existence of β phases of α components of the system where non-boundary work is absent is

$$w = \alpha - \beta + 2 - r, \tag{9.1}$$

where r is the number of independent chemical reactions possibly occurring in the system.

Relationship (9.1) is called the **Gibbs Phase Rule**. In an equilibrium system where macroscopic effects of chemical reactions (change of the system's chemical composition) are negligible and non-boundary work is absent, the Rule reduces to

$$w = \alpha - \beta + 2. \tag{9.2}$$

The w independent *intensive* variables are called the system's **thermodynamic degrees of freedom**.

One should notice that since the value of w in expression (9.2) cannot be negative, one has

$$\alpha - \beta + 2 \geq 0, \tag{9.3}$$

which can be rewritten as

$$\beta \leq \alpha + 2 \tag{9.4}$$

(the number of phases possible to exist in an equilibrium system cannot exceed the number of the system's components by more than two), or as

$$\alpha \geq \beta - 2 \tag{9.5}$$

(an equilibrium system that is expected to have β phases must consist of at least $(\beta - 2)$ components).

Relationship (9.4) tells us that a single-component system in a stable equilibrium (where there are neither chemical reactions nor non-boundary work) cannot have more than three phases.

Consider a single-component ($\alpha = 1$), single-phase ($\beta = 1$) substance in equilibrium where there are neither chemical reactions nor non-boundary work. How many of the state's intensive thermodynamic variables (pressure P, temperature T, specific volume $v = V/M$) can be changed to move the substance to another equilibrium state and still have the substance as a single-component, single-phase system? According to the Gibbs Phase Rule (9.2), the system has $w = 2$ thermodynamic degrees of freedom. Thus, **thermodynamic properties of any single-component, single-phase substance in equilibrium (local equilibrium) can be described in terms of only two intensive thermodynamic variables (P and T, P and v, or T and v)**; the pair P, T is the most convenient in many thermodynamic applications.

Problem 9.1: Consider a pure substance of chemical symbol H_2O where solid, liquid, and gas phases co-exist in a state of stable phase equilibrium where chemical reactions and non-boundary work are negligible. How many of the system's independent intensive thermodynamic variables (temperature, pressure, specific volume) can be changed in order to move the system to another state of equilibrium where all the three phases will still be present?

Solution: The system consists of one component (the substance called H_2O) in the following three phases: solid H_2O (ice), liquid H_2O (water), and gaseous H_2O (water vapor). Thus, the number of the system components is $\alpha = 1$, the number of phases is $\beta = 3$, and, according to relationship (9.2), the number of the thermodynamic degrees of freedom of the system is

$$w = 1 - 3 + 2 = 0. \tag{9.6}$$

Clearly, the system has no thermodynamic 'freedom' at all; the slightest change of any of the intensive properties of the system will destroy the equilibrium co-existence of the three phases (some of the phases will disappear). In other words, the three phases of the system can coexist in stable equilibrium only at some unique values of the system temperature T, pressure P, and specific volume v (or mass density $\rho = 1/v$). The state where the three phases of single-component substance co-exist in a stable phase equilibrium is called the **triple point** of the substance.

The triple point states are marked in P–T diagrams (such as those shown in Figs. 9.3 and 9.4) as states t_1, t_2 and t_3. The pressures P_t and temperatures T_t of these three states are the same but their specific v_t (and, therefore, the corresponding mass densities ρ_t) are different. For example, the pressure and temperature of the three phases co-existing in the triple-point of the pure substance called H_2O are the same ($P_t = 0.006\,$bar and $T_t = 0.01\,°C$) but their mass densities are $916.8\,kg/m^3$, $997\,kg/m^3$, and $0.005\,kg/m^3$.

End of Problem 9.1.

Problem 9.2: Find the number of thermodynamic degrees of freedom of a single-component liquid in a phase equilibrium with its vapor. Assume that chemical reactions and non-boundary work are absent in the system.

Solution: The system has one component ($\alpha = 1$) and two phases ($\beta = 2$). Thus, according to expression (9.2), the number of the system's degrees of thermodynamic freedom is

$$w = 1 - 2 + 2 = 1. \tag{9.7}$$

This means that a change of one intensive variable of the system will thermodynamically move the system to another equilibrium state, and both phases (the liquid and its vapor) will still coexist in the new equilibrium state. This also means that any change of two (or more) intensive properties (say, temperature T and pressure P) will destroy the two-phase equilibrium of the system.

End of Problem 9.2.

9.2   One Component, Two or Three Phases

Let us study thermodynamic properties of a pure substance that can exist in a phase equilibrium, at pressure P and temperature T, of two (liquid+gas, solid+liquid, or solid+gas) or three (solid+liquid+gas) phases. The properties of the equilibrium can be displayed in a three-dimensional P–T–v space which is often called the **phase diagram** (an example of such a diagram is shown in Fig. 9.1). Other common representations, also shown in Fig. 9.1, are the projections of the three-dimensional diagrams on two-dimensional (P–T, P–v,

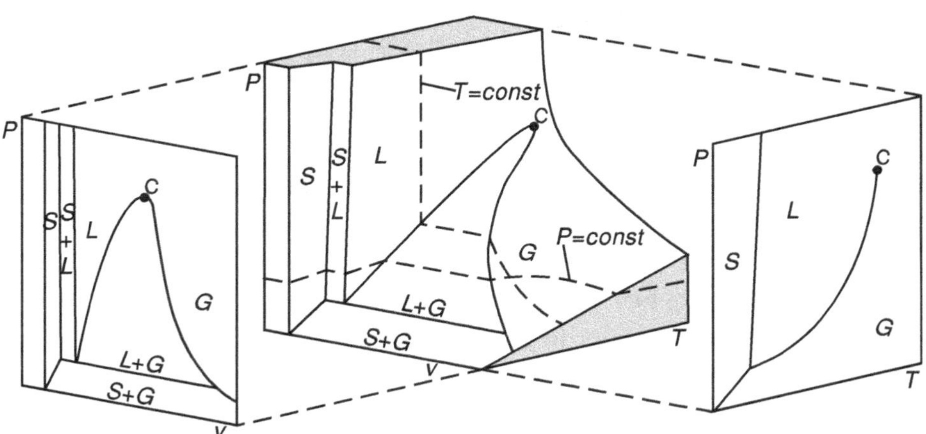

Fig. 9.1: Three-dimensional (pressure P, temperature T, and specific volume v) diagram of properties of a single-component substance. Two-dimensional (P–T, P–v, and T–v) projections of the diagram are very common in studies of phase equilibria. Symbols S, L and G denote the regions where the substance exists only in solid, liquid, and gas phase, respectively.

and $T\!-\!v$) planes. Such projections are very useful in thermodynamic studies because they show in a comprehensive way the **melting curve** (regions of solid-liquid equilibrium co-existence), the **vaporization curve** (regions of liquid-gas equilibrium co-existence), and the **sublimation curve** (regions of solid-gas equilibrium co-existence). The **vaporization curve** is often called **vapor pressure curve**. The melting, vaporization, and sublimation curves are shown in Fig. 9.1 as thick solid lines – also see Fig. 9.3–9.6.

Every single-component liquid-gas, solid-liquid, or solid-gas phase equilibrium shows very interesting and useful features. For example, at a *fixed* (constant) pressure and some *unique* (for this pressure) temperature, the system's temperature also stays constant even when heat is being added to or removed from the system, as long as the component remains a two-phase system (however, the heating or cooling of the system causes changes of the specific volumes of the co-existing phases). This isobaric-isothermal process changing the specific volumes of the two phases is called the **phase equilibrium transition** or just **phase transition**. The pressure and the corresponding unique temperature at which such a transition occurs are called the component's **saturation pressure** and **saturation temperature**, respectively (see below). When enough heat is added to the two-phase system at a given saturation pressure and temperature, the system's matter becomes a single-phase substance of a lower mass density. When enough heat is removed from the two-phase matter at the saturation pressure and temperature, the matter becomes a single-phase substance of a higher mass density.

Changes of state properties (the thermal energy, enthalpy, etc.) of a single-phase component in equilibrium can be calculated in the way discussed in other chapters of this book. The properties of a two-phase component during its phase transition are usually calculated as a weighted average (over the mass fractions of the phases) of the corresponding properties of each of the two phases. For example, in order to calculate this average value for a mixture of liquid and gas, one has to introduce an additional variable called the **quality of a saturated liquid-vapor mixture, phase equilibrium quality,** or just **quality**. The quality x is defined as (similar relationships can be written for solid→liquid and solid→gas phase transitions)

$$x = \frac{\text{mass of gaseous part of the mixture}}{\text{mass of the entire mixture}} = \frac{M_g}{M_g + M_f}, \qquad (9.8)$$

where M_g is the mass of the gas (vapor) phase and M_f is the mass of the liquid phase of the mixture. The phase-equilibrium quality (9.8) has values between zero and one. When $x = 0$ ($M_g = 0$), the considered substance is a pure liquid (the **saturated liquid**) ready to start producing vapor if a heat is added at the (constant) saturation pressure. When $x = 1$ ($M_f = 0$), the last drop of the liquid phase has just evaporated, and the entire substance becomes a pure gas (the **saturated vapor**). As mentioned earlier, the process of phase changing from liquid to vapor is called **vaporization**.

Let us consider thermodynamic properties of pure substance called diatomic nitrogen (N_2) that is heated at a constant pressure and, therefore, can undergo (at temperature equal to the saturated temperature of nitrogen at this pressure) liquid→gas phase transition – see Fig. 9.2. *Qualitative* behavior of an overwhelming majority of pure substances during their phase transitions is very similar to the behavior of nitrogen.

A piston-cylinder device contains diatomic nitrogen in equilibrium at temperature $T_H = 70$ K (STATE H in Fig. 9.2). The outside pressure on the piston and its weight keep the nitrogen pressure constant and equal to $P = P_1 = 1$ bar. Under such conditions, nitrogen exists in liquid phase called the **supercooled liquid** or (somewhat confusingly) **compressed liquid**. As heat is being added *isobarically* (at $P = P_1$) to the compressed liquid but with neither non-boundary work transfer nor chemical reactions, its temperature T and specific volume v increase, causing the piston to move up slightly. The pressure inside the cylinder remains constant during our experiment because the weight of the movable piston and the outside pressure do not change during the experiment. The liquid nitrogen does not boil during the heating until it reaches temperature $T_B = T_1 = 77.24$ K (STATE B in Fig. 9.2). The temperature T_B is called the **saturation temperature** of nitrogen at pressure of 1 bar, and the liquid nitrogen in STATE B is called the **saturated liquid**. Note that in this state, the nitrogen is still liquid (with the quality x equal to zero) but 'saturated' with energy to the point that any further heat addition will keep transforming more and more of the liquid phase into nitrogen gas. As one keeps adding the heat *isobarically* (at $P = P_1$) to the saturated liquid, its (saturation) temperature remains constant ($T_B = T_1$), and the amount of the gaseous phase of the substance increases (that is, the substance's quality x also increases). Such *isobaric-isothermal* process ('phase transition') will continue until the last drop of the liquid disappears and the nitrogen becomes a pure gas which is called **saturated vapor** (STATE F in Fig. 9.2). The quality of STATE F is $x = 1$. The most characteristic feature of the isobaric phase transition from STATE B (the 'saturated liquid') to STATE F

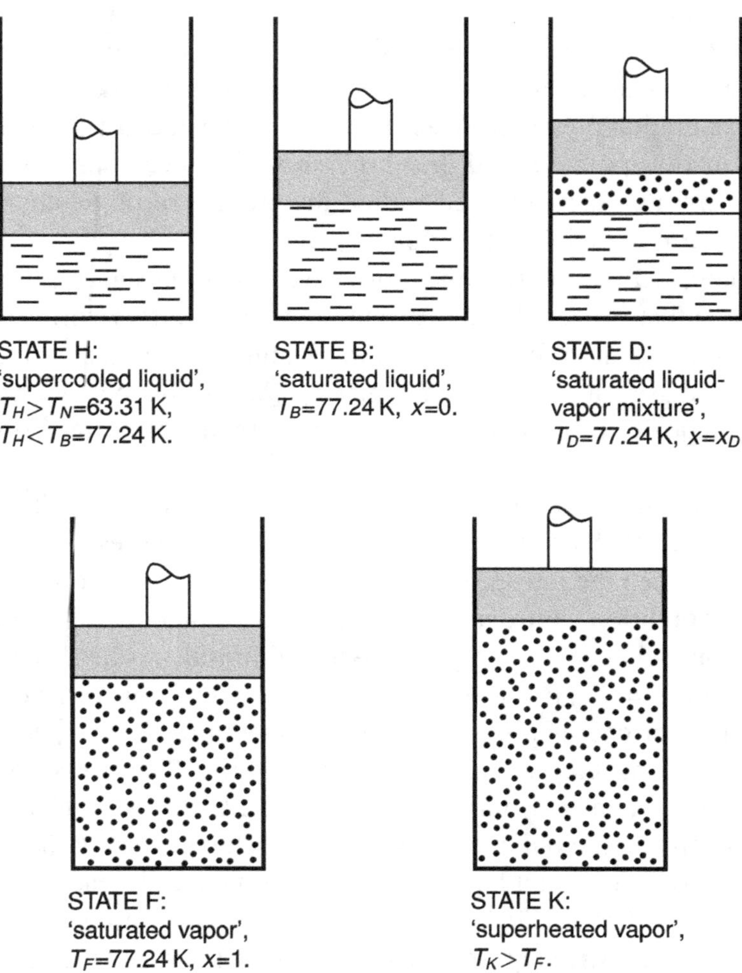

Fig. 9.2: Phase transition, from liquid to vapor, in diatomic nitrogen at a constant pressure of 1 bar. The gas and liquid in STATE D are sometimes mixed in a less orderly manner than it is shown in the figure. Then, one can see many bubbles of gas moving up inside the liquid.

(the 'saturated vapor') is that the temperature of the two-phase component is also constant during the entire transition ($T = T_1 = T_B = T_F$). (It should be noticed that the constancy of both pressure and temperature during phase transition from STATE B to STATE F allows a significant simplification of

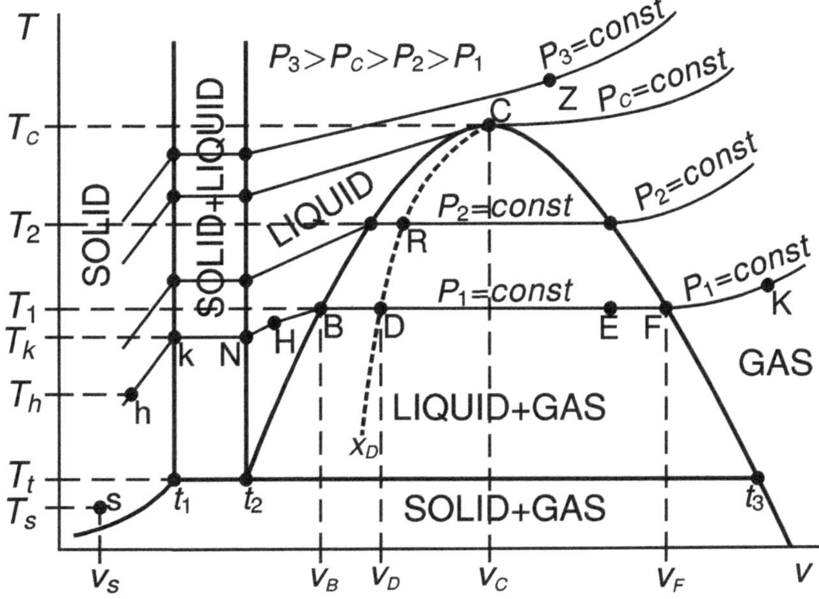

Fig. 9.3: Projection of typical three-dimensional (P–T–v) phase equilibrium diagram shown in Fig. 9.1 on the T–v plane. Meaning of the symbols is discussed in the text.

theoretical description of the isobaric-isothermal process.) The states of the phase transition between STATE B and STATE F (examples of such states are STATE D and STATE E – see Fig. 9.3) have the same pressure (the saturated pressure) and the same temperature (the saturated temperature) but different values of their qualities x, that is, different fractions of the liquid and vapor phases. For example, the quality of STATE D is x_D.

The complete liquid→gas phase transition at constant pressure $P = P_1$ and constant temperature $T = T_1$ begins in STATE B and ends in STATE F. Further isobaric (at $P = P_1$) addition of heat to the pure gas in STATE F increases the gas temperature above $T_F = T_1$. Then the gas is called the superheated gas or **superheated vapor**. An example of a superheated vapor is the gas in STATE K (see Fig. 9.3).

We studied above a phase transition caused by an *isobaric* (at $P = P_1$) addition of heat to a system (with no non-boundary work transfer and with no chemical reactions) moving from a compressed liquid (STATE H) to a superheated vapor (STATE K). The transition can be shown as a surface in a three-dimensional diagram similar to that shown in Fig. 9.1. The projection of the surface on planes $T–v$, $P–T$, or $P–v$ (see Figs. 9.3–9.7) is shown as a solid line connecting STATE B and STATE F. One should notice in Fig. 9.3 that the studied isobaric heating could have started not from state H (compressed liquid nitrogen at temperature T_H) but from state h (solid nitrogen of temperature T_h). Adding heat isobarically (at $P = P_1$) to the solid increases its temperature until the solid reaches STATE k. Further addition of heat causes melting of the solid which produces a two-phase (solid+liquid) mixture. This two-phase isobaric-isothermal co-existence begins at temperature T_k and ends when the solid-liquid mixture reaches STATE N (at temperature T_N) where the last bit of solid melts and the substance becomes a pure liquid. A further isobaric addition of heat increases the liquid's temperature until the liquid reaches STATE H (the state which was the starting point of the H→K process discussed above). It should be emphasized that the $T = T(v)$ dependence for the isobaric h→H (solid→liquid) heating is qualitatively similar to such dependence for the isobaric H→K (liquid→gas) heating; the temperature T_k (of the solid→liquid phase transition k→N) and the temperature T_B (of the liquid→vapor phase transition B→F) are both constant (but not equal) when the pressures of both transitions are the same.

The isobaric (at $P = P_1$) transition from a compressed liquid in STATE H to a superheated vapor in STATE K can be reversed by isobarically cooling the nitrogen from STATE K to STATE H. When the superheated vapor is isobarically cooled down to STATE F, the vapor becomes saturated vapor. Any further cooling causes more condensation of the gas, that is, continuation of the isobaric-isothermal transformation of the vapor into liquid (STATE E, then STATE D, etc.) at constant temperature $T_F = T_E = T_D = T_B = T_1$. The **condensation** process (the gas→liquid phase transition) ends in the saturated-liquid STATE B. Further isobaric cooling (at $P = P_1$) of the liquid will be non-isothermal and will move the liquid to a compressed-liquid state (say, STATE H) of temperature which is lower than the saturation temperature T_B.

One should notice that the saturated liquid states and the saturated vapor states of substances become identical at a point (called the **critical point** of the substance) which is shown in Figs. 9.1, 9.3, and 9.4 as STATE C. The pressure, temperature, and specific volume of a substance at its critical point are called, respectively, the **critical pressure** P_C, the **critical temperature** T_C, and the **critical specific volume** v_C. For example, the critical-point properties of the substance called H_2O are: $P_C = 220.9$ bar, $T_C = 647.3$ K and $v_C = 0.0031$ m^3/kg. The liquid phase and vapor phase of a pure substance never co-exist in phase equilibrium at pressures greater than the critical pressure of the substance.

Phase diagrams such as that shown in Fig. 9.3 are unique for each substance. In other words, the values of temperature T_1 and pressure P_1 at which isobaric-isothermal liquid→vapor phase transition takes place are unique for each substance; as discussed earlier, the temperature and pressure are called the saturation temperature and the saturation pressure, respectively. Similar remarks can be made about the substance's solid→liquid isobaric-isothermal (stable) phase transition, but then the transition's saturation pressure and temperature differ from their values for the liquid-vapor phase transition. Therefore, the thick solid lines t_2–B–C and C–F–t_3 shown in Fig. 9.3 give unique relationships between the saturation temperature and saturation pressure of the substance under consideration. The constant temperature $T_B = T_D = T_E = T_F = T_1$ at which a particular single-component liquid changes isobarically into gas (vapor) is uniquely determined by the phase transition's constant pressure $P_B = P_D = P_E = P_F = P_1$. Reversely, the constant pressure P_1 at which a particular liquid is transformed during phase transition into gas (vapor) is uniquely determined by the transition's constant temperature T_1. The databases of saturation temperature as a function of pressure, and of saturation pressure as a function of temperature, for a large number of pure substances are available in literature of the subject.

As discussed earlier, the saturation temperature of diatomic nitrogen (N_2) at pressure $P = 1$ bar is $T_B = 77.24$ K ($-195.91\,°C$) (see Figs. 9.2 and 6.1), and the saturation pressure of the nitrogen at temperature $T = 120$ K ($-153.15\,°C$) is $P_B = 25.15$ bar. Similarly, the saturation temperature of substance called H_2O at pressure $P = 10$ bar is $T_B = 453.06$ K ($179.91\,°C$) and the saturation pressure of H_2O at temperature $T = 373.15$ K ($100\,°C$) is $P_B = 1.013$ bar $= 1$ Atm. The saturation temperature of H_2O at pressure of 1 bar is 372.78 K ($99.63\,°C$). Therefore, it is often assumed in practical studies that the saturation pressure and saturation temperature of H_2O are 1 bar and 373.15 K, respectively.

The reader should notice in Figs. 9.3 and 9.4 that the states t_1 (solid), t_2 (liquid) and t_3 (gas) (all have the same pressure and temperature but different specific volumes) are the states of the 'triple point' of the pure substance under consideration – see Problem 9.1.

Since STATE B (saturated liquid) and STATE F (saturated vapor) are unique states of every pure substance at a given pressure P_1 (see Fig. 9.3), the changes of the substance properties of state (the thermal energy, enthalpy, entropy, etc.) during the isobaric-isothermal phase transition B→F also have unique values which can be easily tabularized. Also, the change of enthalpy ΔH during any isobaric-isothermal phase transition in a closed nonflow system with neither chemical reactions nor non-boundary work transfer is equal to the process's heat transfer Q (see expression (6.16), and the discussion associated with relationships (7.14) and (7.15)),

$$Q = \Delta H \qquad \text{stable phase transitions.} \qquad (9.9)$$

This heat per unit of mass is called the **specific latent heat** of the phase transition from STATE B to STATE F. Thus, one can speak about the substance's latent heat of **melting** (solid→liquid phase transition k→N) and the latent heat of **vaporization** (liquid→gas phase transition B→F) (see Fig. 9.3). The magnitude of the latent heat of the **freezing** transition N→k is the same as that of the melting transition, but the direction of the heat flow is opposite. Similarly, the magnitude of the latent heat of **condensation** F→B is the same as that of the vaporization (B→F) but the direction of the heat flow is opposite. (Heat must be added to the substance during the melting and vaporization transitions, and it has to be removed from the substance during the freezing and condensation transitions.) In other words, one can write (adopting the common terminology where properties of saturated liquid are marked with subscript f while properties of saturated vapor are marked with subscript g) that

$$q_{fg} = l_{fg} = h_g - h_f \qquad \text{and} \qquad q_{gf} = l_{gf} = h_f - h_g, \qquad (9.10)$$

and

$$l_{fg} = -l_{gf}, \qquad (9.11)$$

where h_f and h_g are specific enthalpies of the saturated-liquid STATE B and the saturated-vapor STATE F, respectively (see Figs. 9.3 and 9.4); l_{fg} and q_{fg} are the specific latent heat and the heat transfer (per unit of mass of the substance), respectively, of the complete vaporization transition of the substance from its saturated-liquid state to its saturated-vapor state; l_{gf} and q_{gf} are the specific latent heat and the heat transfer (per unit of mass of the substance), respectively, of the complete phase transition of the substance from its saturated-vapor state to its saturated-liquid state. Of course, the latent heat l_{fg} is absorbed by the substance while the latent heat l_{gf} is removed from the substance during the corresponding phase transition.

One should mention two other isobaric-isothermal phase transitions (less common in applications) that are similar in character to the phase transitions discussed above: **sublimation** (solid→gas) transition, and its reverse process called **deposition** (gas→solid) transition. These two transitions have the same values of pressure and temperature, and their latent heats are the same in magnitude but move in opposite directions.

The values of saturation temperatures of the solid→liquid, liquid→vapor, and solid→vapor phase transitions at the standard pressure P° are called the substance's **melting point**, the **boiling point**, and the **sublimation point**, respectively. The saturation temperatures of the reverse phase transitions (liquid→solid, vapor→liquid, and vapor→solid) are called, respectively, the **freezing point**, the **condensation point**, and the **deposition point** of the substance.

As already mentioned, the concept of quality x (see expression (9.8)) is used in Thermodynamics to study properties of liquid⇌vapor phase transitions. For example, a *specific* property a (that is, the value of the property per unit of mass of the matter undergoing a liquid→vapor transition) can be given as the following statistical average weighted over the mass fractions of the two phases participating in the process:

$$a(x) = \frac{M_f}{M_g + M_f} a_f + \frac{M_g}{M_g + M_f} a_g$$
$$= (1 - x)a_f + xa_g = a_f + x(a_g - a_f), \qquad (9.12)$$

where a_f and a_g are the values of the specific property a $(u, h, s, ...)$ in the saturated-liquid STATE B and in the saturated-vapor STATE F, respectively,

$$a_f = a(x = 0) \qquad \text{and} \qquad a_g = a(x = 1). \qquad (9.13)$$

An example of a constant-quality line of a saturated liquid-gas mixture is shown in Fig. 9.3 as the dotted curve x_D passing through STATES C, R and D. The STATES D and R belong to two different phase transitions but have the same qualities, $x = x_D = x_R$.

Relationship (9.12) gives the following linear expressions for the specific volume υ, the specific thermal energy u, the specific enthalpy h, and the specific entropy s, respectively, of a single-component, two-phase system of quality x:

$$v(x) = v_f + x(v_g - v_f),\qquad(9.14)$$

$$u(x) = u_f + x(u_g - u_f),\qquad(9.15)$$

$$h(x) = h_f + x(h_g - h_f),\qquad(9.16)$$

and

$$s(x) = s_f + x(s_g - s_f),\qquad(9.17)$$

where, as before, the subscript f denotes the values of the properties of the saturated liquid in the beginning of the considered phase transition (STATE B when $P = P_1$), and the subscript g denotes the values of the properties of the saturated vapor at the transition end (STATE F when $P = P_1$).

Table 9.1 of Appendix B lists values of several thermodynamic state properties (specific thermal energy u, specific enthalpy h, etc.) of liquid H_2O at various saturated pressures and the corresponding saturated temperatures when the substance quality $x = 0$ (STATE B in Fig. 9.3). Table 9.2 of the Appendix lists the values of the properties for gaseous H_2O at the saturated pressures and the corresponding saturated temperatures when the substance quality $x = 1$ (STATE F in Fig. 9.3). The values listed in Tables 9.1 and 9.2 are taken from the *NIST Chemistry WebBook, NIST Standard Reference Database No.69*.

9.2.1 Incompressible Matter

In most thermodynamic applications, specific volumes and mass densities of solids change very little with pressure. The change is usually larger in liquids, but its magnitude is usually small. Such solids and liquids are called **incompressible substances** – see Section 4.2.2. Usually, the thermal properties of liquid at a given pressure and temperature (STATE H in Fig. 9.3) are not much different from its properties at the same pressure and the liquid saturation temperature (STATE B in Fig. 9.3). Therefore, one can write the following relationships valid in most incompressible liquids at pressure P and temperature T:

$$v(P,T) \simeq v_f(P),\quad \rho(P,T) \simeq \rho_f(P) = 1/v_f(P)\qquad(9.18)$$

and

$$u(P,T) \simeq u_f(P),\quad h(P,T) \simeq h_f(P) = u_f(P) + Pv_f(P),\qquad(9.19)$$

where υ, $\rho = 1/\upsilon$, u, and h are the liquid's specific volume, mass density, specific thermal energy, and specific enthalpy at pressure P and temperature T, respectively, and υ_f, ρ_f, u_f, and h_f are the values of the properties at pressure P and the corresponding saturation temperature of the liquid.

9.2.2 Water Expansion on Freezing

According to the Gibbs Phase Rule (see Section 9.1), two different phases of a pure substance without chemical reactions and non-boundary work can co-exist as a stable phase equilibrium if both phases have the same pressure P and the same temperature T, and if the values of these two thermodynamic variables are located on one of the substance phase transition curves (the **melting curve**, the **vaporization curve** (usually called **vapor pressure curve**), or the **sublimation curve**) – see Figs. 9.4–9.7.

The change of volume of a single-component system during a *complete* liquid→gas transition at a (constant) saturation pressure P and the corresponding (constant) saturation temperature T can be found from the so-called **Clausius-Clapeyron Equation**,

$$\frac{dP}{dT} = \frac{\Delta H_{\text{trans}}}{T \Delta V_{\text{trans}}}, \tag{9.20}$$

where ΔH_{trans} and ΔV_{trans} are the changes of the component's enthalpy and volume, respectively, during the phase transition from STATE B to STATE F (see Figs. 9.5a and 9.5b). The enthalpy change ΔH_{trans} is easy to measure since it is equal to the heat transfer of this isobaric-isothermal transition – see relationship (9.9). The derivative dP/dT is obtainable from the substance's *melting curve* (in the case of solid⇌liquid transition), *vaporization curve* (in the case of liquid⇌gas transition), or *sublimation curve* (in the case of solid⇌gas transition) – see Figs. 9.3–9.5.

One should notice that $\Delta H_{\text{trans}} > 0$ in all solid→gas, solid→liquid and liquid→gas phase transitions (because heat is added to the system during the transitions) while the opposite is true for gas→solid, liquid→solid, and gas→liquid transitions where heat is removed from the system. Therefore, the sign of ΔV_{trans} for any freezing (liquid→solid) phase transition is the same as the sign of the derivative dP/dT on the left-hand side of expression (9.20). Subsequently, solids contract ($\Delta V_{\text{trans}} < 0$) during freezing processes where $dP/dT > 0$ (see Fig. 9.6a), but they expand ($\Delta V_{\text{trans}} > 0$) during freezing where $dP/dT < 0$ (see Fig. 9.6b).

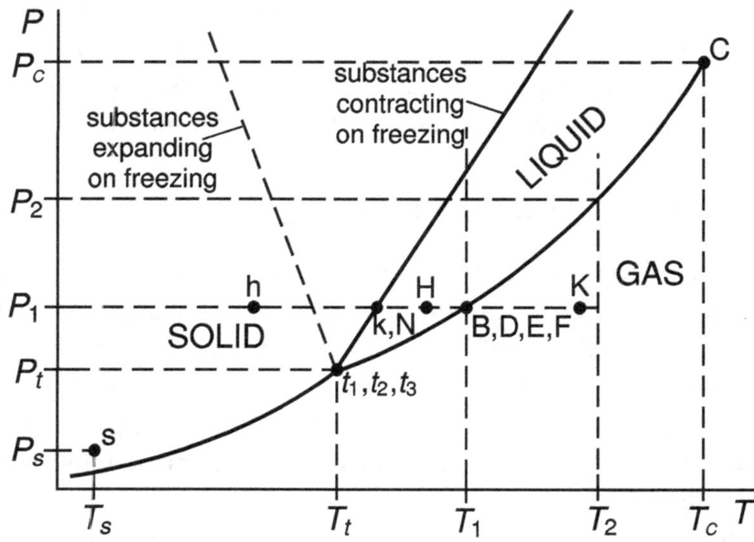

Fig. 9.4: Projection of the three-dimensional (P–T–v) phase equilibrium diagram shown in Fig. 9.1 on the P–T plane. Meaning of the symbols is the same as in Fig. 9.3. The derivative of the melting curve that separates the solid and liquid phases can be positive (the case represented by the solid line originating at point (t_1, t_2, t_3) and passing through point (k,N)), or negative (the case represented by the dashed line originating at point (t_1, t_2, t_3)). The former situation occurs in substances that contract on freezing, the latter in substances that expand on freezing (see Fig. 9.6).

Almost all substances contract on freezing, that is, their P–T phase diagrams are similar to those shown in Fig. 9.6a. The few substances that expand on freezing are Bi (bismuth), Ga (gallium), PVC-vinyl, and H_2O; their P–T diagrams are similar to that shown in Fig. 9.6b.

One can see in Fig. 9.7 that skating is possible only on solid substances where the melting curve's derivative $dP/dT < 0$, that is, on substances like Bi, Ga, PVC-vinyl, and H_2O. Only then can a substantial increase of a substance's pressure (which is caused by a skater's weight distributed over a very small area of the part of the skate that touches the substance) transforms some of the solid into a liquid which allows the skater to 'slide'. A detailed physical picture of the melting phenomenon also requires considering some molecular processes, but the main overall mechanism is the macroscopic process described here.

The fact that H_2O expands upon freezing (that is, that its mass density de-

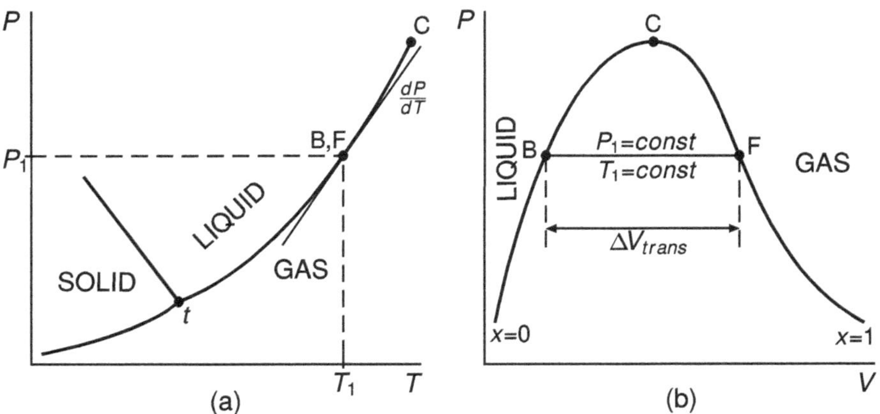

Fig. 9.5: Explanation of Clausius-Clapeyron Equation (9.20): (a) the P–T plane projection, (b) the P–V plane projection.

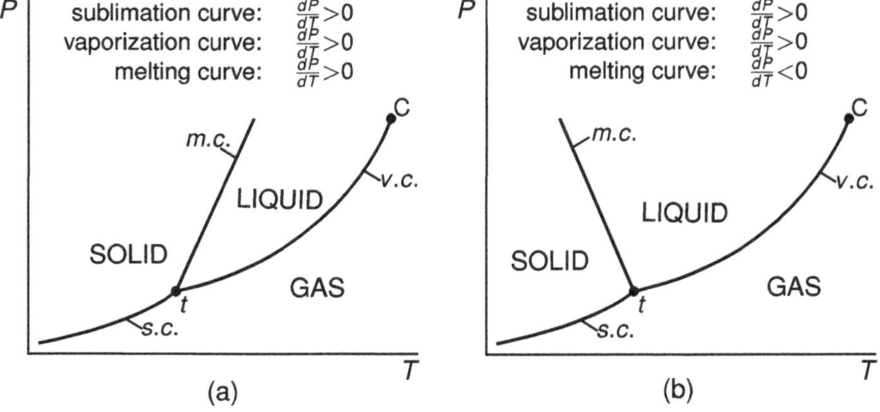

Fig. 9.6: The P–T diagram of stable coexistence of two phases of single-component substance when: (a) the derivative dP/dT is positive everywhere on the substance melting curve (*m.c.*), vaporization curve (*v.c.*), and sublimation curve (*s.c.*); (b) the derivative dP/dT is positive everywhere on the vaporization curve and sublimation curve, but it is negative everywhere on the melting curve.

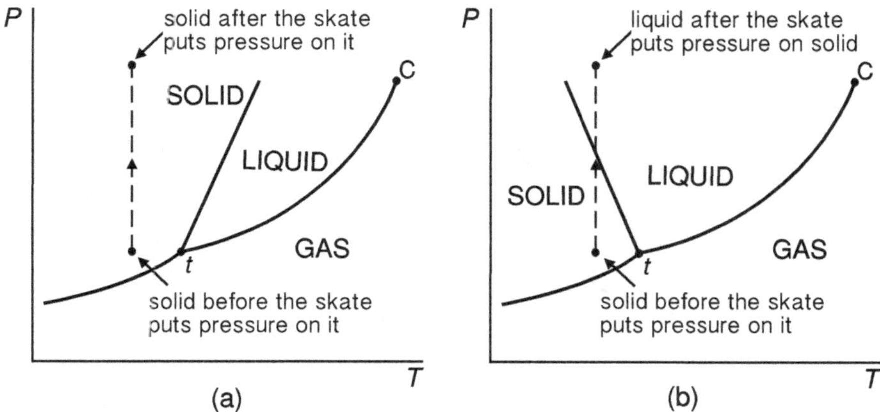

Fig. 9.7: The P–T diagram of a stable coexistence of two phases of single-component substances: (a) skating is impossible on most substances because the derivatives dP/dT for their melting curves are positive, (b) skating is possible on some substances (H_2O is one of them) because the derivatives dP/dT for their melting curves are negative.

creases during the process) has been important for the evolutionary development. Since the density of ice produced in freezing process is smaller than the density of water, the buoyancy force acting on ice immersed in water is greater than the gravity force acting on the ice. As a result, icebergs float on the surface of oceans instead of sinking. This has prevented accumulation of ice (over millions of years) at the bottom of the oceans making them shallow. If this accumulation had happened, only very simple forms of life (much simpler than a human) could have been formed.

Problem 9.3: The vaporization curve $P = P(T)$ ('vapor pressure curve') of pure substance called H_2O in temperature range $300\,\text{K} < T < 400\,\text{K}$ is often approximated by the so-called **Antoine Equation**,

$$P = \exp\left[a_1 - \frac{a_2}{a_3 + T}\right], \tag{9.21}$$

where temperature T is given in kelvins, pressure P in pascals, $a_1 = 23.19$, $a_2 = 3814$, and $a_3 = -46.29$. Calculate the latent heat needed to completely transform 1 mol of the substance from saturated liquid into saturated vapor at constant pressure of 1 bar. Non-boundary work and chemical reactions are absent in the system.

<u>*Solution*</u>: The constant temperature of the isobaric-isothermal phase transition at pressure $P = 10^5$ Pa can be obtained from relationship (9.21) as

$$T = \frac{a_2}{a_1 - \ln P} - a_3 = 372.91 \text{ K.} \tag{9.22}$$

The derivative of vapor pressure (9.21) with respect to temperature is

$$\frac{dP}{dT} = \frac{a_2}{(a_3 + T)^2} \exp\left[a_1 - \frac{a_2}{a_3 + T}\right], \tag{9.23}$$

so that Clausius-Clapeyron Equation (9.20), and expressions (9.9) and (9.23) lead to

$$\Delta \bar{H}_{\text{trans}} = \frac{a_2}{(a_3 + T)^2} T \Delta \bar{V}_{\text{trans}} \exp\left[a_1 - \frac{a_2}{a_3 + T}\right] = \bar{Q}_{\text{trans}}, \tag{9.24}$$

where $\Delta \bar{H}_{\text{trans}}$ is the change of the enthalpy of 1 mol of saturated water in STATE B during its complete isobaric-isothermal phase transition into the saturated vapor in STATE F (see Figs. 9.2–9.5), and $\bar{Q}_{\text{trans}}$ is the heat transfer needed to complete the transition. This heat is the latent heat of the transition – see relationship (9.10).

Since the volume of one mol of saturated vapor of a substance is much greater than the molar volume of its saturated liquid, the change of the volume of the substance during its completed vaporization can be given as

$$\Delta \bar{V}_{\text{trans}} = \bar{V}_g - \bar{V}_f \simeq \bar{V}_g, \tag{9.25}$$

where $\bar{V}_f$ and $\bar{V}_g$ are molar volumes of the saturated liquid and the saturated vapor, respectively. If, in addition, one assumes that the saturated vapor is close to being an ideal gas (for which the equation of state is $\bar{V}_g = RT/P$, where the molar gas constant $R = 8.314$ J/mol·K – see Section 5.1), then, according to relationship (9.24), the latent heat of the complete isobaric-isothermal (at pressure of 1 bar) phase transition of 1 mol of H_2O from saturated liquid to saturated vapor is

$$\bar{Q}_{\text{trans}} = \Delta \bar{H}_{\text{trans}} = R \frac{a_2}{(1 + a_3/T)^2} = +41.33 \text{ kJ/mol.} \tag{9.26}$$

End of Problem 9.3.

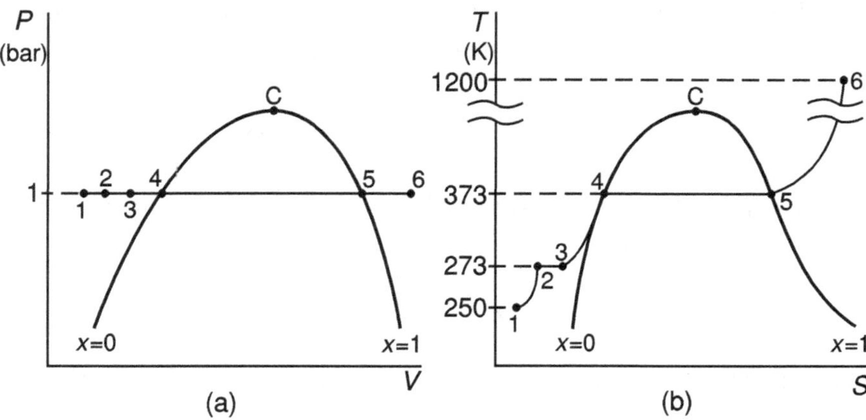

Fig. 9.8: Isobaric transformation (at constant pressure $P = 1$ bar, with no chemical reactions and non-boundary work transfer) of solid H_2O (ice) in state 1 into superheated vapor in state 6. The transformation is forced by adding some heat to the substance. The 2→3 part of the transformation is melting of the ice into water, and the 4→5 part is evaporating of the water into gas; (a) the 'work diagram', and (b) the 'heat diagram' of the transformation.

Problem 9.4: How much heat has to be added at constant pressure $P = 1$ bar to $M = 3$ kg of ice at $T = 250$ K to transform it entirely into gas (vapor) at $T = 1200$ K? Find the change of the system entropy during the heating process with no chemical reactions and with no non-boundary work transfer. Assume that 1 bar = 1 Atm (in fact, 1 Atm = 1.013 bar), that the saturation temperature for H_2O melting at pressure of 1 bar is 273.15 K, and that the saturation temperature for H_2O evaporating at pressure of 1 bar is 373.15 K (in fact, the saturation temperature of H_2O evaporating at 1 bar is 372.78 K). The entire heating process is shown in Fig. 9.8.

State 1: the initial state of the ice (solid phase of H_2O at pressure $P_1 = 1$ bar, and temperature $T_1 = 250$ K).

Process 1→2: isobaric heating of the ice.

State 2: saturated ice ready to start melting ($P_2 = 1$ bar, $T_2 = 273.15$ K).

Process 2→3: isobaric-isothermal phase transition (melting) of the ice into water (liquid H_2O); $P_3 = P_2 = 1$ bar, $T_3 = T_2 = 273.15$ K.

State 3: liquid H_2O (water); the entire ice has just melted ($P_3 = 1$ bar, $T_3 = 273.15$ K).

Process 3→4: isobaric heating of the water.

State 4: saturated water; the beginning of production of vapor (P_4 = 1 bar, T_4 = 373.15 K).

Process 4→5: isobaric-isothermal transition (vaporization) of the saturated liquid (water) into the saturated gas (vapor); the quality x of the mixture of the two phases changes during the process from 0 to 1 (P_5 = P_4 = 1 bar, T_5 = T_4 = 373.15 K).

State 5: saturated vapor; the entire liquid has just 'boiled out' (P_5 = 1 bar, T_5 = 373.15 K).

Process 5→6: isobaric heating ('superheating') of the gas.

State 6: the final state of superheated gas (P_6 = 1 bar, T_6 = 1200 K).
   The specific latent heats of H_2O at P = 1 bar are:
l_{23} = 335 kJ/kg (solid→liquid transition (melting) at 273.15 K);
l_{45} = 2256 kJ/kg (liquid→vapor transition (vaporization) at 373.15 K).

We assume that the specific heats of the ice, water, and vapor are different, but each of them does not change much within the temperature ranges of the subprocesses considered here. Subsequently, we use the following constant values of the constant-pressure specific heats: $c_P^{(i)}$(ice) = 2.11 kJ/kg·K, $c_P^{(w)}$(water) = 4.18 kJ/kg·K, and $c_P^{(g)}$(vapor) = 1.99 kJ/kg·K.

Solution: Heat transfers q_{mn} and changes of entropy Δs_{mn} per 1 kg of H_2O during the five processes making the entire **1→6** transformation are:

Process 1→2 (heating of 1 kg of ice); since this process is an isobaric transition of a single-component and single-phase substance between two equilibrium states, the change of enthalpy during the process is given by expression (7.31). There are no non-boundary work, no chemical reactions, no radiation transfer during the transition, and the gradients of pressure and temperature in the isobaric-isothermal process are zero. Therefore, the Energy Balance Equation for the process is (see (9.9) and expression (6.17))

$$q_{12} = c_P^{(i)}(T_2 - T_1) = 2.11(273.15 - 250)\,\text{K} = 48.85\,\text{kJ/kg}, \tag{9.27}$$

where $c_P^{(i)}$ is assumed constant.
   The change of the specific entropy during the 1→2 transition can be obtained from expression (7.33), assuming that the specific heat of ice is constant,

$$\Delta s_{12} = c_P^{(i)} \ln(T_2/T_1) = 2.11\ln(273.15/250) = 0.18\,\text{kJ/kg·K}. \tag{9.28}$$

Process 2→3 (isobaric-isothermal melting of 1 kg of ice); according to expression (9.10), the heat transfer per 1 kg of the system substance during this *complete* isobaric-isothermal phase transition is

$$q_{23} = l_{23} = 335\,\text{kJ/kg}, \tag{9.29}$$

and the change of the specific entropy of this (reversible) transition is (see expressions (2.13) and (2.17))

$$\Delta s_{23} = q_{23}/T_2 = l_{23}/T_2 = 335/273.15 = 1.23\,\text{kJ/kg·K}. \tag{9.30}$$

Heat and entropy transfers during processes 3→4 and 5→6 are isobaric transitions of single-component and single-phase matter (liquid H_2O and gaseous H_2O, respectively). Therefore, they can be calculated in the same way as the corresponding changes in process 1→2. Similarly, the heat and entropy transfers during the (isobaric-isothermal) phase transition 4→5 can be calculated in the same way as similar transfers in phase transition 2→3. In other words,

Process 3→4 (isobaric heating of 1 kg of water):

$$q_{34} = c_P^{(w)}(T_4 - T_3) = 4.18(373.15 - 273.15)\,\text{K} = 418\,\text{kJ/kg}, \tag{9.31}$$

$$\Delta s_{34} = c_P^{(w)} \ln(T_4/T_3) = 4.18\ln(373.15/273.15) = 1.30\,\text{kJ/kg·K}. \tag{9.32}$$

Process 4→5 (isobaric-isothermal evaporating of 1 kg of water):

$$q_{45} = l_{45} = 2256\,\text{kJ/kg}, \tag{9.33}$$

$$\Delta s_{45} = l_{45}/T_4 = 2256/373.15 = 6.04\,\text{kJ/kg·K}. \tag{9.34}$$

Process 5→6 (isobaric heating of 1 kg of vapor):

$$q_{56} = c_P^{(g)}(T_6 - T_5) = 1.99(1200 - 373.15)\,\text{K} = 1645\,\text{kJ/kg}, \tag{9.35}$$

$$\Delta s_{56} = c_P^{(g)} \ln(T_6/T_5) = 1.99\ln(1200/373.15) = 2.32\,\text{kJ/kg·K}. \tag{9.36}$$

The total heat added to the system during the entire process **1→6** is

$$Q_{16} = M(q_{12} + q_{23} + q_{34} + q_{45} + q_{56}) = +14.11\,\text{MJ}, \tag{9.37}$$

and the change of the system entropy during the entire process is

$$\Delta S_{16} = M(\Delta s_{12} + \Delta s_{23} + \Delta s_{34} + \Delta s_{45} + \Delta s_{56}) = +33.49\,\text{kJ/K}. \tag{9.38}$$

The specific entropy changes Δs_{mn} in processes $m \rightarrow n$ calculated above can also be calculated as differences $s_n - s_m$ if accurate values of s_m and s_n are known. The total change of the system entropy during the entire heating process can be given as $\Delta S_{16} = M(s_6 - s_1)$. However, ΔS_{16} calculated as $M(s_6 - s_1)$ would be slightly different from that obtained from expression (9.38). This is because we assumed in our calculations that the specific heat capacities $c_P^{(i)}$, $c_P^{(w)}$, and $c_P^{(g)}$ are constant (temperature-independent) while in fact they vary (but not much) with temperature. Similar remarks can be made about the changes of specific enthalpies (Δh_{mn}) and the total change of the system enthalpy $\Delta H_{16} = M(h_6 - h_1)$.

End of Problem 9.4.

9.3 One Phase, Many Components

9.3.1 Properties of Ideal-Gas Mixtures

Consider a non-reactive mixture of J components that form a thermodynamic system in a complete (mechanical+thermal+chemical) equilibrium at pressure P and temperature T. The mass of the ith component of the mixture is M_i, and the number of mols of the component is v_i. When the mixture is a closed system, one can write

$$\sum_{i=1}^{J} M_i = M \qquad \text{and} \qquad \sum_{i=1}^{J} v_i = v, \tag{9.39}$$

where M and v are the mass and number of mols of the entire mixture, respectively.

One can define the **mass fraction** g_i and the **mol fraction** z_i of the ith component of the mixture as

$$g_i = \frac{M_i}{M} \qquad \text{and} \qquad z_i = \frac{v_i}{v}, \tag{9.40}$$

so that

$$\sum_{i=1}^{J} g_i = 1 \qquad \text{and} \qquad \sum_{i=1}^{J} z_i = 1. \tag{9.41}$$

Since $M_i = v_i \mu_i$ and $M = v \mu$ (μ_i and μ are molar masses of the ith component and the entire mixture, respectively – see below), the relationship between the

ith component mass fraction and the component's mol fraction can be written as

$$g_i = z_i \frac{\mu_i}{\mu}. \tag{9.42}$$

The **Dalton model** of an ideal-gas mixture in equilibrium assumes that the energies of the particle-particle interactions in the gas are negligible – see Section 5.1. As a result, behavior of one component of the mixture has no impact on behavior of the other components, and the particles of each component move freely everywhere inside the entire volume accessible to the mixture. Therefore, the volume V_i of the mixture's ith component is the same as the volume V of the entire mixture. The model also assumes that temperature T_i of the ith component is the same as the temperature T of the entire mixture,

$$T_1 = T_2 = \cdots = T_J = T \quad \text{and} \quad V_1 = V_2 = \cdots = V_J = V, \tag{9.43}$$

but the pressures of the mixture components are assumed to be different. Since the pressure of the ith component is (see expression (5.4))

$$P_i = \frac{v_i R T_i}{V_i} = \frac{v_i R T}{V}, \tag{9.44}$$

while the total pressure of the entire mixture is

$$P = \frac{v R T}{V}, \tag{9.45}$$

the Dalton model gives the **partial pressure** of the ith component of the ideal mixture as

$$P_i = z_i P. \tag{9.46}$$

Expressions (9.41) and (9.46) lead to the so-called **Dalton's Law of partial pressures: the sum of the partial pressures of all components of an ideal mixture in equilibrium equals the total pressure of the mixture,**

$$\sum_{i=1}^{J} P_i = P. \tag{9.47}$$

The ratio of the total mass M of an ideal mixture of J components to its molar mass μ can be given as a sum of the ratios of the components' masses to their molar masses,

$$\frac{M}{\mu} = \frac{M_1}{\mu_1} + \frac{M_2}{\mu_2} + \cdots + \frac{M_J}{\mu_J}, \tag{9.48}$$

or

$$\frac{1}{\mu} = \frac{g_1}{\mu_1} + \frac{g_2}{\mu_2} + \cdots + \frac{g_J}{\mu_J}. \tag{9.49}$$

Thus, the equation of state for any ideal mixture is (see expression (9.47))

$$(P_1 + P_2 + \cdots + P_J)V = \left(\frac{M_1}{\mu_1} + \frac{M_2}{\mu_2} + \cdots + \frac{M_J}{\mu_J}\right)RT, \tag{9.50}$$

or

$$PV = \frac{M}{\mu}RT = MR_gT, \tag{9.51}$$

or

$$\rho = \mu P/RT = P/R_gT, \tag{9.52}$$

where, as before, P, V, T, and ρ are the mixture pressure, volume, temperature, and mass density, respectively, R is the universal (molar) gas constant of the mixture, and

$$R_g = R/\mu \tag{9.53}$$

is the mixture's **individual gas constant**. Since $M_i = v_i\mu_i$, the molar mass of the mixture can be given as (see expression (9.48))

$$\mu = \frac{M}{v_1 + v_2 + \cdots + v_J} = \frac{M}{v}. \tag{9.54}$$

When composition of an ideal-gas mixture is defined by a set of mass fractions of the mixture components, relationship (9.49) can be written as

$$\mu = 1/\sum_{i=1}^{J}(g_i/\mu_i), \tag{9.55}$$

which reduces, in the common case of two-component mixture, to

$$\mu = \mu_1\mu_2/(g_1\mu_2 + (1 - g_1)\mu_1). \tag{9.56}$$

Since components of ideal-gas mixture do not interact, the mixture mass fractions g_i and mol fractions z_i can be considered 'statistical weights' representing the contributions of the components to the physical properties of the entire mixture. Therefore, any *specific physical property* of the mixture can be given as a linear weighted sum. (*Specific property* of a thermodynamic system is the amount of the property carried by a unit of the system's mass.) This sum, often called the **mixture rule**, is

$$p_{\text{mix}} = \sum_{i=1}^{J} g_i p_i, \tag{9.57}$$

where the mass fractions $\{g_i\}$ are the statistical weights of the sum ($\sum_i g_i = 1$ is the 'normalization condition' for the weights), and p_i is the specific property (enthalpy, entropy, heat capacity, etc.) carried by the ith component. For example, the specific enthalpy h_{mix} (in kJ/kg) of an ideal-gas mixture consisting of J gases can be calculated as

$$h_{\text{mix}} = \sum_{i=1}^{J} g_i h_i, \tag{9.58}$$

where h_i (in kJ/kg) is the specific enthalpy of the ith component of the mixture. Similarly, the mixture's constant-pressure specific heat (in kJ/kg·K) can be given as

$$c_{P,\text{mix}} = \sum_{i=1}^{J} g_i c_{P,i}, \tag{9.59}$$

where $c_{P,i}$ is the constant-pressure specific heat of the ith component of the mixture.

Problem 9.5: Assume that air at some pressure and temperature near the earth surface is a calorically perfect ideal mixture (see Section 6.4) consisting only of nitrogen and oxygen gases with their mass fractions $g_{N_2} = 0.77$ (77%) and $g_{O_2} = 0.23$ (23%), respectively. Calculate the molar mass of the air, the individual constant of the gas and its components, the constant-volume specific heat capacity of the air, and the mol fractions of the components.

Solution: According to the *Periodic Chart of Elements*, the molar masses of diatomic nitrogen and diatomic oxygen are

$$\mu_{N_2} = 28\,\text{kg/kmol} \qquad \text{and} \qquad \mu_{O_2} = 32\,\text{kg/kmol}, \tag{9.60}$$

so that the molar mass of the air is (see expression (9.57))

$$\mu_{\text{air}} = \frac{\mu_{N_2}\mu_{O_2}}{g_{N_2}\mu_{O_2} + g_{O_2}\mu_{N_2}} = \frac{(28\,\text{kg/kmol})(32\,\text{kg/kmol})}{0.77(32\,\text{kg/kmol}) + 0.23(28\,\text{kg/kmol})}$$
$$= 28.83\,\text{kg/kmol}. \tag{9.61}$$

One can see here the reason why molar masses of mixtures are sometimes called 'apparent molar masses' – they represent molar masses of fictitious gases of particles that do not exist in Nature.

The individual gas constant of the air is (see expressions (5.5) and (9.53))

$$(R_g)_\text{air} = \frac{R}{\mu_\text{air}} = \frac{8.314\,\text{kJ/kmol·K}}{28.83\,\text{kg/kmol}} = 0.288\,\text{kJ/kg·K}, \tag{9.62}$$

and the individual gas constants of the air components are

$$(R_g)_\text{N}_2 = \frac{R}{\mu_{\text{N}_2}} = \frac{8.314\,\text{kJ/kmol·K}}{28\,\text{kg/kmol}} = 0.297\,\text{kJ/kg·K}, \tag{9.63}$$

and

$$(R_g)_{\text{O}_2} = \frac{R}{\mu_{\text{O}_2}} = \frac{8.314\,\text{kJ/kmol·K}}{32\,\text{kg/kmol}} = 0.260\,\text{kJ/kg·K}. \tag{9.64}$$

Since the air can be considered a calorically perfect ideal gas, the specific heat capacities of the air components are temperature-independent and equal to (see Section 6.4)

$$c_{V,\text{N}_2} = \frac{5}{2}R_{\text{N}_2} = 0.742\,\frac{\text{kJ}}{\text{kg·K}} \quad\text{and}\quad c_{V,\text{O}_2} = \frac{5}{2}R_{\text{O}_2} = 0.650\,\frac{\text{kJ}}{\text{kg·K}}, \tag{9.65}$$

and the constant-volume specific heat capacity of the air is

$$\begin{aligned} c_{V,\text{air}} &= g_{\text{N}_2}c_{V,\text{N}_2} + g_{\text{O}_2}c_{V,\text{O}_2} \\ &= 0.77 \times 0.742\,\text{kJ/kg·K} + 0.23 \times 0.650\,\text{kJ/kg·K} \\ &= 0.720\,\text{kJ/kg·K}. \end{aligned} \tag{9.66}$$

Using relationship (9.42), one obtains the mol fractions of nitrogen and oxygen, respectively, as

$$z_{\text{N}_2} = g_{\text{N}_2}\frac{\mu_\text{air}}{\mu_{\text{N}_2}} = 0.77\frac{28.83\,\text{kg/kmol}}{28\,\text{kg/kmol}} = 0.79, \tag{9.67}$$

and

$$z_{\text{O}_2} = g_{\text{O}_2}\frac{\mu_\text{air}}{\mu_{\text{O}_2}} = 0.23\frac{28.83\,\text{kg/kmol}}{32\,\text{kg/kmol}} = 0.21. \tag{9.68}$$

End of Problem 9.5.

9.3.2   Properties of Real-Gas Mixtures

Studying the thermodynamic properties of mixtures of non-ideal gases ('real gases' where particle-particle interactions cannot be ignored) is difficult. The main problem is lack of reliable 'mixture rules' for superposing (analogously to the ideal-mixture weighted sum (9.57)) the contributions of the individual components of the mixtures. The difficulty increases when the number of components is large and when the contributions vary within a wide range. Mathematically, a mixture rule for a real-gas mixture can be similar to expression (9.57), but it must include some additional terms (called the 'cross terms') accounting for the contribution of the particle-particle interactions.

A useful method for studying the thermodynamic properties of real gases and real-gas mixtures is discussed in Section 8.3. It is based on the so-called **thermodynamic departure functions** (or **thermodynamic departure factors**) which give the differences between the values of the thermodynamic properties of real gas and the values of the corresponding properties of its ideal-gas model.

Chapter 10

Thermodynamic Cycles

10.1 Introduction

A **thermodynamic cycle** is a process in which a thermodynamic system in a given initial state A moves through a number of different states to return to the initial state.

There are several types of energy-processing devices that operate as thermodynamic cycles. These are devices that can cyclically transfer heat into work (such devices are called **heat engines**), work into heat (**heaters**), or move heat from one place to another (**refrigerators** and **heat pumps**) – see Fig. 10.1. The refrigerator and heat pump 'follow' the same thermodynamic cycles but have different objectives. The function of a refrigerator is to extract heat from an object in order to keep it at some lower temperature, while the function of a heat pump is to deliver heat to an object in order to keep it at some higher temperature.

Energy-processing devices interact with their environment, which may include other technological devices. Parts of the environment (or the entire environment) that exchange heat (but not work) with the energy-processing device can often be modeled as the so-called **heat reservoirs** (**heat baths**). The thermal energy of every heat reservoir is much, much greater than the amount of heat transferred between the energy-processing device and the reservoir. Therefore, one can assume that the temperature and pressure of the reservoir (but not necessarily of the energy-processing device) are constant during the heat transfer. As a result, the transfer is 'seen' *by the reservoir* as a reversible

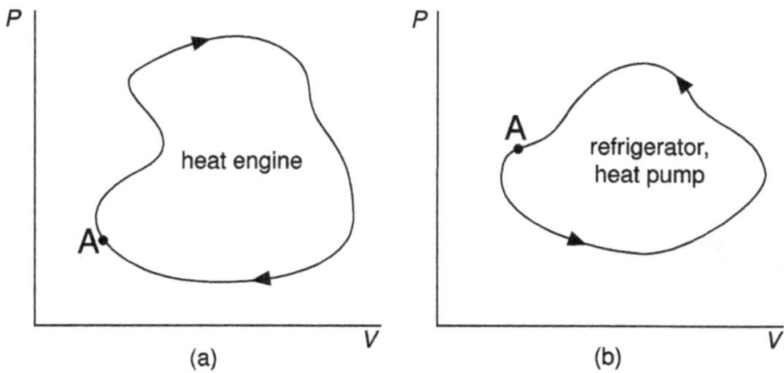

Fig. 10.1: Thermodynamic cycles in the work plane P–V: (a) the heat engine cycle, (b) the refrigerator cycle or heat pump cycle.

isobaric-isothermal process at the reservoir's temperature T, and the change of the reservoir's entropy during the process can be given as (see expression (2.13))

$$\Delta S_{\text{res}} = Q/T, \tag{10.1}$$

where, in accordance with the International Convention for the Sign of Energy, the transferred heat Q is a positive number when it is absorbed by the reservoir), but a negative number when it leaves the reservoir.

Let us study energy-processing devices operating in a *cyclic* manner while exchanging heat with surroundings that can be considered heat reservoirs. In order to simplify our considerations (without changing their main conclusions) we assume here that each of the devices interacts with no more than two heat reservoirs: no more than one reservoir supplies heat to the device, and no more than one reservoir receives heat from the device. (Various energy-processing devices will be further studied in Chapter 13.) We will use below the fact that the change of every property of state during any (reversible or not) complete thermodynamic cycle is always zero (see Section 2.8.1). In other words, the changes of the thermal energy U, enthalpy H, entropy S, the Helmholtz free energy A, and the Gibbs free energy G of any energy-processing device during one complete thermodynamic cycle (or during an integral number of complete cycles) are

$$\Delta U = 0, \quad \Delta H = 0, \quad \Delta S = 0, \quad \Delta A = 0, \quad \Delta G = 0. \tag{10.2}$$

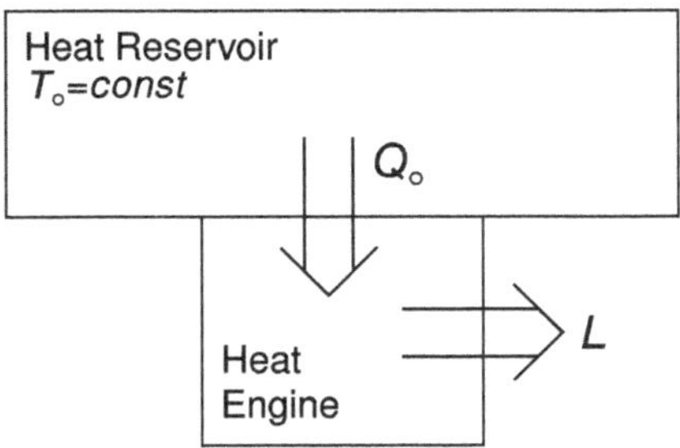

Fig. 10.2: A heat engine producing, during one thermodynamic cycle, work L while receiving heat Q_o from a single heat reservoir of constant temperature T_o.

10.2 Heat Engine using One Heat Reservoir

Consider a heat engine which produces, in a cyclic and steady-state way, some work L while receiving heat Q_o from a heat reservoir at a constant temperature T_o (see Fig. 10.2). In order to check whether such an engine can operate in practice, we must test its operation against the Second Law of Thermodynamics (see Section 3.5.3). The Law says that the total entropy of every *isolated* physical system undergoing any thermodynamic process increases during the process. (The reader may recall that the total entropy of every isolated system is just the system's natural entropy, which always increases in all processes in such system.) Thus, the heat engine plus the part of the environment (the heat reservoir) that supplies the heat Q_o to the engine, plus the part of the environment receiving the work produced by the engine, can be considered an *isolated* system called hereafter the 'overall system'.

The change of entropy of the overall system during one operational cycle of the engine is

$$\Delta S_{\text{sys}} = \Delta S_{\text{eng}} + \Delta S_{\text{res}}, \tag{10.3}$$

where ΔS_{eng} and ΔS_{res} are the changes of the entropies of the engine and the heat reservoir, respectively, and where we took the change of the entropy of the part of the environment that receives the work L as equal to zero because work does not carry entropy – see Section 2.3.

During any *complete* cycle of the engine, its entropy (property of state) does not change,

$$\Delta S_{\text{eng}} = 0. \tag{10.4}$$

Since the heat Q_o leaves the reservoir at its constant temperature T_o, the change of the entropy of the reservoir during one thermodynamic cycle is (see expression (2.13))

$$\Delta S_{res} = \frac{Q_o}{T_o} = -\frac{|Q_o|}{T_o},\tag{10.5}$$

where the heat Q_o is 'seen' by the reservoir as a negative number, in agreement with the International Convention for the Sign of Energy, because the heat leaves the reservoir. Thus, according to relationships (10.3)–(10.5), the change of the total entropy of the (isolated) overall system is

$$\Delta S_{sys} = 0 - \frac{|Q_o|}{T_o} < 0 \quad \text{always!}\tag{10.6}$$

This relationship violates the Second Law of Thermodynamics which requires that the change of the total entropy in any process taking place in any *isolated* system (such as the overall system discussed here) cannot be negative. Conclusion: **it is impossible to build a heat engine that would cyclically produce work while receiving heat from a single heat reservoir.** One should notice that the discussed heat engine could produce work if it operated in a non-cyclic way. In such a case, ΔS_{eng} can be greater than zero, and possibly large enough to make ΔS_{sys} also greater than zero, in agreement with the Second Law of Thermodynamics.

10.3 Heater using One Heat Reservoir

Consider a simple heater, a device transforming cyclically some work L into heat Q_o, which is then moved to a heat reservoir that maintains constant temperature T_o (see Fig. 10.3). The heater, the heat reservoir, and the part of the environment that supplies the work to the heater form an *isolated* thermodynamic system (the overall system). The change of the total entropy of such a system during its one-cycle operation is (see expression (2.13))

$$\Delta S_{sys} = \Delta S_{res} + \Delta S_{hea} = \frac{Q_o}{T_o} + 0 = \frac{|Q_o|}{T_o} > 0 \quad \text{always!},\tag{10.7}$$

where ΔS_{res} and ΔS_{hea} are the changes of entropies of the reservoir and the heater, respectively. The heat Q_o is 'seen' by the reservoir as a positive quantity

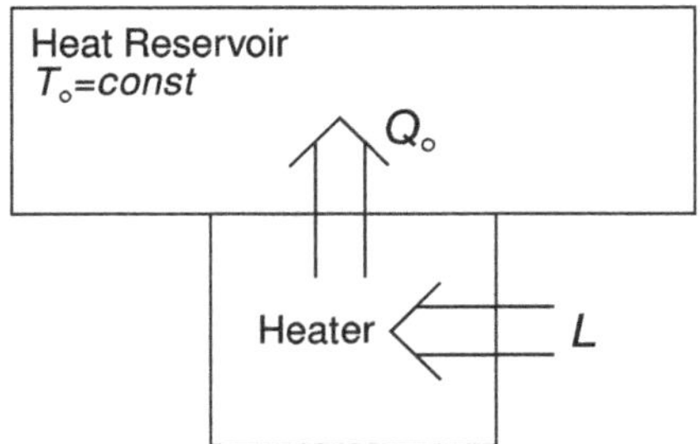

Fig. 10.3: Work L supplied to a heater during a thermodynamic cycle is converted into heat Q_o, which moves to a single heat reservoir of constant temperature T_o.

because it is being received by the reservoir. The change of entropy of the heater during one cycle of its operation is zero because the change of any thermodynamic function of state in any thermodynamic cycle is zero. As before, we did not consider in relationship (10.7) the change of the entropy of the part of the environment that supplies work L to the heater because work does not carry entropy. Conclusion: **It is possible to operate a heater that transforms (in a cyclic manner) work into heat which is absorbed by a single heat reservoir.**

10.4   Heat Engine using Two Heat Reservoirs

A heat engine using two heat reservoirs is shown in Fig. 10.4. During an operational cycle, some heat Q_1 is taken from reservoir 1 at the reservoir's constant temperature T_1 and processed in the engine in order to produce some work L. In the same cycle, the engine rejects heat Q_2, which is taken in by heat reservoir 2 at its constant temperature T_2. The temperature of reservoir 1 is higher than the temperature of reservoir 2, $T_1 > T_2$. The engine, the two heat reservoirs, and the part of the environment that receives the work produced by the engine form an *isolated* thermodynamic system. Because work does not carry entropy, the change of the total entropy of the overall system during its one-cycle operation is (see expression (2.13))

$$\Delta S_{\text{sys}} = \Delta S_{\text{res1}} + \Delta S_{\text{res2}} + \Delta S_{\text{eng}} = -\frac{|Q_1|}{T_1} + \frac{|Q_2|}{T_2} + 0, \qquad (10.8)$$

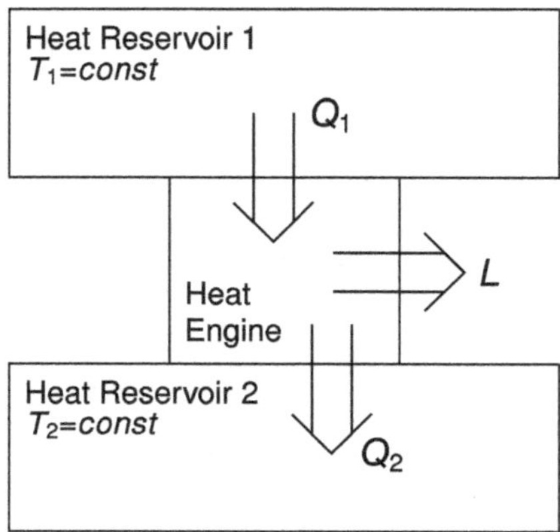

Fig. 10.4: An engine operating in a cyclic way receives heat Q_1 from heat reservoir 1 (of temperature T_1), produces work L, and rejects heat Q_2 to heat reservoir 2 (of temperature T_2).

where ΔS_{eng} is the change (equal to zero) of the entropy of the engine, and ΔS_{res1} and ΔS_{res2} are the changes of the entropies of reservoirs 1 and 2, respectively. Heat Q_1 is 'seen' by reservoir 1 as a negative quantity (because it leaves the reservoir) while heat Q_2 is 'seen' by reservoir 2 as a positive quantity (because it is received by the reservoir). ΔS_{sys} in relationship (10.8) has a positive value (so that operation of the considered overall system does not violate the Second Law of Thermodynamics) when

$$\frac{|Q_2|}{T_2} > \frac{|Q_1|}{T_1}.$$
(10.9)

Conclusion: **It is possible to build a heat engine operating cyclically between two heat reservoirs. However, such an engine must interact with the two heat reservoirs in a way that does not violate requirement (10.9).**

10.5 Refrigerator or Heat Pump using Two Heat Reservoirs

Refrigerators and heat pumps follow similar (counterclockwise in P–V plane) thermodynamic cycles but have different objectives. A refrigerator removes

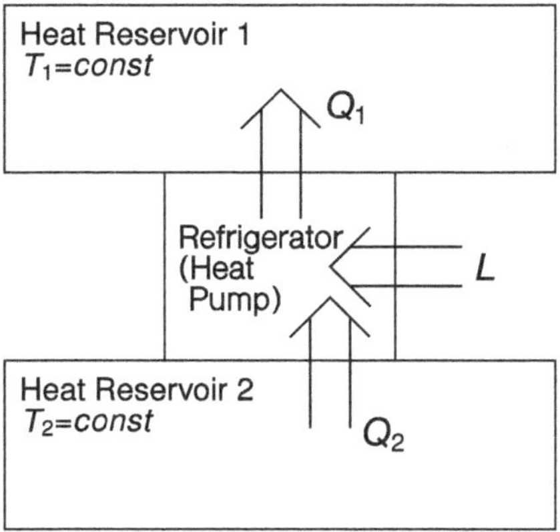

Fig. 10.5: A refrigerator (or heat pump) operating in a cyclic way receives heat Q_2 from heat reservoir 2 (of temperature T_2), uses some work L, and moves heat Q_1 to heat reservoir 1 (of temperature T_1).

a desired amount of heat from an object while a heat pump delivers a desired amount of heat to an object.

A refrigerator (or a heat pump) using two heat reservoirs is shown in Fig. 10.5. Work L is supplied to the refrigerator (heat pump) from its surroundings in order to move heat from the lower reservoir 2 to the upper reservoir 1. Heat Q_2 is moved to the refrigerator (heat pump) from the heat reservoir 2 that has constant temperature T_2. Heat Q_1 is moved to the upper heat reservoir 1 that has constant temperature T_1. The temperature of reservoir 1 is greater than the temperature of reservoir 2, $T_1 > T_2$.

The refrigerator (heat pump), heat reservoirs 1 and 2, and the part of the environment that supplies the work L form an isolated overall system. The change of the entropy of the system's part that supplies work L is not considered because work itself does not carry entropy. Heat Q_1 is 'seen' by reservoir 1 as a positive quantity because it enters the reservoir, while heat Q_2 is 'seen' by reservoir 2 as a negative quantity because it leaves the reservoir. The change of the total entropy of the overall system is (see expression (2.13))

$$\Delta S_{\text{sys}} = \Delta S_{\text{res1}} + \Delta S_{\text{res2}} + \Delta S_{r,p} = \frac{|Q_1|}{T_1} - \frac{|Q_2|}{T_2}, \tag{10.10}$$

where $\Delta S_{r,p} = 0$ is the change of the entropy of the refrigerator (heat pump) during one complete cycle of its operation, and ΔS_{res1} and ΔS_{res2} are the changes of the entropies of reservoirs 1 and 2, respectively. ΔS_{sys} in relationship (10.10) is greater than zero (as required by the Second Law of Thermodynamics) only when

$$\frac{|Q_1|}{|Q_2|} > \frac{T_1}{T_2}.$$ (10.11)

Conclusion: **Heat can be moved in a cyclic way from a heat reservoir to another heat reservoir if the requirement (10.11) is not violated during the operation.**

10.6 Performance of Cycle-Based Devices

Each overall system studied in the previous sections had a central unit (an energy-processing device such as a heat engine, refrigerator, heater or heat pump) and one or two heat reservoirs exchanging energy (in the form of heat) with the central unit. The reservoirs were releasing or receiving heat during isothermal processes at the reservoirs' constant temperatures. The central units operated as thermodynamic cycles, so that the change of the entropy (and the other thermodynamic functions of state) of each of the units during one cycle was always zero. In general, *theoretical* measures of operational performance of a heat engine, refrigerator, or heat pump shown in Figs. 10.4–10.5 are defined as (recall the International Convention for the Sign of Energy):

1) the thermal efficiency of a heat engine:

$$\eta = \frac{\text{work output}}{\text{heat input}} = \frac{|L|}{|Q_1|},$$ (10.12)

2) the coefficient of performance of a refrigerator:

$$\beta = \frac{\text{heat output}}{\text{work input}} = \frac{|Q_2|}{|L|},$$ (10.13)

and

3) the coefficient of performance of a heat pump:

$$\gamma = \frac{\text{heat output}}{\text{work input}} = \frac{|Q_1|}{|L|}.$$ (10.14)

The energy balance of a heat engine, refrigerator, or heat pump operating *reversibly* between two heat reservoirs can be given as (see Figs. 10.4 and 10.5)

$$|L_{\text{rev}}| = |Q_1| - |Q_2|, \tag{10.15}$$

where L_{rev} is the work done on or done by the reversibly-operating central unit, Q_1 is the heat exchanged between reservoir 1 and the unit, and Q_2 is the heat exchanged between reservoir 2 and the unit.

The *theoretical* measures of the performance of the discussed central units operating in *reversible* ways while exchanging heat with two heat reservoirs can be given as

1) thermal efficiency of a heat engine operating in a reversible way:

$$\eta_{\text{rev}} = \frac{|L_{\text{rev}}|}{|Q_1|} = \frac{|Q_1| - |Q_2|}{|Q_1|} = 1 - \frac{|Q_2|}{|Q_1|}, \tag{10.16}$$

2) coefficient of performance of a refrigerator operating in a reversible way:

$$\beta_{\text{rev}} = \frac{|Q_2|}{|L_{\text{rev}}|} = \frac{|Q_2|}{|Q_1| - |Q_2|}, \tag{10.17}$$

and

3) coefficient of performance of a heat pump operating in a reversible way:

$$\gamma_{\text{rev}} = \frac{|Q_1|}{|L_{\text{rev}}|} = \frac{|Q_1|}{|Q_1| - |Q_2|}. \tag{10.18}$$

When the central unit operates *irreversibly*, the balance of its energy when the unit is a heat engine (work leaves the unit) is

$$|L_{\text{irr}}| = |L_{\text{rev}}| - |L_{\text{diss}}| = (|Q_1| - |Q_2|) - |Q_{\text{diss}}|, \tag{10.19}$$

while the energy balances in the case of a refrigerator or a heat pump (work enters the unit) are

$$|L_{\text{irr}}| = |L_{\text{rev}}| + |L_{\text{diss}}| = (|Q_1| - |Q_2|) + |Q_{\text{diss}}|, \tag{10.20}$$

where L_{diss} is the amount of the unit's work dissipated into heat Q_{diss} (see Chapter 13). Of course, $|Q_{\text{diss}}| = |L_{\text{diss}}|$ because energy can be dissipated but cannot disappear.

Taking the above into account, the theoretical measures of the efficiencies of the discussed central units operating irreversibly while exchanging heat with two heat reservoirs can be given as

1) thermal efficiency of a heat engine operating in an irreversible way:

$$\eta_{\text{irr}} = \frac{|L_{\text{irr}}|}{|Q_1|} = \frac{|Q_1| - |Q_2| - |Q_{\text{diss}}|}{|Q_1|} = 1 - \frac{|Q_2| + |Q_{\text{diss}}|}{|Q_1|}, \qquad (10.21)$$

2) coefficient of performance of a refrigerator operating in an irreversible way:

$$\beta_{\text{irr}} = \frac{|Q_2|}{|L_{\text{irr}}|} = \frac{|Q_2|}{|Q_1| - |Q_2| + |Q_{\text{diss}}|}, \qquad (10.22)$$

and

3) coefficient of performance of a heat pump operating in an irreversible way:

$$\gamma_{\text{irr}} = \frac{|Q_1|}{|L_{\text{irr}}|} = \frac{|Q_1|}{|Q_1| - |Q_2| + |Q_{\text{diss}}|}. \qquad (10.23)$$

Comparison of relationships (10.16)–(10.18) with relationships (10.21)–(10.23) shows that the **theoretical performance of irreversible heat engines, refrigerators, and heat pumps is always worse than the performance of the corresponding devices operating reversibly.**

Irreversibilities present in practical thermodynamic cycles are further discussed in Chapter 13.

10.7 The Carnot Cycle

We studied above a heat engine, a refrigerator, and a heat pump that receive some heat from a heat reservoir at a constant temperature and releases some heat to another heat reservoir at a constant temperature. However, in processes without a phase change, isothermal heat transfers require large-surface heat exchangers and/or large amounts of time to transfer a meaningful amount of heat between the central units and the reservoirs. The sizes of the surface areas of the heat exchangers put constraints on the size and costs of the devices, while the long durations of the heat transfers substantially limit their actual power outputs. Therefore, in practice, the heat transfers are done by processes other than isotherms.

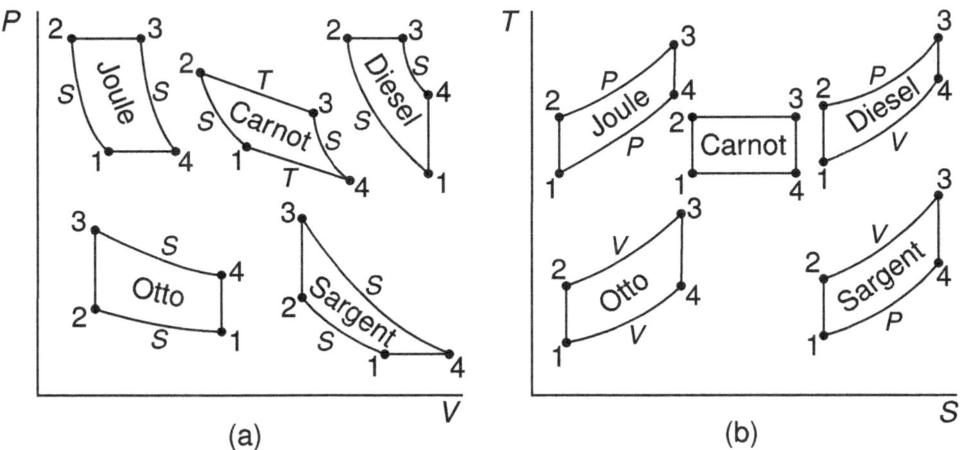

Fig. 10.6: Examples of thermodynamic cycles. The letters next to the processes making the cycles indicate the thermodynamic properties which are kept constant during the processes – the letters P, V, T, and S stand for pressure, volume, temperature, and entropy, respectively; (a) the work plane, (b) the heat plane.

Use of non-isothermal heat transfers in the energy-processing devices leads to many different thermodynamic cycles along which the devices operate. Several such cycles are shown in Fig. 10.6. Two of them, the Otto cycle and the Diesel cycle, are discussed in Sections 10.8 and 10.9, respectively.

Reversible thermodynamic cycles are easy-to-deal-with models of operations of thermodynamic devices discussed here because the models ignore the irreversibility-related dissipation effects that limit the operations. One should recall that the changes of properties of state in any real thermodynamic cycle are zero regardless of whether the cycle is reversible or not. However, the transfers of work and heat (both functions of path) between the cycle-based system and the environment depend on the degree of irreversibility of the cycle. This will be discussed in Chapter 13.

Different thermodynamic cycles process energy differently. Therefore, it is necessary to ask what cycle is the most efficient in the processing. It is shown below that this is the so-called **Carnot cycle**. This fact is often summarized in **Carnot's Theorem: No heat engine operating cyclically between two heat reservoirs can be more efficient than a Carnot engine operating between**

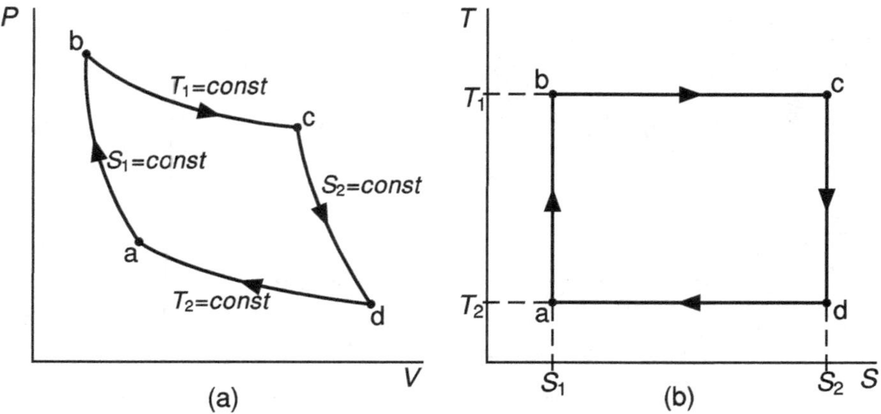

Fig. 10.7: The Carnot cycle for a heat engine: (a) the work plane, (b) the heat plane.

the two reservoirs. The theorem can be extended on refrigerator and heat pump cycles.

The Carnot cycle for a heat engine consists of two reversible isotherms (each representing interaction of one heat reservoir with the engine) and two adiabat-isentrops – see Fig. 10.7. One can see in the figure that the amount of heat released from the upper heat reservoir (reservoir 1 of temperature T_1) and taken in by the heat engine operating as a (clockwise) Carnot cycle is equal to the heat of the isotherm $b \rightarrow c$. This heat is (see expression (10.1))

$$Q_1 = T_1(S_2 - S_1). \tag{10.24}$$

The heat Q_2 released by the engine and received by the lower heat reservoir (reservoir 2 of temperature T_2) can be calculated as the heat of the reversible isotherm $d \rightarrow a$,

$$Q_2 = T_2(S_1 - S_2). \tag{10.25}$$

Relationships (10.24)–(10.25) are also valid in the counterclockwise Carnot cycles for refrigerators and heat pumps, but there the directions of the heat flows Q_1 and Q_2 are opposite to their directions in the clockwise Carnot cycles of heat engines (see Figs. 10.4 and 10.5).

Using definitions (10.12)–(10.14) and relationships (10.24)–(10.25), the thermal efficiencies of the Carnot's engine, Carnot's refrigerator, and Carnot's heat pump can be given, respectively, as

$$\eta_c = 1 - \frac{T_2}{T_1} \qquad \text{Carnot heat engine,} \qquad (10.26)$$

$$\beta_c = \frac{1}{T_1/T_2 - 1} \qquad \text{Carnot refrigerator,} \qquad (10.27)$$

and

$$\gamma_c = \frac{1}{1 - T_2/T_1} \qquad \text{Carnot heat pump.} \qquad (10.28)$$

One should notice that the performance measures (10.26)–(10.28) depend neither on the kind of the substance nor on the design details of the central units and the reservoirs used in the Carnot cycles.

It is interesting to compare the thermal efficiencies and coefficients of performance of non-Carnot cycles with the corresponding measures of the Carnot cycle. To do so, we compare (see Fig. 10.8) a Carnot cycle c and a reversible cycle r, each having the same highest temperature $T_{\max}$ and the same lowest temperature $T_{\min}$. The minimum and maximum entropies (S_1 and S_2, respectively) of the power cycle r are the same as the corresponding entropies of the Carnot power cycle c.

The thermal efficiency of the thermodynamic cycle r is (see expression (10.16))

$$\eta_r = \frac{|L_{r,\text{out}}|}{|Q_{r,\text{in}}|} = \frac{|Q_{r,\text{in}}| - |Q_{r,\text{out}}|}{|Q_{r,\text{in}}|}, \qquad (10.29)$$

and the thermal efficiency of the Carnot cycle c is (see expression (10.26))

$$\eta_c = \frac{|L_{c,\text{out}}|}{|Q_{c,\text{in}}|} = \frac{|Q_{c,\text{in}}| - |Q_{c,\text{out}}|}{|Q_{c,\text{in}}|} = 1 - \frac{T_{\min}}{T_{\max}}, \qquad (10.30)$$

where $L_{r,\text{out}}$ and $L_{c,\text{out}}$ are the works of the engine operating reversibly along the cycles r and c, respectively; $Q_{r,\text{in}}$ and $Q_{r,\text{out}}$ are heats taken in and released, respectively, by the engine performing the cycle r, and $Q_{c,\text{in}}$ and $Q_{c,\text{out}}$ are the heats taken in and released, respectively, by the engine performing the Carnot cycle c.

In the T–S diagram shown in Fig. 10.8, the heat $Q_{r,\text{in}}$ is equal to the area under the upper dashed curve connecting the states 1 and 2, while the heat $Q_{r,\text{out}}$ is equal to the area under the lower dashed curve connecting the states.

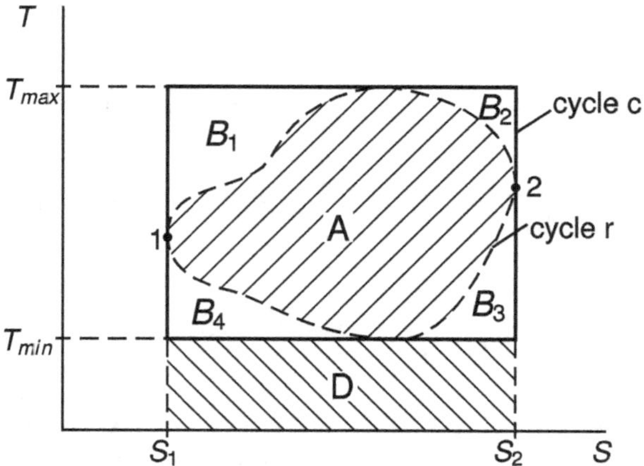

Fig. 10.8: Comparison of efficiencies of a Carnot cycle c (solid line) and a reversible cycle r (dashed line) when the cycles' minimum temperatures $T_{\min}$ are the same, the maximum temperatures $T_{\max}$ are the same, the minimum entropies S_1 are the same, and the maximum entropies S_2 are the same.

Therefore, the work produced by the cycle r is equal to the area A inside the cycle (this area is hatched diagonally upwards to the right). Subsequently,

$$\eta_r = \frac{A}{D + A + B_3 + B_4},$$
(10.31)

where D is the area hatched diagonally upwards to the left, and

$$\eta_c = \frac{A + B_1 + B_2 + B_3 + B_4}{D + A + B_1 + B_2 + B_3 + B_4},$$
(10.32)

so that one always has

$$\frac{\eta_r}{\eta_c} < 1.$$
(10.33)

One can see that the efficiency of the Carnot engine is the upper limit of the efficiency of any engine, and that this highest efficiency is obtained when the ratio T_2/T_1 is the smallest. In other words, heat should be added to the engine at a temperature as high as possible and removed from it at a temperature as low as possible. The typical maximum operational temperature common in heat engines is about 1200 K, while the minimum temperature is usually that of the outdoor air (about 300 K). Therefore, according to expression (10.30), the thermal efficiency of a typical heat engine working along the Carnot cycle is

$$\eta_c \approx 1 - \frac{300}{1200} = 75\%. \qquad (10.34)$$

It should be emphasized that the thermal efficiencies of the heat engines considered above are the *theoretical* efficiencies that are controlled only by constraints imposed by the fundamental laws of Nature (the First Law of Thermodynamics and the Second Law of Thermodynamics) on the interplay between the work and heat transfers during the engines' operations. The efficiencies of real engines are always smaller than their theoretical efficiencies because of the weaknesses of the engine design, manufactoring, the quality of lubrication, etc.

Problem 10.1: Find which of the following two processes is more effective in maintaining the desired temperature inside a house (see Fig. 10.9): 1) burning a fuel in the combustion chamber and supplying the obtained heat directly to the house ('direct' heating of the house), or 2) first using the heat from the combustion chamber as input to a heat engine and then using the work produced by the engine to move heat from the outdoors to the house using a heat pump ('indirect' heating of the house). Assume that the fuel combustion chamber and the outdoors maintain different but constant temperatures, and that both the heat engine and the heat pump operate as Carnot cycles.

Solution: Let us introduce a 'multiplication factor' m defined as the ratio of the number of heat-equivalent tons of fuel used for heating of the house to the number of heat-equivalent tons of fuel burned in the combustion chamber. As can be seen in Fig. 10.9, the multiplication factors for the direct and indirect heatings of the house are, respectively,

$$m_{\mathrm{dir}} = \frac{|Q_1| - |Q_1'|}{|Q_1|} = 1 - \frac{|Q_1'|}{|Q_1|}, \qquad (10.35)$$

and

$$m_{\mathrm{ind}} = \frac{|Q_2| + |Q_2'|}{|Q_1|}. \qquad (10.36)$$

Note that m_{dir} is always smaller than one while m_{ind} can be greater than one.

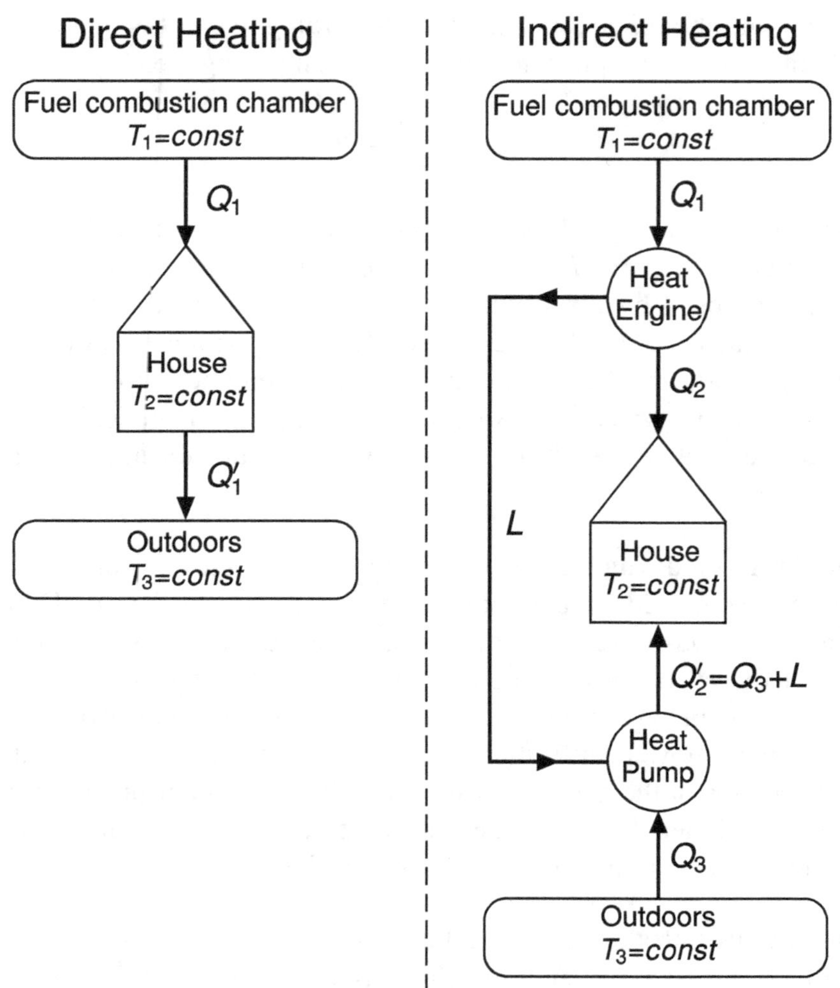

Fig. 10.9: Warming of a house by direct and indirect heating.

During the indirect heating of the house, the efficiency of the Carnot engine is (see expression (10.26))

$$\frac{|L|}{|Q_1|} = 1 - \frac{|Q_2|}{|Q_1|} = 1 - \frac{T_2}{T_1}, \tag{10.37}$$

so that

$$|L| = |Q_1|\left(1 - \frac{T_2}{T_1}\right) \quad \text{and} \quad |Q_2| = |Q_1|\frac{T_2}{T_1}. \tag{10.38}$$

The performance coefficient of the Carnot heat pump in the process of the indirect heating of the house is (see expression (10.28))

$$\frac{|Q_2'|}{|L|} = \frac{|Q_2'|}{|Q_2'| - |Q_3|} = \frac{T_2}{T_2 - T_3}, \tag{10.39}$$

so that relationships (10.37)–(10.39) lead to

$$|Q_2'| = |L|\frac{T_2}{T_2 - T_3} = |Q_1|\left(1 - \frac{T_2}{T_1}\right)\frac{T_2}{T_2 - T_3}, \tag{10.40}$$

and

$$m_{\text{ind}} = \frac{|Q_2|}{|Q_1|} + \frac{|Q_2'|}{|Q_1|} = \frac{T_2}{T_1}\left(1 + \frac{T_1 - T_2}{T_2 - T_3}\right). \tag{10.41}$$

In order to make some numerical estimates of the multiplication factors, we assume the following values of the system temperature:

$T_1 = 473$ K (temperature in the fuel combustion chamber),
$T_2 = 293$ K (temperature of the indoor air),
$T_3 = 273$ K (temperature of the outdoor air).

Use of these values in expression (10.41) gives

$$m_{\text{ind}} = \frac{293\,\text{K}}{473\,\text{K}}\left(1 + \frac{180\,\text{K}}{20\,\text{K}}\right) \approx 6. \tag{10.42}$$

This means that every ton of fuel burned in the combustion chamber in indirect heating of the house delivers a heat equivalent of about six tons of fuel burned in the chamber. Since m_{dir} is always less than one, the indirect heating of the house is *thermodynamically* much more efficient than the direct heating. But heating the house indirectly may be too expensive and/or inconvenient to operate. Still, one can see from the above discussion that there is a lot of flexibility in ways of using the cyclic devices studied in this chapter.

End of Problem 10.1.

10.8   The Otto Cycle

The Otto cycle is a reversible thermodynamic cycle of an ideal gas working in a closed (cylinder-piston) system that was used to build the first passenger-car engine which is called the 'Otto engine', 'gasoline engine', 'spark-ignition engine', or 'internal combustion engine'. The cycle is shown in P–V and T–S diagrams in Fig. 10.10.

The $1{\rightarrow}2{\rightarrow}3{\rightarrow}4{\rightarrow}1$ cycle consists of four consecutive reversible transitions where states 1, 2, 3, and 4 are equilibrium states. The only kind of work transferred during the cycle is boundary work. The four thermodynamic transitions that make the Otto cycle are:

(a) adiabatic-isentropic compression from state 1 to state 2,
(b) isochoric heat addition $2{\rightarrow}3$,
(c) adiabatic-isentropic expansion $3{\rightarrow}4$,
(d) isochoric heat removal $4{\rightarrow}1$.

The changes of motion properties of the working gas (its kinetic and potential energies) are insignificant and are therefore ignored. Thus, in this problem, the 'system internal energy' also means the 'system thermal energy'.

Assuming that the mass of the working gas is M, that its individual gas constant is R_g, and that the heat capacities $(C_V)_{i{\rightarrow}j}$ and $(C_P)_{i{\rightarrow}j}$ (see expressions (6.3)–(6.6)) are constant during each of the four transitions (a)–(d), the properties of the entire Otto cycle can be calculated as follows:

(a) *the adiabatic-isentropic compression* $1{\rightarrow}2$:

The thermodynamic properties of equilibrium state 2 can be 'connected' with the thermodynamic properties of equilibrium state 1 using the ideal gas equation of state and the equation of adiabat-isentrop (see expressions (7.43) and (7.88))

$$P_1 = MR_gT_1/V_1 \qquad \text{and} \qquad P_2 = MR_gT_2/V_2, \qquad (10.43)$$

and

$$T_2 = T_1\left(\frac{P_2}{P_1}\right)^{(\kappa-1)/\kappa} = T_1\left(\frac{V_1}{V_2}\right)^{\kappa-1}, \qquad (10.44)$$

where T_1, P_1, and V_1 are temperature, pressure, and volume of the gas, respectively, in the beginning of the compression, and T_2, P_2, and V_2 are temperature,

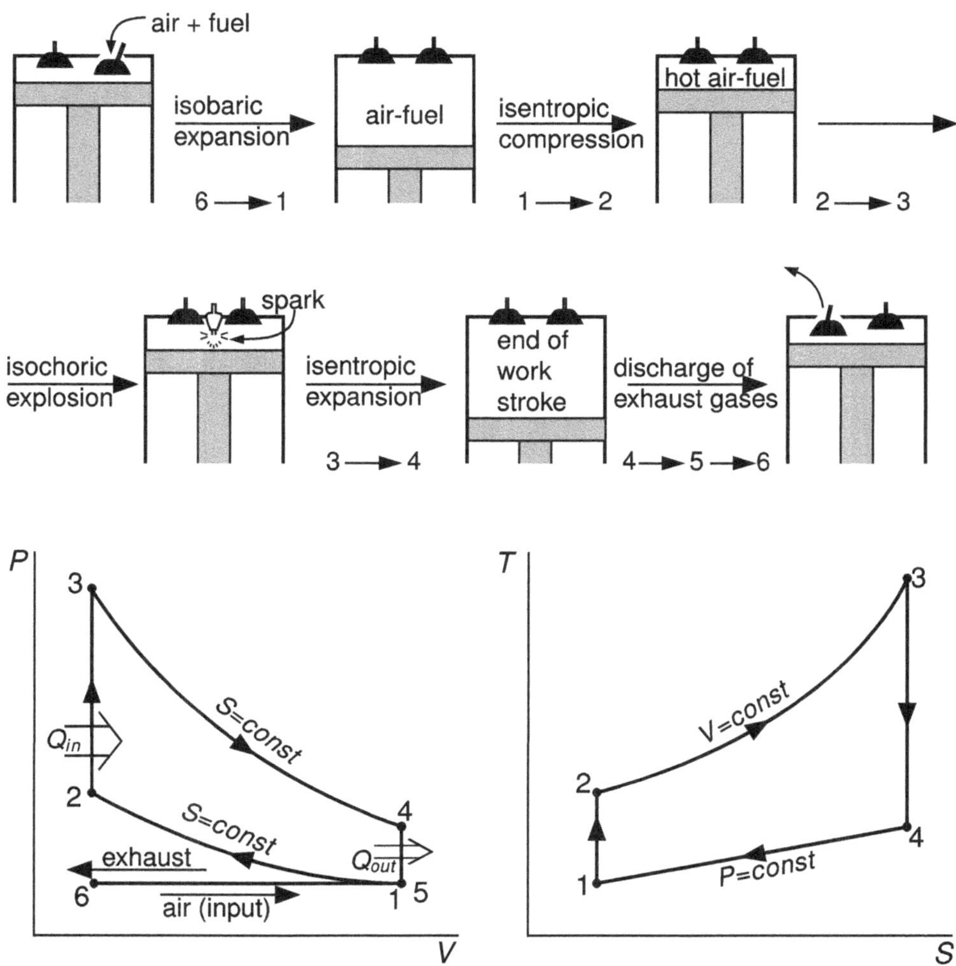

Fig. 10.10: The Otto cycle of an internal combustion engine.

pressure, and volume, respectively, of the compression final state, and κ is the isentropic exponent.

The change of the gas entropy during the adiabatic-isentropic transition is

$$\Delta S_{1\rightarrow 2} = 0. \tag{10.45}$$

Since there is no exchange of heat between the system and the environment during the transition, one has

$$Q_{1\rightarrow 2} = 0, \tag{10.46}$$

and, according to the First Law of Thermodynamics (see expression (7.59)), the boundary work of the process is

$$W_{1\to2} = \Delta U_{1\to2} - Q_{1\to2} = \Delta U_{1\to2}. \qquad (10.47)$$

Assuming that heat capacity $(C_V)_{1\to2}$ is constant during the compression, the change of the internal energy during the isentropic process between equilibrium states 1 and 2 is (see expression (7.55))

$$\Delta U_{1\to2} = (C_V)_{1\to2}(T_2 - T_1), \qquad (10.48)$$

or, when U_1 and U_2 are known, as

$$\Delta U_{1\to2} = U_2 - U_1; \qquad (10.49)$$

the values of U_1 and U_2 within a broad range of thermodynamic conditions can be found in thermodynamic databases discussed in Chapter 8.

Expression (10.48) is less accurate than expression (10.49) because of the assumption used in the former expression that the heat capacity $(C_V)_{1\to2}$ has a constant, temperature-independent value. Usually, the difference is very small in single-phase transitions where the ranges of temperature are not large.

(b) *the isochoric heat addition* 2→3:

The thermodynamic properties of the gas at the beginning and at the end of the process can be found from expression (7.43),

$$V_3 = V_2, \quad P_2 = MR_gT_2/V_2, \quad P_3 = MR_gT_3/V_3. \qquad (10.50)$$

The boundary work of the isochoric ($\Delta V_{2\to3}=0$) transition is

$$W_{2\to3} = 0, \qquad (10.51)$$

so the heat transfer between the closed nonflow system and the environment during the process is (see expression (7.59))

$$Q_{2\to3} = \Delta U_{2\to3}, \qquad (10.52)$$

where

$$\Delta U_{2\to3} = (C_V)_{2\to3}(T_3 - T_2) \quad \text{or} \quad \Delta U_{2\to3} = U_3 - U_2, \qquad (10.53)$$

and the values of U_2 and U_3 can be obtained from the thermochemical databases discussed in Chapter 8.

The change of the gas entropy during the transition is (see expression (7.57))

$$\Delta S_{2\to3} = (C_V)_{2\to3}\ln(T_3/T_2) \qquad \text{or} \qquad \Delta S_{2\to3} = S_3 - S_2, \qquad (10.54)$$

where the values of S_2 and S_3 can be found in the thermochemical databases discussed in Chapter 8.

(c) *the adiabatic-isentropic expansion* $3\to4$:

The thermodynamic properties of equilibrium state 3 and equilibrium state 4 are (see expressions (7.43) and (7.88)),

$$P_3 = MR_gT_3/V_3 \qquad \text{and} \qquad P_4 = MR_gT_4/V_4, \qquad (10.55)$$

and

$$T_3 = T_4\left(\frac{P_3}{P_4}\right)^{(\kappa-1)/\kappa} = T_4\left(\frac{V_4}{V_3}\right)^{\kappa-1}. \qquad (10.56)$$

The change of the thermal energy during the transition is (see expression (7.55)),

$$\Delta U_{3\to4} = (C_V)_{3\to4}(T_4 - T_3) \qquad \text{or} \qquad \Delta U_{3\to4} = U_4 - U_3, \qquad (10.57)$$

and the change of the gas entropy is

$$\Delta S_{3\to4} = 0. \qquad (10.58)$$

There is no heat transfer between the system and the environment during the transition,

$$Q_{3\to4} = 0, \qquad (10.59)$$

so the boundary work of the transition is (see expression (7.59))

$$W_{3\to4} = \Delta U_{3\to4}. \qquad (10.60)$$

(d) *the isochoric heat removal* $4\to1$:

The thermodynamic properties of the gas at the beginning and at the end of the removal can be found from expression (7.43),

$$V_4 = V_1, \qquad P_4 = MR_gT_4/V_4, \qquad P_1 = MR_gT_1/V_1. \qquad (10.61)$$

The boundary work of the isochor is

$$W_{4\to1} = 0, \tag{10.62}$$

and the heat transfer between the system and the environment is (see expression (7.59))

$$Q_{4\to1} = \Delta U_{4\to1}, \tag{10.63}$$

where (see expression (7.55))

$$\Delta U_{4\to1} = (C_V)_{4\to1}(T_1 - T_4) \qquad \text{or} \qquad \Delta U_{4\to1} = U_1 - U_4. \tag{10.64}$$

The change of the system entropy during the transition is (see expression (7.57))

$$\Delta S_{4\to1} = (C_V)_{4\to1}\ln(T_1/T_4) \qquad \text{or} \qquad \Delta S_{4\to1} = S_1 - S_4. \tag{10.65}$$

One should notice that the changes of properties of path (work and heat) of the processes discussed in this section were obtained from the Energy Balance Equations for the processes.

(e) *the entire Otto cycle* $1\to2\to3\to4\to1$:

The change of the gas internal energy during the cycle is

$$\Delta U_{1\to2\to3\to4\to1} = \Delta U_{1\to2} + \Delta U_{2\to3} + \Delta U_{3\to4} + \Delta U_{4\to1} = 0, \tag{10.66}$$

and the change of the system entropy is

$$\Delta S_{1\to2\to3\to4\to1} = \Delta S_{1\to2} + \Delta S_{2\to3} + \Delta S_{3\to4} + \Delta S_{4\to1} = 0, \tag{10.67}$$

because there are no changes of state properties during thermodynamic cycles – see Section 2.8.1.

It is a significant computational convenience to realize that relationships (10.66) and (10.67) are *identities* resulting from the definition of a thermodynamic cycle (*the total change of every property of state in every thermodynamic cycle is zero*). Therefore, the identities can be used to validate their parts calculated above.

The *net* heat transfer of the entire Otto cycle is

$$Q_{1\to2\to3\to4\to1} = Q_{1\to2} + Q_{2\to3} + Q_{3\to4} + Q_{4\to1}, \tag{10.68}$$

and the *net* boundary work of the cycle is

$$W_{1\to2\to3\to4\to1} = W_{1\to2} + W_{2\to3} + W_{3\to4} + W_{4\to1}. \qquad (10.69)$$

Since the Otto cycle is a reversible (no dissipation) process, the *net* heat of the cycle is equal to the cycle's *net* boundary work,

$$W_{1\to2\to3\to4\to1} = -Q_{1\to2\to3\to4\to1}, \qquad (10.70)$$

and the thermal efficiency of the cycle is (see expression (10.12))

$$\eta_{\text{Otto}} = \frac{|\text{net work}|}{|\text{heat in}|} = 1 - \frac{|Q_{4\to1}|}{|Q_{2\to3}|} = 1 - \frac{T_4 - T_1}{T_3 - T_2} = 1 - \frac{1}{r^{\kappa-1}}, \qquad (10.71)$$

where r is the so-called **compression ratio**,

$$r = V_{\text{max}}/V_{\text{min}} = V_1/V_2. \qquad (10.72)$$

If the Otto cycle was treated as an irreversible cycle, one would not be able to calculate its thermal efficiency because one would not know how to calculate the amount of the energy dissipated during the cycle (see expression (10.21) and Chapter 13).

One can see that the thermal efficiency of the reversible Otto cycle depends on the compression ratio r (the ratio represents the size of the engine's cylinder) and on the isentropic exponent κ which represents the ratio of the working gas heat capacities C_P/C_V during the two adiabat-isentrops of the cycle. The efficiency of the Otto engine increases with both r and κ.

10.9   The Diesel Cycle

The Diesel cycle is a reversible thermodynamic cycle of an ideal gas working in a closed (cylinder-piston) system often called the 'Diesel engine' or 'compression-ignition engine'. The cycle is shown in the P–V and T–S diagrams in Fig. 10.11. The $1\to2\to3\to4\to1$ cycle consists of four consecutive transitions where states 1, 2, 3, and 4 are equilibrium states. The only kind of work transferred during the cycle is boundary work.

The four transitions that make the Diesel cycle are:

(a) adiabatic-isentropic compression from state 1 to state 2,

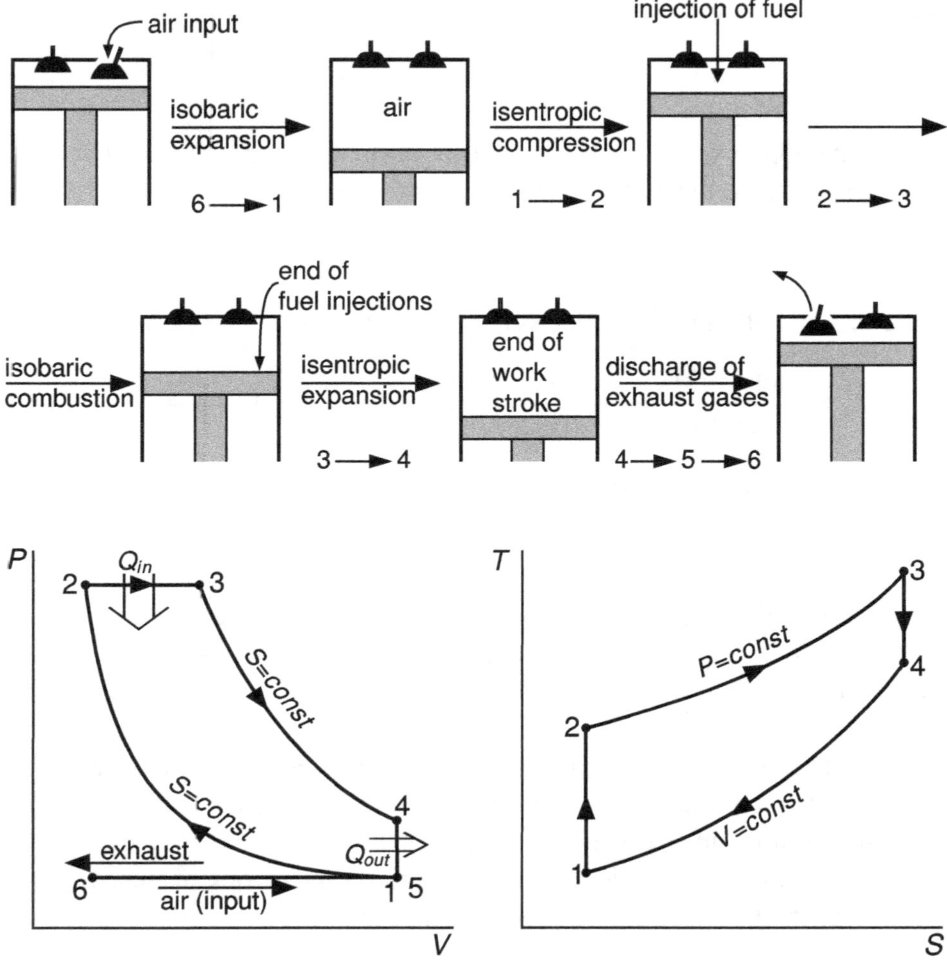

Fig. 10.11: The Diesel cycle of a compression-ignition engine.

(b) isobaric heat addition 2→3,
(c) adiabatic-isentropic expansion 3→4,
(d) isochoric heat removal 4→1.

The changes of the working gas's kinetic and potential (gravitational) energies are negligible. Thus, in this problem, the 'system internal energy' also means the 'system thermal energy'.

Assuming that the gas heat capacities $(C_V)_{i\to j}$ and $(C_P)_{i\to j}$ are constant during each of the transitions (a)–(d) (also, see epressions (6.3)–(6.6)), the thermodynamic properties of the cycle can be calculated as follows:

(a) *the adiabatic-isentropic compression* $1 \rightarrow 2$:

The thermodynamic properties of the gas equilibrium states 1 and 2 can be obtained from expressions (7.43) and (7.88), respectively,

$$P_1 = MR_g T_1/V_1 \qquad \text{and} \qquad P_2 = MR_g T_2/V_2, \tag{10.73}$$

and

$$T_2 = T_1 \left(\frac{P_2}{P_1}\right)^{(\kappa-1)/\kappa} = T_1 \left(\frac{V_1}{V_2}\right)^{\kappa-1}, \tag{10.74}$$

where T_1, P_1, and V_1 are the temperature, pressure, and volume of the gas in the beginning of the compression, respectively, and T_2, P_2, and V_2 are the temperature, pressure, and volume of the compression final state, respectively, M is the mass of the gas, R_g is its individual gas constant, and κ is the isentropic exponent of the process.

The change of entropy of the gas during the transition is

$$\Delta S_{1 \rightarrow 2} = 0. \tag{10.75}$$

There is no heat transfer between the system and the environment,

$$Q_{1 \rightarrow 2} = 0, \tag{10.76}$$

so the (boundary) work of the transition is (see expression (7.59))

$$W_{1 \rightarrow 2} = \Delta U_{1 \rightarrow 2}. \tag{10.77}$$

The thermal energy change during the adiabatic-isentropic process can be found from relationship (7.55) or as the difference between U_2 (the internal energy of state 2), and U_1 (the internal energy of state 1) if U_1 and U_2 are known (say, from the thermochemical databases discussed in Chapter 8). In other words,

$$\Delta U_{1 \rightarrow 2} = (C_V)_{1 \rightarrow 2}(T_2 - T_1) \qquad \text{or} \qquad \Delta U_{1 \rightarrow 2} = U_2 - U_1; \tag{10.78}$$

see the paragraph that follows relationship (10.49).

(b) *the isobaric heat addition* 2→3:

The thermodynamic properties of equilibrium states 2 and 3 can be obtained from expression (7.43),

$$P_3 = P_2, \quad T_2 = P_2 V_2/M R_g, \quad T_3 = P_3 V_3/M R_g, \tag{10.79}$$

and the boundary work of the reversible isobaric transition is (see expression (2.71))

$$W_{2\to3} = -P_2(V_3 - V_2) = -P_3(V_3 - V_2). \tag{10.80}$$

Therefore, the heat transfer during the transition is (see expressions (7.60) and (7.56))

$$Q_{2\to3} = \Delta H_{2\to3} \tag{10.81}$$

where

$$\Delta H_{2\to3} = (C_P)_{2\to3}(T_3 - T_2) \quad \text{or} \quad \Delta H_{2\to3} = H_3 - H_2, \tag{10.82}$$

and $(C_P)_{2\to3}$ is the constant-pressure heat capacity of the matter participating in the process. Thus, the change of the internal energy of the gas during the process is (see expression (7.59))

$$\Delta U_{2\to3} = Q_{2\to3} + W_{2\to3} = (C_P)_{2\to3}(T_3 - T_2) - P_2(V_3 - V_2), \tag{10.83}$$

or

$$\Delta U_{2\to3} = U_3 - U_2. \tag{10.84}$$

The change of the gas entropy during the transition is (see expression (7.58))

$$\Delta S_{2\to3} = (C_P)_{2\to3} \ln(T_3/T_2) \quad \text{or} \quad \Delta S_{2\to3} = S_3 - S_2, \tag{10.85}$$

where the values of S_2 and S_3 can be found in the thermochemical databases.

(c) *the adiabatic-isentropic expansion* 3→4:

The thermodynamic properties of equilibrium states 3 and 4 are given by expressions (7.43) and (7.88),

$$P_3 = M R_g T_3/V_3 \quad \text{and} \quad P_4 = M R_g T_4/V_4, \tag{10.86}$$

and

$$T_3 = T_4 \left(\frac{P_3}{P_4}\right)^{(\kappa-1)/\kappa} = T_4 \left(\frac{V_4}{V_3}\right)^{\kappa-1}. \tag{10.87}$$

Also,

$$\Delta U_{3\to4} = (C_V)_{3\to4}(T_4 - T_3) \qquad \text{or} \qquad \Delta U_{3\to4} = U_4 - U_3, \tag{10.88}$$

and the change of the gas entropy is

$$\Delta S_{3\to4} = 0. \tag{10.89}$$

There is no heat transfer between the system and the environment,

$$Q_{3\to4} = 0, \tag{10.90}$$

so, according to the First Law of Thermodynamics, the boundary work transfer during the process is

$$W_{3\to4} = \Delta U_{3\to4}. \tag{10.91}$$

(d) *the isochoric heat removal* $4\to1$:

The thermodynamic properties of the gas at the beginning and at the end of the process can be found from expression (7.43),

$$V_4 = V_1, \quad P_4 = MR_g T_4/V_4, \quad P_1 = MR_g T_1/V_1, \tag{10.92}$$

and the boundary work of the process is

$$W_{4\to1} = 0, \tag{10.93}$$

so the heat transfer between the system and the environment is (see expression (7.59))

$$Q_{4\to1} = \Delta U_{4\to1}, \tag{10.94}$$

where

$$\Delta U_{4\to1} = (C_V)_{4\to1}(T_1 - T_4) \qquad \text{or} \qquad \Delta U_{4\to1} = U_1 - U_4. \tag{10.95}$$

The change of the gas entropy during the transition is (see expression (7.57))

$$\Delta S_{4\to1} = (C_V)_{4\to1} \ln(T_1/T_4) \qquad \text{or} \qquad \Delta S_{4\to1} = S_1 - S_4. \tag{10.96}$$

One should notice that most of the path properties (work and heat) of the Diesel cycle were calculated above from the Energy Balance Equations. Only the isobaric boundary work $W_{2\to3}$ was obtained from expression (10.80) which was derived independently under the assumption of reversibility of the isobaric transition.

(e) *the entire Diesel cycle* $1\to2\to3\to4\to1$:

The change of the internal energy of a complete Diesel cycle (changes of all thermodynamic properties of state during any cycle are zero – see Section 2.8.1),

$$\Delta U_{1\to2\to3\to4\to1} = \Delta U_{1\to2} + \Delta U_{2\to3} + \Delta U_{3\to4} + \Delta U_{4\to1} = 0, \qquad (10.97)$$

and the change of the system entropy during the cycle is

$$\Delta S_{1\to2\to3\to4\to1} = \Delta S_{1\to2} + \Delta S_{2\to3} + \Delta S_{3\to4} + \Delta S_{4\to1} = 0. \qquad (10.98)$$

It is worthy to realize that relationships (10.97) and (10.98) are *identities* resulting from the definition of a thermodynamic cycle – see the discussion that follows relationships (10.66) and (10.67). Therefore, the identities can be used to validate their parts calculated above.

The *net* heat transfer of the entire Diesel cycle is

$$Q_{1\to2\to3\to4\to1} = Q_{1\to2} + Q_{2\to3} + Q_{3\to4} + Q_{4\to1}, \qquad (10.99)$$

and the *net* boundary work of the cycle is

$$W_{1\to2\to3\to4\to1} = W_{1\to2} + W_{2\to3} + W_{3\to4} + W_{4\to1}. \qquad (10.100)$$

Since the Diesel cycle is a *reversible* process, there is no dissipation of energy during the process. Therefore,

$$W_{1\to2\to3\to4\to1} = -Q_{1\to2\to3\to4\to1}, \qquad (10.101)$$

that is, the entire (net) heat of the Diesel cycle is transformed into (net) work. If the cycle were an irreversible one, the changes of the system's thermal energy and its entropy during the cycle would still be zero because both of these properties are functions of state. However, the net work of an irreversible Diesel cycle would not be equal to the cycle's net heat $Q_{1\to2\to3\to4\to1}$ because of the dissipation effects which are always present in irreversible processes (see expression (10.21) and Chapter 13).

The thermal efficiency of the reversible Diesel cycle is (see expression (10.12))

$$\eta_{\text{Diesel}} = \frac{|\text{net work}|}{|\text{heat in}|} = 1 - \frac{|Q_{4\to1}|}{|Q_{2\to3}|} = 1 - \frac{(C_V)_{1\to4}(T_4 - T_1)}{(C_P)_{2\to3}(T_3 - T_2)}, \qquad (10.102)$$

or, using relationships (10.74) and (10.87),

$$\eta_{\text{Diesel}} = 1 - \frac{1}{r^{\kappa-1}}\left[\frac{r_c^{\kappa} - 1}{\kappa(r_c - 1)}\right], \qquad (10.103)$$

where, as before, $r = V_1/V_2$, and the so-called 'cutoff ratio' is

$$r_c = V_3/V_2. \qquad (10.104)$$

Note that $r_c = 1$ for the Otto cycle.

10.10 Another Example

Problem 10.2: One mol of an ideal diatomic gas undergoes a reversible thermodynamic cycle shown in the *P–V* diagram in Fig. 10.12. The cycle consists of four transitions: an isochoric heat addition 1→2, isothermal expansion 2→3, adiabatic-isentropic expansion 3→4, and isobaric heat removal 4→1. The states 1, 2, 3, and 4 are equilibrium states. The working gas heat capacities are assumed constant in the considered ranges of temperature, that is, the gas is a 'calorically perfect ideal gas' in every transition making the cycle. The temperature and pressure of the gas in state 1 are $T_1 = 300$ K and $P_1 = 1$ bar, respectively. The pressure of the gas in state 2 is $P_2 = 3$ bar, and the gas volume in state 3 is $V_3 = 0.04$ m^3. There is no non-boundary work done on or done by the engine. The changes of the kinetic and potential energies of the working gas are negligible. Calculate: 1) the changes of the thermal energy, entropy, and the work and heat transfers during each of the four transitions making the cycle, 2) the same properties for the entire 1→2→3→4→1 cycle, and 3) the thermal efficiency of the cycle.

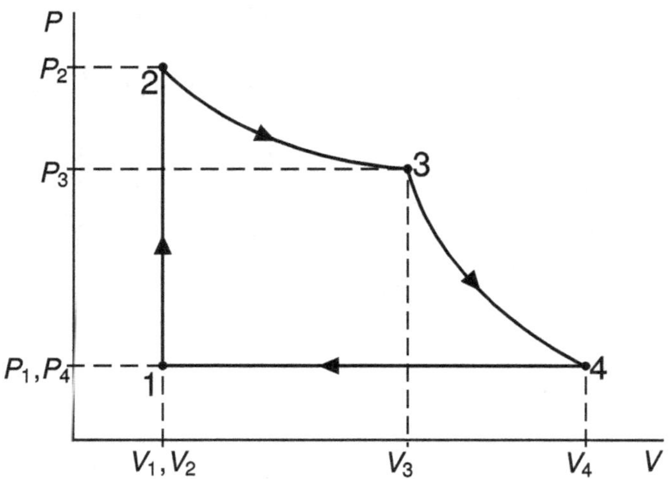

Fig. 10.12: The work diagram of the thermodynamic cycle discussed in Problem 10.2.

Solution: The fact that the cycle operates on *one mol* of gas is emphasized below by marking the thermodynamic properties of the cycle with overbars.

The known thermodynamic variables of equilibrium states 1, 2, 3, and 4 are:

$$T_1 = 300\,\text{K,} \tag{10.105}$$

$$P_1 = P_4 = 1\,\text{bar,} \tag{10.106}$$

$$P_2 = 3\,\text{bar,} \tag{10.107}$$

and

$$\bar{V}_3 = 0.04\,\text{m}^3/\text{mol.} \tag{10.108}$$

Since the constant-volume and constant-pressure heat capacities of the working ideal diatomic gas are assumed to be constant during the transitions making the cycle, their molar values can be approximated by (see expressions (6.34), (6.35), and (6.3)–(6.6))

$$\bar{c}_V = 5R/2 = 5(8.3143\,\text{J/mol·K})/2 = 20.78\,\text{J/mol·K,} \tag{10.109}$$

and

$$\bar{c}_P = 7R/2 = 7(8.3143\,\text{J/mol·K})/2 = 29.10\,\text{J/mol·K,} \tag{10.110}$$

where, as before, $R = 8.3143\,\text{J/mol·K}$ is the universal (molar) gas constant. Thus, the isentropic exponent of the calorically perfect ideal diatomic gas is (see expression (7.84)):

$$\kappa = \bar{c}_P/\bar{c}_V = 1.4. \tag{10.111}$$

The working gas is an ideal gas. Therefore, the unknown thermodynamic variables (pressure P, volume $\bar{V}$, and temperature T) of equilibrium states 1, 2, 3, and 4 can be obtained from the ideal-gas equation of state for one mol of the gas,

$$\bar{V}_1 = \bar{V}_2 = \frac{RT_1}{P_1} = \frac{(8.3143\,\text{J/mol·K})(300\,\text{K})}{10^5\,\text{N/m}^2} = 0.025\,\text{m}^3/\text{mol}, \tag{10.112}$$

$$T_2 = T_3 = \frac{T_1 P_2}{P_1} = \frac{(300\,\text{K})(3 \times 10^5\,\text{N/m}^2)}{10^5\,\text{N/m}^2} = 900\,\text{K}, \tag{10.113}$$

and

$$P_3 = \frac{P_2 \bar{V}_2}{\bar{V}_3} = \frac{(3 \cdot 10^5\,\text{N/m}^2)(0.025\,\text{m}^3/\text{mol}1920)}{0.04\,\text{m}^3/\text{mol}} = 1.87\,\text{bar}. \tag{10.114}$$

Use of isentropic relationship (7.88) gives

$$T_4 = T_3 \left(\frac{P_4}{P_3}\right)^{(\kappa-1)/\kappa} = (900\,\text{K})\left(\frac{1\,\text{bar}}{1.87\,\text{bar}}\right)^{(1.4-1)/1.4} = 752\,\text{K}, \tag{10.115}$$

and

$$\bar{V}_4 = \frac{RT_4}{P_4} = \frac{(8.3143\,\text{J/mol·K})(752\,\text{K})}{10^5\,\text{N/m}^2} = 0.062\,\text{m}^3/\text{mol}. \tag{10.116}$$

The other thermodynamic properties of the cycle can be obtained from the following:

(a) *the isochoric heat addition* $1 \rightarrow 2$:

The change of the internal energy of one mol of the calorically perfect ideal gas during this transition is (see expression (7.55))

$$\Delta \bar{U}_{1 \rightarrow 2} = \bar{c}_V(T_2 - T_1) = 12{,}471\,\text{J/mol}, \tag{10.117}$$

while the boundary work transfer is

$$\bar{W}_{1 \rightarrow 2} = 0. \tag{10.118}$$

Subsequently, the heat transferred between the system and the environment is (see expression (7.59))

$$\bar{Q}_{1\to2} = \Delta\bar{U}_{1\to2} = 12,471 \text{ J/mol}, \qquad (10.119)$$

and the change of the gas entropy during the process is (see expression (7.57))

$$\Delta\bar{S}_{1\to2} = \bar{c}_V \ln(T_2/T_1) = 22.83 \text{ J/mol·K}. \qquad (10.120)$$

One can see that the entire heat supplied to the system in the $1\to2$ isochoric process is used by the system to increase its internal energy while neither producing nor receiving work.

(b) *the isothermal expansion* $2\to3$:

The change of the internal (thermal) energy in the closed, nonflow system consisting of one mol of the ideal gas that undergoes the reversible isothermal transition $2\to3$ is

$$\Delta\bar{U}_{2\to3} = 0, \qquad (10.121)$$

because the thermal energy of ideal gas is a function of temperature only (see Section 7.5).

The change of the system entropy during the transition is (see expression (7.63))

$$\Delta\bar{S}_{2\to3} = R \ln(\bar{V}_3/\bar{V}_2) = 3.92 \text{ J/mol·K}. \qquad (10.122)$$

The heat exchanged isothermally between the system and the environment is (see expression (2.58))

$$\bar{Q}_{2\to3} = T_2\Delta\bar{S}_{2\to3} = (900 \text{ K})(3.92 \text{ J/mol·K}) = 3534 \text{ J/mol}, \qquad (10.123)$$

and, according to the First Law of Thermodynamics, the transition's boundary work is (see expression (7.59))

$$\bar{W}_{2\to3} = -\bar{Q}_{2\to3} = -3534 \text{ J/mol}. \qquad (10.124)$$

Thus, during the $2\to3$ isothermal transition, the system does not change its internal energy but produces boundary work that is equal to the amount of heat taken in during the transition.

(c) *the adiabatic-isentropic expansion* 3→4:

The change of the internal energy during the adiabatic-isentropic transition between (equilibrium) states 3 and 4 is (see relationship (7.55))

$$\Delta \bar{U}_{3\rightarrow 4} = \bar{c}_V(T_4 - T_3) = -3065 \, \text{J/mol}, \tag{10.125}$$

and the change of the gas entropy during the transition is

$$\Delta \bar{S}_{3\rightarrow 4} = 0. \tag{10.126}$$

Since there is no heat transfer between the system and the environment during the process, one can write

$$\bar{Q}_{3\rightarrow 4} = 0, \tag{10.127}$$

so that the process's boundary work is (see expression (7.59))

$$\bar{W}_{3\rightarrow 4} = \Delta \bar{U}_{3\rightarrow 4} = -3065 \, \text{J/mol}. \tag{10.128}$$

Thus, during the expansion, the system produces some work without exchanging heat with the environment – the work is produced by the decrease of the system thermal energy.

(d) *the isobaric heat removal* 4→1:

The boundary work of the removal is (see expression (2.71))

$$\bar{W}_{4\rightarrow 1} = -P_4(\bar{V}_1 - \bar{V}_4) = 3762 \, \text{J/mol}, \tag{10.129}$$

and the heat transfer between the gas and the environment is (see expressions (7.56) and (7.60))

$$\bar{Q}_{4\rightarrow 1} = \bar{c}_P(T_1 - T_4) = -13{,}168 \, \text{J/mol}. \tag{10.130}$$

The change of the internal energy of the gas during the isobaric process is (see expression (7.55))

$$\Delta \bar{U}_{4\rightarrow 1} = \bar{c}_V(T_1 - T_4) = -9406 \, \text{J/mol}, \tag{10.131}$$

and the change of the system's entropy during the process is (see expression (7.58))

$$\Delta \bar{S}_{4 \to 1} = \bar{c}_P \ln(T_1/T_4) = -26.76 \, \text{J/mol·K}. \tag{10.132}$$

Thus, the isobaric process $4 \to 1$ has decreased the system's internal energy while some boundary work has been done on the system and some heat left the system.

(e) *the entire cycle* $1 \to 2 \to 3 \to 4 \to 1$:

The change of the internal energy of one mol of the working gas during one cycle $1 \to 2 \to 3 \to 4 \to 1$ is

$$\Delta \bar{U}_{1 \to 2 \to 3 \to 4 \to 1} = \Delta \bar{U}_{1 \to 2} + \Delta \bar{U}_{2 \to 3} + \Delta \bar{U}_{3 \to 4} + \Delta \bar{U}_{4 \to 1} = 0, \tag{10.133}$$

and the change of the system entropy during the cycle is

$$\Delta \bar{S}_{1 \to 2 \to 3 \to 4 \to 1} = \Delta \bar{S}_{1 \to 2} + \Delta \bar{S}_{2 \to 3} + \Delta \bar{S}_{3 \to 4} + \Delta \bar{S}_{4 \to 1} = 0. \tag{10.134}$$

Expressions (10.133) and (10.134) were expected because the internal energy and entropy are functions of state. As discussed after relationships (10.66)–(10.67) and (10.97)–(10.98), the expressions are identities that can be used to validate their parts calculated above.

The *net* heat transferred between the environment and the working gas during the cycle $1 \to 2 \to 3 \to 4 \to 1$ is

$$\bar{Q}_{1 \to 2 \to 3 \to 4 \to 1} = \bar{Q}_{1 \to 2} + \bar{Q}_{2 \to 3} + \bar{Q}_{3 \to 4} + \bar{Q}_{4 \to 1} = 2836 \, \text{J/mol}, \tag{10.135}$$

and the *net* boundary work produced by the cycle is

$$\bar{W}_{1 \to 2 \to 3 \to 4 \to 1} = \bar{W}_{1 \to 2} + \bar{W}_{2 \to 3} + \bar{W}_{3 \to 4} + \bar{W}_{4 \to 1} = -2836 \, \text{J/mol}. \tag{10.136}$$

Thus, the net heat transferred between the system and the environment in the reversible cycle $1 \to 2 \to 3 \to 4 \to 1$ is used by the system to produce work. If the cycle was an irreversible process, then the changes of the system's internal energy and entropy during the cycle would still be zero because both of these properties are functions of state. However, the net work $\bar{W}_{1 \to 2 \to 3 \to 4 \to 1}$ of such an irreversible cycle would not be equal to the amount of the cycle's net heat $\bar{Q}_{1 \to 2 \to 3 \to 4 \to 1}$ because of the energy dissipation effects which are always present in irreversible processes – see Chapter 13.

The heat added to and the heat removed from the system during the cycle are, respectively,

$$\bar{Q}_{\text{in}} = \bar{Q}_{1\rightarrow2} + \bar{Q}_{2\rightarrow3} = 16{,}005\,\text{J/mol}, \tag{10.137}$$

and

$$\bar{Q}_{\text{out}} = \bar{Q}_{4\rightarrow1} = -13{,}168\,\text{J/mol}. \tag{10.138}$$

Thus, the thermal efficiency of the cycle is

$$\eta = \frac{\bar{Q}_{\text{in}} - |\bar{Q}_{\text{out}}|}{\bar{Q}_{\text{in}}} = 17.7\%. \tag{10.139}$$

End of Problem 10.2.

Chapter 11

Open Systems (Flows)

11.1 The Speed of Sound

A weak perturbation propagating in a medium is called the **sound wave** (named so because the vibrations caused by the wave can sometimes be heard as a sound). The speed of the sound wave is called the **speed of sound**. Typical sound wave has very little energy. Therefore, one can say, when observing the movement of the wave in a gas in a coordinate frame attached to the wave, that the gas is moving with speed a in front of the wave and with speed $a + da$ behind the wave. The changes in the gas pressure, temperature, and mass density across the wave (dP, dT, and $d\rho$, respectively) are also very small. Therefore, conservation of matter across the wave when there is no transfer of heat and work between the wave and the environment requires that

$$\rho a = (\rho + d\rho)(a + da) \rightarrow a = -\rho\frac{da}{d\rho}, \tag{11.1}$$

where we ignored the product $d\rho\, da$ since it is a very small second-order term. Relationship (11.1) allows one to write

$$P + \rho a^2 = (P + dP) + (\rho + d\rho)(a + da)^2 \rightarrow da = \frac{dP + a^2 d\rho}{-2a\rho}, \tag{11.2}$$

where we again ignored the very small products of differentials. Combining expressions (11.1) and (11.2) leads to

$$a^2 = \frac{dP}{d\rho}. \tag{11.3}$$

Expression (11.3) was obtained assuming that neither work nor heat are exchanged between the sound wave and its surroundings, and that the irreversibilities present in the wave are locally negligible. Subsequently, the moving wave can be treated as an **adiabatic-isentropic system**. To emphasize this fact, we rewrite relationship (11.3) as

$$a = \sqrt{\left(\frac{\partial P}{\partial \rho}\right)_S}, \tag{11.4}$$

where the subscript S reminds us that the local change of the gas pressure with density takes place at a constant value of the local entropy S. Using expression (7.84), one can write for adiabatic-isentropic flows with constant heat capacities that

$$P\rho^{-\kappa} = \text{const}, \tag{11.5}$$

where κ is the isentropic exponent for the flowing gas. As a result, the speed of sound (11.4) can be given as

$$a = \sqrt{\kappa P/\rho}. \tag{11.6}$$

If the sound wave is moving in an ideal gas in equilibrium, then the gas equation of state is $P = \rho R_g T$, and relationship (11.6) gives

$$a = \sqrt{\kappa R_g T}. \tag{11.7}$$

This relationship tells us that the **speed of sound in an ideal gas in equilibrium (local equilibrium) depends only on the gas temperature**.

An easy-to-memorize approximate formula for the speed of sound (11.7) in air in equilibrium at temperature T and $\kappa = 1.4$ (air is a mixture of mainly diatomic gases) can be given as

$$a\,(\text{m/s}) \simeq 20[T\,(\text{K})]^{1/2}. \tag{11.8}$$

It is a common and convenient practice to use, at any location in any flow, the so-called **Mach number** M as a measure of the ratio of the speed of flow ω to the speed of sound a at the location,

$$M = \omega/a. \tag{11.9}$$

The concept of Mach number is often used to define the following six categories of gas flows:

$$
\begin{aligned}
M &\ll 1 \quad &\text{incompressible flow,} \\
M &< 1 \quad &\text{subsonic flow,} \\
M &= 1 \quad &\text{sonic flow,} \\
M &\simeq 1 \quad &\text{transonic flow,} \\
M &> 1 \quad &\text{supersonic flow,} \\
M &\gg 1 \quad &\text{hypersonic flow.}
\end{aligned}
$$

The intensity of the weakest audible sound wave (that is, the flux of the wave acoustic energy) at frequency 1000 Hz (the approximate value of the lower threshold of hearing for an average person) is

$$
I_{\text{sound}}^{\text{thr}} = 10^{-12}\,\text{W}/\text{m}^2. \tag{11.10}
$$

Sound wave intensity I_{sound} is usually measured on the so-called **decibel scale** where the intensity measured in decibels (dB) is

$$
\Delta = 10\log_{10}(I_{\text{sound}}/I_{\text{sound}}^{\text{thr}}). \tag{11.11}
$$

Thus, the intensity of the weakest sound wave still audible to an average person is $\Delta = 0\,\text{dB}$. Whispers, conversations, and loud rock music concerts produce sounds of intensity of about 20 dB, 60 dB, and 110 dB, respectively. Ear drum rupture of an average person is usually done at $\Delta = 140\,\text{dB}$ or $100\,\text{W}/\text{m}^2$.

11.2 The Adiabatic Steady-Flow Ellipse

Consider a steady flow that exchanges neither heat nor work with the surroundings, and where the impact of gravitational forces on properties of the flow can be ignored – such flows are common in applications. The Energy Balance Equation for one kilogram of the flowing gas at any location is (see expression (3.46))

$$
\frac{\omega^2}{2} + h = \text{const} = \frac{\omega_\circ^2}{2} + h_\circ = h_\circ, \tag{11.12}
$$

where $\omega^2/2$ and h are the specific kinetic energy and specific enthalpy, respectively, of the flowing gas at this location (the reader may recall that a physical property of one kilogram of a substance is called the substance's *specific* property), and $\omega_o^2/2$ and h_o are the specific kinetic energy and specific enthalpy of the gas under its 'stagnation conditions' when the gas stops (then $\omega = \omega_o = 0$ – see expression (3.36)). The gas's properties under the stagnation conditions are marked hereafter with the subscript o.

The concept of the 'stagnation conditions' is useful in studies of flows discussed here because, according to expression (11.12), a flow's specific stagnation enthalpy h_o is just the sum of the specific enthalpy h and specific kinetic energy $\omega^2/2$ of the flow. Thus, the **flow stagnation enthalpy represents the total energy of the flowing gas. The total energy has the same value at every location in the flow with no heat and work transfers**.

At the ideal-gas flow's location where the speed of the flow is ω and the speed of sound is a, one can write, assuming that the constant-pressure specific heat c_P is temperature-independent, that

$$h - h_o = c_P(T - T_o) = \frac{c_P}{\kappa R_g}(a^2 - a_o^2), \tag{11.13}$$

because $dh = c_P dT$ (see expression (7.47)), $T = a^2/\kappa R_g$, and $T_o = a_o^2/\kappa R_g$ (see expression (11.7)), where T_o and a_o are the stagnation temperature and the stagnation speed of sound, respectively. Therefore, the flow's Energy Balance Equation (11.12) can be written as

$$\omega^2 + \frac{2}{\kappa - 1}a^2 = \frac{2}{\kappa - 1}a_o^2 = \text{const}, \tag{11.14}$$

which is an equation of ellipse in a–ω coordinate plane (see Fig. 11.1). The ellipse is called the **adiabatic-isentropic steady-flow ellipse**. Different fragments of the ellipse represent different steady adiabatic-isentropic flows of ideal gases that exchange neither heat nor work with their surroundings. Depending on the value of the Mach number of the flow at a given location, the flow at this location can be categorized as:

1) *Incompressible flow* ($M \ll 1$): the flow speed ω is small when compared to the speed of sound a in the flow. Any change of the sound speed in such a flow is very small in comparison with the change of the flow speed.
2) *Subsonic flow* ($M < 1$): the changes in the flow Mach number occur primarily because of the changes in the speed of the flow.

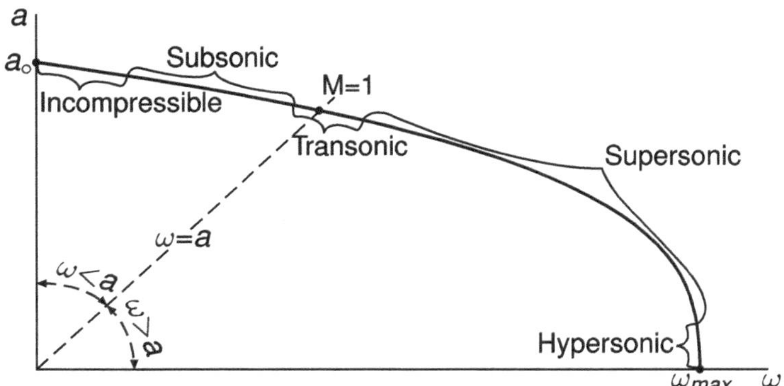

Fig. 11.1: Relationship between the speed of sound a and the speed of flow ω in a broad range of steady adiabatic-isentropic flows; $a_\circ$ is the stagnation speed of sound, $\omega_{\max}$ is the flow maximum speed, and M is the Mach number.

3) *Transonic flow* ($M \simeq 1$): the flow speed is close to the speed of sound, and the changes of ω and a in the flow are close to one another.

4) *Supersonic flow* ($M > 1$): the speed of the flow and the speed of sound are of similar order, but the former is larger than the latter. The changes in Mach number during the flow take place through variations in both ω and a.

5) *Hypersonic flow* ($M \gg 1$): the flow speed is much larger than the speed of sound. The changes in the flow's speed are small, and the changes in the flow's Mach number result almost exclusively from the changes in the speed of sound.

11.3 Flow without Gasdynamic Shock

Let us consider a steady axially-symmetric flow of a compressible gas through a channel that changes its geometry along the channel's axis x – see Fig. 11.2. The changes of the flow's potential energy are neglected. The changes of the flow properties ($\rho, T, P, ...$) in directions normal to the axis x also are neglected as small in comparison with the corresponding changes along the axis. Such a flow is called a **quasi-one-dimensional flow**. The flowing gas properties at the inlet and at the outlet of the channel are marked with subscripts 1 and 2, respectively.

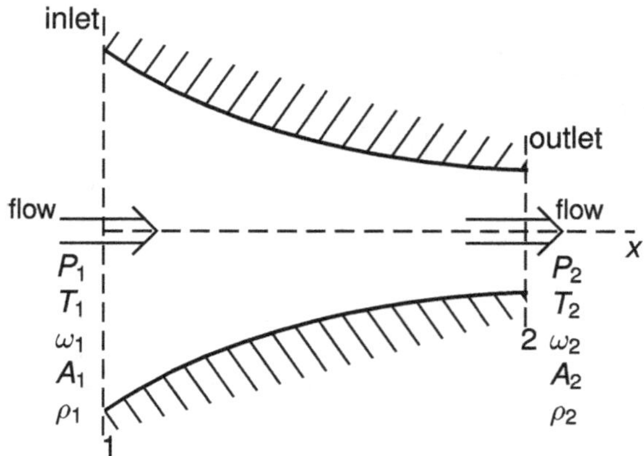

Fig. 11.2: Notation for the properties of a quasi-one-dimensional flow moving between the inlet and the outlet of a converging channel.

The dissipative effects are small, and neither heat nor work are transferred between the flow and the environment. Thus, the flow can be considered an adiabatic-isentropic flow.

The Principle of Conservation of Matter for such *steadily* flowing gas requires that (see expression (3.23))

$$\dot{m}_1 = \dot{m}_2 = \dot{m}(x) = \text{const}, \tag{11.15}$$

where $\dot{m}_1$ and $\dot{m}_2$ are the mass flow rates (in kg/s) passing through the channel inlet and outlet, respectively, and the mass flow rate at a location x is (see expression (3.19))

$$\dot{m}(x) = \rho(x)A(x)\omega(x), \tag{11.16}$$

where $\rho(x)$, $A(x)$, and $\omega(x)$ are the mass density, the cross section of the channel, and the speed of the flow, respectively, at location x. Therefore, relationship (11.15) is often written as

$$\rho_1 A_1 \omega_1 = \rho_2 A_2 \omega_2 = \rho(x)A(x)\omega(x) = \text{const}, \tag{11.17}$$

and called the **steady flow continuity equation**.

The Energy Conservation Equation for the steady flow moving between the system's inlet 1 and outlet 2 is (see expression (11.12))

$$h_1 + \frac{1}{2}\omega_1^2 = h_2 + \frac{1}{2}\omega_2^2 = h(x) + \frac{\omega^2(x)}{2} = h_\circ = \text{const}, \qquad (11.18)$$

where, as before, h_1 and h_2 are the flow specific enthalpies (in J/kg) at the inlet and at the outlet, respectively, and $h_\circ$ is the flow's specific enthalpy of stagnation. Equation (11.18) is called the **steady flow energy equation**.

Conservation of linear momentum of the discussed flow along the x axis (the angular momentum change about the axis x is negligible in most applications) can be described by the following (somewhat simplified) equation:

$$P_1 + \rho_1\omega_1^2 = P_2 + \rho_2\omega_2^2 = \text{const}. \qquad (11.19)$$

This equation is called the **steady flow momentum equation**.

The above conservation equations can also be written in their differential forms when the nozzle (differential) length in the x direction is dx. Denoting the flow properties at x as P, h, ω, A, and ρ, and the flow properties at $x + dx$ as $P + dP$, $h + dh$, $\omega + d\omega$, $A + dA$, and $\rho + d\rho$, allows one to derive the following differential equations for conservation of matter, energy, and linear momentum in the flow:

(a) the flow continuity equation:

$$\frac{d\rho}{\rho} + \frac{dA}{A} + \frac{d\omega}{\omega} = 0, \qquad (11.20)$$

(b) the flow energy equation:

$$dh + \omega d\omega = 0, \qquad (11.21)$$

(c) the flow momentum equation:

$$dP + \rho\omega d\omega = 0, \qquad (11.22)$$

where we ignored the small terms such as products of differentials.

It should be noticed that the mass density ρ was not assumed constant in the above discussion. Therefore, the entire discussion is valid for *compressible* flows.

Combining expressions (11.20) and (11.22) with definitions of the speed of sound (expression (11.3)) and Mach number (expression (11.9)) leads to the following relationships valid at any (local-equilibrium) location x:

$$\frac{d\omega}{\omega} = -\frac{1}{M^2}\frac{d\rho}{\rho}, \tag{11.23}$$

and

$$\frac{d\omega}{\omega} = \frac{1}{M^2 - 1}\frac{dA}{A}. \tag{11.24}$$

Expressions (11.23)–(11.24) tell us a lot about the adiabatic-isentropic flows discussed in this Chapter. One can see that:

1) in order to accelerate ($d\omega > 0$) a subsonic flow ($M < 1$, $M^2 - 1 < 0$), one has to use a channel that is converging ($dA < 0$) along the flow direction x; the decrease of the channel's cross section causes a decrease of the flow pressure ($dP < 0$) along the direction x (see expression (11.22)). Such channel is called the **subsonic nozzle** or **converging nozzle** – see Fig. 11.3a. However, acceleration of a supersonic flow ($M > 1$, $M^2 - 1 > 0$) requires a channel that diverges ($dA > 0$) along the flow direction; the flow pressure then decreases ($dP < 0$) with x. Such channel is called the **supersonic nozzle** or **diverging nozzle** – see Fig. 11.3b.

2) in order to slow down ($d\omega < 0$) a subsonic flow ($M < 1$, $M^2 - 1 < 0$) one has to use a diverging channel ($dA > 0$); the flow pressure in the channel increases ($dP > 0$) along the flow direction. A channel of this kind is called the **subsonic diffuser** – see Fig. 11.3c. However, slowing down of a supersonic flow ($M > 1$, $M^2 - 1 > 0$) requires a converging channel ($dA < 0$); pressure of the flow in such a channel increases ($dP > 0$) along the direction x. This kind of channel is called the **supersonic diffuser** – see Fig. 11.3d.

3) subsonic flow in a converging channel can reach the critical speed (then $M = 1$, $\omega = a$, $M^2 - 1 = 0$) only at the location where $dA = 0$, that is, where the channel cross section is the smallest (the so-called **throat of the nozzle**). Therefore, in order to accelerate a subsonic flow to a supersonic speed, one has to use the so-called **converging-diverging nozzle (de Laval nozzle)** – see Fig. 11.3e. In such a nozzle, the subsonic flow accelerates in the converging part (if the difference of pressures at the inlet and at the outlet of the nozzle is

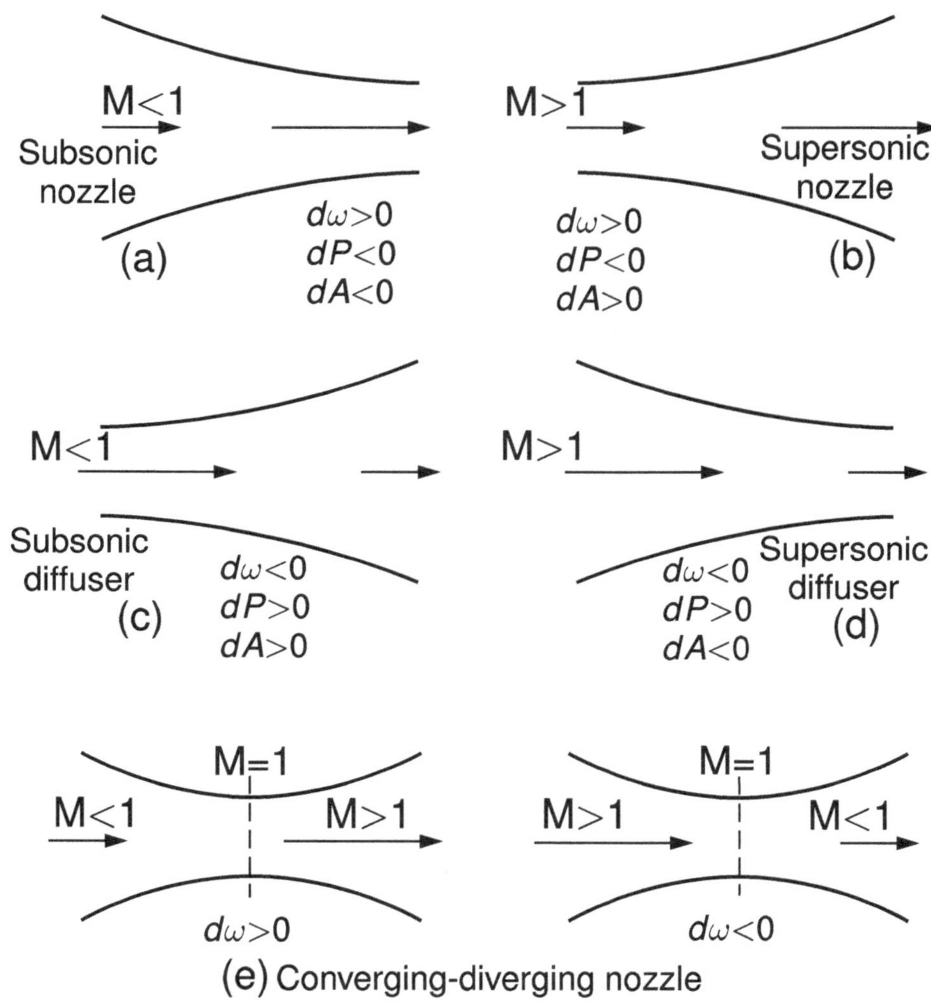

Fig. 11.3: Changes of properties of a steady flow moving through different nozzles.

large enough – see below) until it reaches $M = 1$ in the nozzle throat. After passing the throat, the flow keeps increasing its (now supersonic) speed in the diverging part of the nozzle. Conversely, a flow can decelerate from supersonic to subsonic speeds only in a **diverging-converging nozzle**, becoming a sonic flow ($M = 1$) at the channel throat (see Fig. 11.3e).

In an adiabatic-isentropic flow of an ideal gas of constant specific heat c_P (calorically perfect ideal gas), one has (see expression (7.47))

$$h - h_\circ = c_P(T - T_\circ), \tag{11.25}$$

so that relationship (11.18) can be written as

$$c_P T + \frac{\omega^2}{2} = h_\circ = c_P T_\circ, \tag{11.26}$$

where, as before, $h_\circ$ and $T_\circ$ are the flow's specific stagnation enthalpy and stagnation temperature, respectively.

As can be seen in expression (11.26), speed of the adiabatic-isentropic flow of an ideal gas with constant specific heat c_P cannot be higher than

$$\omega_{\max} = (2h_\circ)^{1/2} = (2c_P T_\circ)^{1/2}. \tag{11.27}$$

Relationship (11.26) can be rewritten as

$$1 + \frac{\omega^2}{2c_P T} = \frac{T_\circ}{T}, \tag{11.28}$$

or

$$\frac{T}{T_\circ} = \left(1 + \frac{\kappa - 1}{2} M^2\right)^{-1}, \tag{11.29}$$

because in a calorically perfect ideal gas (see expression (11.13))

$$c_P T = \frac{\kappa R_g}{\kappa - 1} T = \frac{a^2}{\kappa - 1}. \tag{11.30}$$

Since the speed of sound in an ideal gas is proportional to $\sqrt{T}$ (see expression (11.7)), relationship (11.29) leads to

$$\frac{a}{a_\circ} = \left(\frac{T}{T_\circ}\right)^{1/2} = \left(1 + \frac{\kappa - 1}{2} M^2\right)^{-1/2}, \tag{11.31}$$

and, using the equation of adiabat-isentrop ($P\rho^{-\kappa} = \text{const}$ – see expression (11.5)), to

$$\frac{P}{P_\circ} = \left(\frac{T}{T_\circ}\right)^{\kappa/(\kappa-1)} = \left(1 + \frac{\kappa - 1}{2} M^2\right)^{\kappa/(1-\kappa)}, \tag{11.32}$$

and

$$\frac{\rho}{\rho_o} = \left(\frac{T}{T_o}\right)^{1/(\kappa-1)} = \left(1 + \frac{\kappa-1}{2}M^2\right)^{1/(1-\kappa)}. \qquad (11.33)$$

It should be noticed that relationships (11.29)–(11.33) do not depend explicitly on the geometry of the channel through which the gas flows. However, such dependence can easily be obtained by using the continuity equation (11.17) in the following form:

$$\frac{A(x)}{A_*} = \frac{\rho_* \omega_*}{\rho(x)\omega(x)}, \qquad (11.34)$$

where, as before, $A(x)$ is the cross section of the channel at a location x, and the asterisk indicates that the corresponding property is for the location where A is the *smallest* cross section of the channel. As already mentioned, the location of the converging-diverging channel where $A = A_*$ is called the **channel throat** or the **nozzle throat**. The flow properties at the throat are called the **critical properties of the flow** (not to be confused with the critical thermodynamic properties of substances that were discussed in Section 5.4).

Taking the above into account, one can write the following expressions, valid at any location in a steady, adiabatic-isentropic, and quasi-one-dimensional flow of a calorically perfect ideal gas:

$$\frac{\omega}{\omega_*} = \frac{\omega}{(\kappa R_g T_*)^{1/2}} = M\left[\frac{2}{\kappa+1}\left(1 + \frac{\kappa-1}{2}M^2\right)\right]^{-1/2}, \qquad (11.35)$$

$$\frac{\rho}{\rho_*} = \left[\frac{2}{\kappa+1}\left(1 + \frac{\kappa-1}{2}M^2\right)\right]^{1/(1-\kappa)}, \qquad (11.36)$$

$$\frac{T}{T_*} = \frac{\kappa+1}{2}\left(1 + \frac{\kappa-1}{2}M^2\right)^{-1}, \qquad (11.37)$$

$$\frac{P}{P_*} = \left[\frac{2}{\kappa+1}\left(1 + \frac{\kappa-1}{2}M^2\right)\right]^{\kappa/(1-\kappa)}, \qquad (11.38)$$

$$\frac{a}{a_*} = \left[\frac{2}{\kappa+1}\left(1 + \frac{\kappa-1}{2}M^2\right)\right]^{-1/2}, \qquad (11.39)$$

and

$$\frac{A}{A_*} = \frac{1}{M}\left[\frac{2 + (\kappa-1)M^2}{\kappa+1}\right]^{(\kappa+1)/[2(\kappa-1)]}. \qquad (11.40)$$

The flow's critical speed ω_* (the speed of the flow at the nozzle throat) can be obtained from relationship (11.35) as

$$\omega_* = a_* = (\kappa R_g T_*)^{1/2} = \left(\frac{2\kappa}{\kappa + 1} R_g T_\circ\right)^{1/2}, \tag{11.41}$$

so that

$$\dot{m} = \rho(x)A(x)\omega(x) = \rho_* A_* \omega_*, \tag{11.42}$$

and

$$\dot{m} = \rho_\circ \left(\frac{2}{\kappa + 1}\right)^{1/(\kappa-1)} \left(\frac{2\kappa}{\kappa + 1} R_g T_\circ\right)^{1/2} A_*. \tag{11.43}$$

When the flow under consideration is sonic at the channel throat ($M_* = 1$), then the ratios $T_*/T_\circ$, $a_*/a_\circ$, $P_*/P_\circ$ and $\rho_*/\rho_\circ$ can be obtained by assuming $M = 1$ in relationships (11.29) and (11.31)–(11.33). As a result, one has

$$\frac{T_*}{T_\circ} = \frac{2}{\kappa + 1}, \tag{11.44}$$

$$\frac{a_*}{a_\circ} = \left(\frac{2}{\kappa + 1}\right)^{1/2}, \tag{11.45}$$

$$\frac{P_*}{P_\circ} = \left(\frac{2}{\kappa + 1}\right)^{\kappa/(\kappa-1)}, \tag{11.46}$$

and

$$\frac{\rho_*}{\rho_\circ} = \left(\frac{2}{\kappa + 1}\right)^{1/(\kappa-1)}. \tag{11.47}$$

Expressions (11.29)–(11.47) are valid only when the flowing matter in *both* locations considered in each of the expressions is in some (different) local-equilibrium state. Ratios (11.29)–(11.33) for various nozzle locations x in a steady, quasi-one-dimensional, and adiabatic-isentropic flow of a calorically perfect ideal gas are shown in Fig. 11.4. One can see there that when ratio $P_e/P_\circ$ for a flow is greater than the ratio β,

$$\beta = P_*/P_\circ, \tag{11.48}$$

the flow Mach number is smaller than one everywhere in the nozzle. Thus, such a flow cannot become supersonic and, therefore, no shock can be formed in the nozzle. In a flow in a converging-diverging nozzle where the ratio $P_e/P_\circ$ falls below β (but is unequal to some 'special' ratio $P_{e,6}/P_\circ$ – see below), the flow

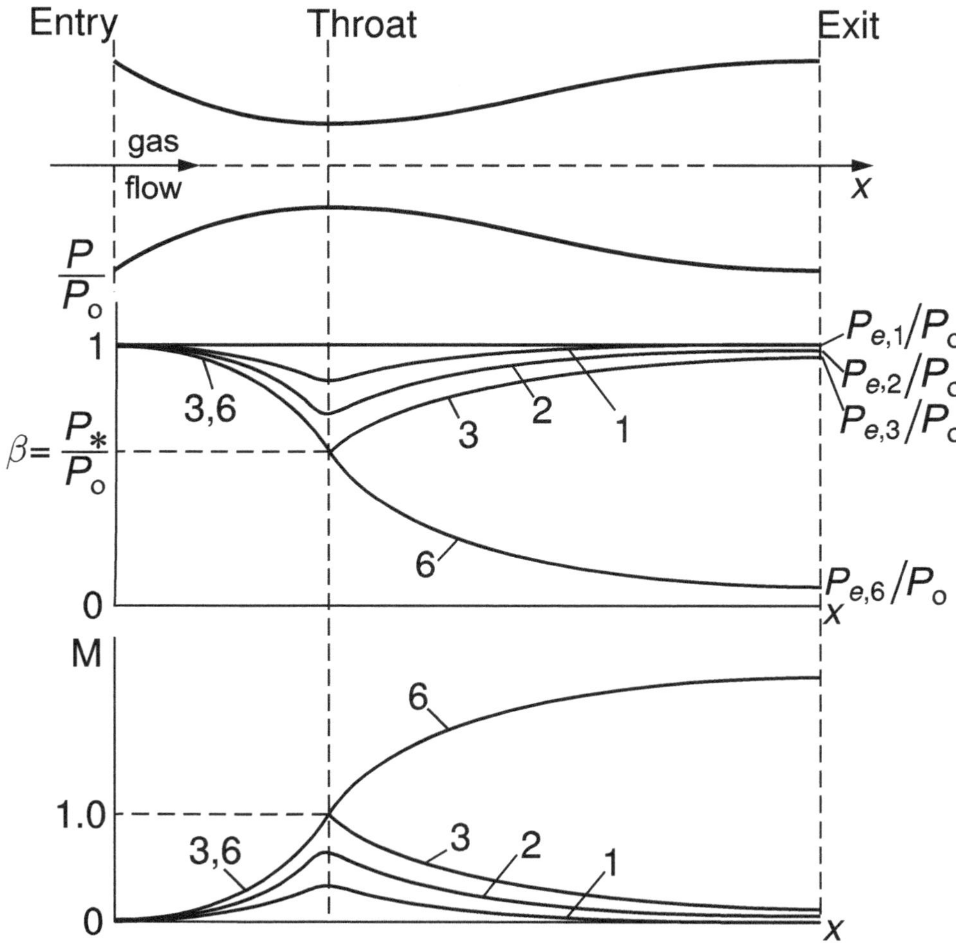

Fig. 11.4: Steady, shockless, adiabatic-isentropic, quasi-one-dimensional flow in a converging-diverging nozzle. P is the pressure of the flow at a given location x, and P_o is the flow's stagnation pressure. Some exit pressures ($P_{e,1}, P_{e,2}, P_{e,3}$) are not small enough, when compared to the pressure at the nozzle entry, to produce a supersonic flow ($M > 1$) in the nozzle diverging part. Therefore, flows 1, 2, and 3 cannot produce a standing shock. At some 'special' value of the exit pressure $P_e = P_{e,6}$, the flow becomes supersonic in the nozzle's diverging part but with no shock there (also see Figs. 11.5 and 11.6).

becomes supersonic in the diverging part of the nozzle, and a standing normal shock can occur in this part – see Section 11.4 and Fig. 11.5.

The above discussion shows the usefulness of the concept of stagnation pressure P_o in studies of flow dynamics. It is a general measure of possible changes taking place in flowing gas. It also has a direct practical meaning in applications where a flow starts from a vessel filled with a gas at rest. Then, the static pressure of the gas is just the stagnation pressure of the entire flow originating from the vessel.

Values of the ratios P/P_o (expression (11.32)) and $\beta = P_*/P_\text{o}$ (expressions (11.46) and (11.48)) depend on the value of the isentropic exponent κ for the flowing gas; $\kappa = 1.67$ (monatomic gas), 1.40 (diatomic gas), or about 1.33 (gas of polyatomic molecules). The corresponding values of β are 0.490, 0.528, and 0.541, respectively. In addition, the change of the ratio P/P_o along the direction of the flow depends on the value of pressure P_e, that is, the pressure of the gas at the exit of the nozzle.

Flow 1 (with the exit pressure $P_{e,1}$) and Flow 2 (with the exit pressure $P_{e,2}$) shown in Fig. 11.4 are examples of the flows where decrease of the exit pressure to or below some 'special' value $P_{e,3}$ causes acceleration of the flows at the nozzle throat to sonic speed ($\omega_* = a_*$). One should notice that $M_* < 1$ in Flows 1 and 2, while $M_* = 1$ in Flows 3 and 6. Since no shock occurs in Flows 1, 2, 3, and 6, they can be treated as isentropic flows.

Flows where the nozzle exit pressure P_e is smaller than the 'special' pressure $P_{e,3}$ but greater than the 'special' pressure $P_{e,6}$ will be discussed in Section 11.4. Here we only mention that even when $P_e < P_{e,3}$, the flow Mach number at the throat of converging-diverging nozzle is always $M_* = 1$ (see relationship (11.24)). Then the mass flow rate does not increase even when P_e decreases; such a flow is called the 'choked' flow.

Problem 11.1: Find the change of the kinetic energy of 1 kg of a fluid moving through a channel where: (A) the fluid enters the channel with speed $\omega_{A,1} = 1\,\text{m/s}$ and leaves it with speed $\omega_{A,2} = 45\,\text{m/s}$, and (B) the fluid enters the channel with speed $\omega_{B,1} = 500\,\text{m/s}$ and leaves it with speed $\omega_{B,2} = 502\,\text{m/s}$.

Solution: The kinetic energy of matter of mass M that moves with speed ω is $M\omega^2/2$. Thus, the changes of the kinetic energies of 1 kg of the matter moving through the channel are

$$\Delta e_{A,\text{kin}} = \frac{\omega_{A,2}^2 - \omega_{A,1}^2}{2} = \frac{(45\,\text{m/s})^2 - (1\,\text{m/s})^2}{2} \simeq 1\,\text{kJ/kg}, \qquad (11.49)$$

and

$$\Delta e_{B,\text{kin}} = \frac{\omega_{B,2}^2 - \omega_{B,1}^2}{2} = \frac{(502 \text{ m/s})^2 - (500 \text{ m/s})^2}{2} \simeq 1 \text{ kJ/kg}. \qquad (11.50)$$

End of Problem 11.1.

Problem 11.2: A shuttle flies at altitude of 20 km with Mach number $M = 6$. What would the shuttle speed be if it was flying close to the ground with the same Mach number?

Solution: Both the air close to the ground and at the altitude 20 km can be considered ideal gases in states of equilibrium. Therefore, the speeds of sound at the two altitudes can be given as $a = (\kappa R_g T)^{1/2}$ (see expression (11.7)), where T is the absolute temperature of the gas, $R_g = 287$ J/kg·K is the air individual gas constant, and κ is the isentropic exponent equal to 1.4 because the air is dominated by diatomic nitrogen and diatomic oxygen. Thus, the speeds of sound close to the ground and at altitude 20 km are, respectively,

$$a_g = [1.4(287 \text{ J/kg·K})(300 \text{ K})]^{1/2} = 347 \text{ m/s} \qquad (11.51)$$

and

$$a_u = [1.4(287 \text{ J/kg·K})(217 \text{ K})]^{1/2} = 295 \text{ m/s}, \qquad (11.52)$$

where we assumed, in agreement with the *NASA Standard Atmosphere*, that on a typical California day, the air temperatures at the ground and at altitude 20 km are about 300 K and 217 K, respectively. Subsequently, the speeds of the shuttle at these two locations are, respectively,

$$\omega_g = M a_g = 6(347 \text{ m/s}) = 2082 \text{ m/s} \qquad (11.53)$$

and

$$\omega_u = M a_u = 6(295 \text{ m/s}) = 1770 \text{ m/s}. \qquad (11.54)$$

One can see that it is easier to break the 'sound barrier' ($M = 1$) at higher altitudes than at lower altitudes.

End of Problem 11.2.

Problem 11.3: Ideal diatomic oxygen is flowing isentropically and adiabatically through a pipe. The temperature and the speed of the gas measured 20 cm from the pipe's inlet (the gas is in a local equilibrium there) are $t = 10\,°C$ and $\omega = 10\,m/s$, respectively ($x = 0$ at the pipe inlet). Is the flow at $x = 20\,cm$ subsonic, sonic, or supersonic?

Solution: The isentropic exponent for diatomic oxygen is $\kappa = 1.4$. Thus, the speed of sound in the local-equilibrium gas at $x = 20\,cm$ is (see expression (11.7)),

$$a(\text{at } x = 20\,cm) = [\kappa (R/\mu)T]^{1/2}$$

$$= \left[1.4 \frac{8.314\,J/mol\cdot K}{32\cdot 10^{-3}\,kg/mol} (283.15\,K) \right]^{1/2} = 321\,m/s, \qquad (11.55)$$

where $R = 8.314\,J/mol\cdot K$ is the universal (molar) gas constant, and $\mu = 32 \times 10^{-3}\,kg/mol$ is the molecular (molar) mass of diatomic oxygen.

The Mach number of the flow at $x = 20\,cm$ is

$$M(\text{at } x = 20\,cm) = \frac{\omega(\text{at } x = 20\,cm)}{a(\text{at } x = 20\,cm)} = \frac{10\,m/s}{321\,m/s} = 0.031, \qquad (11.56)$$

which means that this is a subsonic flow.

End of Problem 11.3.

Problem 11.4: A small amount of air flows steadily out of a large tank, through a small converging nozzle. The adiabatic-isentropic flow is sonic ($M = 1$) at the exit of the nozzle where the pressure of the flowing air is 1 bar. Find the minimum value of the pressure of the air in the center of the tank that is necessary to produce the sonic flow at the nozzle exit. The air can be treated as a calorically perfect ideal gas, that is, a gas that has constant-pressure heat capacity independent of temperature. Assume that the air at rest in the tank is in a state of equilibrium, and that it is in local equilibrium at the nozzle exit.

Solution: The air in the tank regions away from the nozzle is at rest because the amount of the air leaving the nozzle is small in comparison with the amount of the gas in the tank. Therefore, the pressure of the air in the regions is practically the same as the flow's stagnation pressure P_o. The (critical) pressure of the sonic flow leaving the converging nozzle is $P_* = 1$ bar. Subsequently (see expressions (11.46) and (11.48)),

$$\frac{P_o}{P_*} = \beta^{-1} = \left(\frac{2}{\kappa + 1}\right)^{\kappa/(1-\kappa)} = 1.893 = (0.528)^{-1}, \qquad (11.57)$$

where we used $\kappa = 1.4$ for the flow isentropic exponent since the air consists mainly of diatomic molecules N_2 and O_2 (see expression (6.40)). Thus, in order to have sonic flow at the exit of the converging nozzle, the pressure P_o of the air in the central regions of the tank should be equal to or greater than 1.893 bar. When this pressure is smaller than 1.893 bar, the flow at the nozzle exit will be a subsonic ($M < 1$) flow. If the pressure is greater than 1.893 bar, then the flow will *always* be sonic ($M = 1$) at the nozzle exit because the discussed adiabatic-isentropic flows in converging channels cannot be accelerated to supersonic speeds.

End of Problem 11.4.

Problem 11.5: An adiabatic-isentropic flow leaves a large tank (filled with calorically perfect ideal nitrogen gas) through a converging nozzle attached to the tank. Find the physical conditions (pressure P_t, temperature T_t, and density ρ_t) of the gas in the tank necessary for the flow to reach speed $\omega_e = 100\,\text{m/s}$ at the nozzle exit where the gas pressure and temperature are $P_e = 1$ bar and $T_e = 300\,\text{K}$, respectively. Assume that the gas is in a state of equilibrium in the tank's regions away from the nozzle, and that the flowing gas is in a local equilibrium at the nozzle exit.

Solution: The isentropic exponent for ideal diatomic nitrogen is $\kappa = 1.4$. The speed of sound in the flow at the nozzle exit is (see expression (11.7))

$$a_e = [\kappa R_g T_e]^{1/2} = \left[1.4\frac{8.314\,\text{J/mol·K}}{28\cdot10^{-3}\,\text{kg/mol}}(300\,\text{K})\right]^{1/2} = 353\,\text{m/s}, \qquad (11.58)$$

where the individual gas constant of the nitrogen gas is $R_g = R/\mu = 297\,\text{J/kg·K}$, $\mu = 28\,\text{g/mol}$ is the molar mass of nitrogen, and $R = 8.314\,\text{J/mol·K}$ is the universal gas constant.

The Mach number of the flow at the nozzle exit is

$$M_e = \frac{\omega_e}{a_e} = \frac{100\,\text{m/s}}{353\,\text{m/s}} = 0.28, \qquad (11.59)$$

which means that the flow is subsonic there.

The density of the flowing gas at the nozzle exit can be obtained from the gas equation of state because the gas is in a local equilibrium at the exit,

$$\rho_e = \frac{P_e}{R_g T_e} = \frac{10^5 \, \text{N/m}^2}{(297 \, \text{J/kg·K})(300 \, \text{K})} = 1.12 \, \text{kg/m}^3. \tag{11.60}$$

The reader should recall that use of the ideal-gas equation of state at a flow locality (as we did in expression (11.60)) is justified only when the flowing gas is in a local-equilibrium state at this locality – see discussion in Section 1.2.2.

The gas in most of the volume of the tank is practically at rest. The gas is moving with noticeable speed only in the region close to the entry of the nozzle. Therefore, the thermodynamic properties of most of the gas in the tank are the stagnation properties of the flow. Thus, pressure P_t, temperature T_t, and density ρ_t of the gas in the tank can be found from the following ratios given in expressions (11.29)–(11.33):

$$\frac{P_e}{P_t} = 0.947, \qquad \frac{T_e}{T_t} = 0.985, \qquad \frac{\rho_e}{\rho_t} = 0.962, \tag{11.61}$$

so that

$$P_t = \frac{1 \, \text{bar}}{0.947} = 1.06 \, \text{bar}, \qquad T_t = \frac{300 \, \text{K}}{0.985} = 305 \, \text{K}, \tag{11.62}$$

and

$$\rho_t = \frac{1.12 \, \text{kg/m}^3}{0.962} = 1.16 \, \text{kg/m}^3. \tag{11.63}$$

End of Problem 11.5.

11.4 Flow with Gasdynamic Shock

A very interesting phenomenon occurs in supersonic flows, that is, flows with the Mach number greater than one. This phenomenon, called the **gasdynamic shock wave** (or just **shock**), is characterized by a very significant change of the flow's properties (pressure, temperature, entropy, etc.) across a very short distance (usually on the order of a millimeter). In this section we consider only **standing normal shocks** that occur in steady supersonic flows without

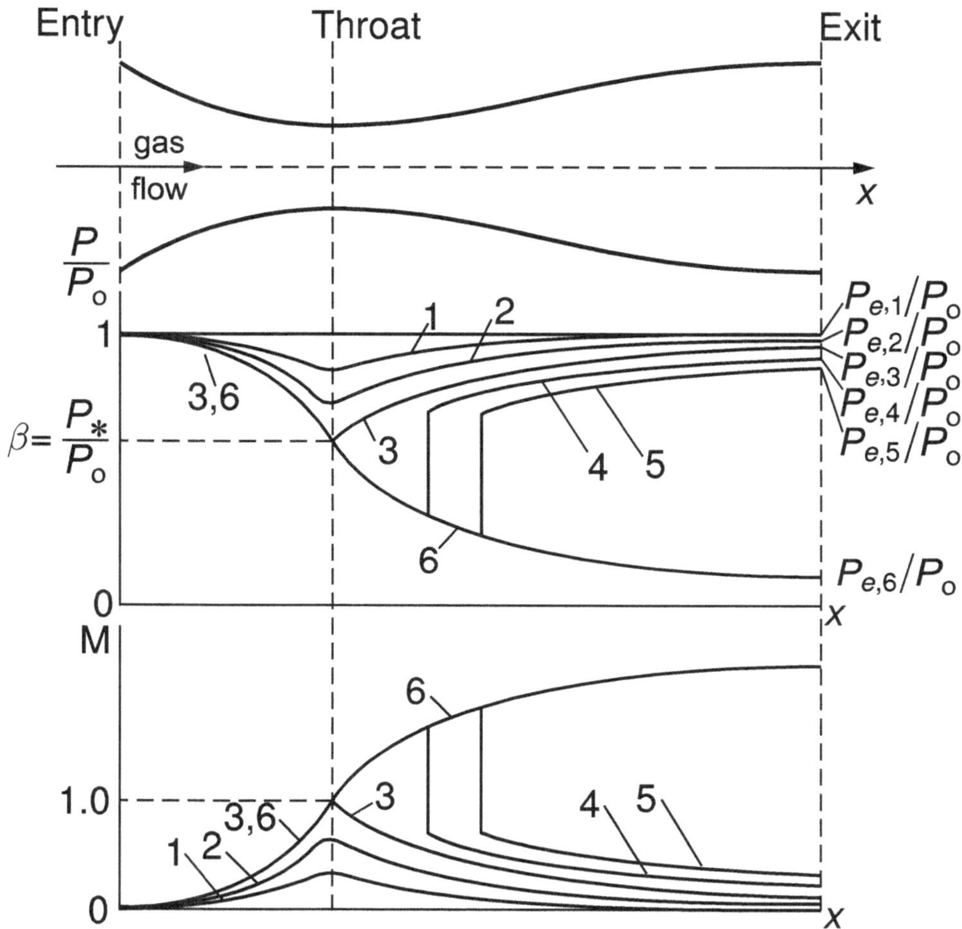

Fig. 11.5: Examples of steady flows moving in direction x through a converging-diverging nozzle with the exit's pressures $P_{e,i}$ ($i = 1 - 6$). Flows 1, 2, 3, and 6 do not produce standing normal shocks (also see Fig. 11.4) but Flow 4 (with exit pressure $P_{e,6} < P_{e,4} < P_{e,3}$) and Flow 5 (with exit pressure $P_{e,6} < P_{e,5} < P_{e,3}$) produce standing shocks in the diverging part of the nozzle.

work and heat transfers. (As mentioned earlier, **gasdynamic shocks do not occur in subsonic flows.**) The standing normal shock does not move in space and is normal to the flow direction x – see Figs. 11.5 and 11.6. **The steady flow before the standing normal shock is always supersonic ($M > 1$) but it**

becomes a subsonic flow ($M < 1$) behind the shock. The flow speed and its Mach number decrease while the gas pressure, temperature, and density, increase across the shock. Since the increase of entropy across a gasdynamic shock is usually significant, flows with shocks cannot be considered isentropic flows. However, in studies of such flows it is often justified to divide the entire flow into three parts: the part from the nozzle entry up to the location *just before* the shock (the so-called 'upstream flow'), the shock itself, and the part from the location *just after* the shock up to the nozzle exit (the so-called 'downstream flow'). Then, the 'upstream' and the 'downstream' parts can be considered as (different) adiabatic-isentropic flows discussed in the previous sections but the middle part (the shock) must be treated as a non-isentropic system. This, however, does not cause much of a problem if the flowing gas just before the shock and just after the shock are in some states of local equilibrium because then the changes of all state properties across the shock can be calculated using the methods discussed in Section 7.2 (see below).

In some steady flows in converging-diverging nozzles with neither heat nor work transfers, the pressure P_e at the nozzle exit can be smaller than the 'special' value $P_{e,3}$ – see the previous section. When $P_e = P_{e,3}$, the subsonic flow, marked in Figs. 11.4 and 11.5 as Flow 3, accelerates in the converging part of the nozzle, reaches the sonic speed ($M_* = 1$) at the nozzle's throat, and keeps moving (with a subsonic speed) through the nozzle's divergent part until it exits the nozzle in a 'smooth' manner, that is, without showing any behavior indicating a significant production of entropy inside the nozzle. Therefore, such a flow can be considered an adiabatic-isentropic flow.

When pressure P_e is smaller than $P_{e,3}$, then the following can happen in the flow (see Fig. 11.5):

1) when $P_{e,6} < P_e < P_{e,3}$, a standing normal shock occurs in the diverging part of the nozzle – Flows 4 and 5 in Fig. 11.5 are examples of such a situation. These flows cannot be considered isentropic because the significant changes of their thermodynamic properties during crossing the standing shocks generate significant amounts of entropy. However, as already mentioned, the part of each of the flows from the nozzle entry up to the beginning of the shock and the part from the end of the shock to the nozzle exit can be treated as isentropic flows.

2) when $P_e = P_{e,6}$, no shock occurs inside the nozzle. Thus, Flow 6 in Fig. 11.5 can be considered an isentropic flow.

3) when $P_e < P_{e,6}$, no shock occurs inside the nozzle, but multiple oblique shocks can be formed outside the nozzle exit.

The conservation laws for a steady quasi-one-dimensional flow, with neither heat nor work transfers but with a normal shock standing in the diverging part of

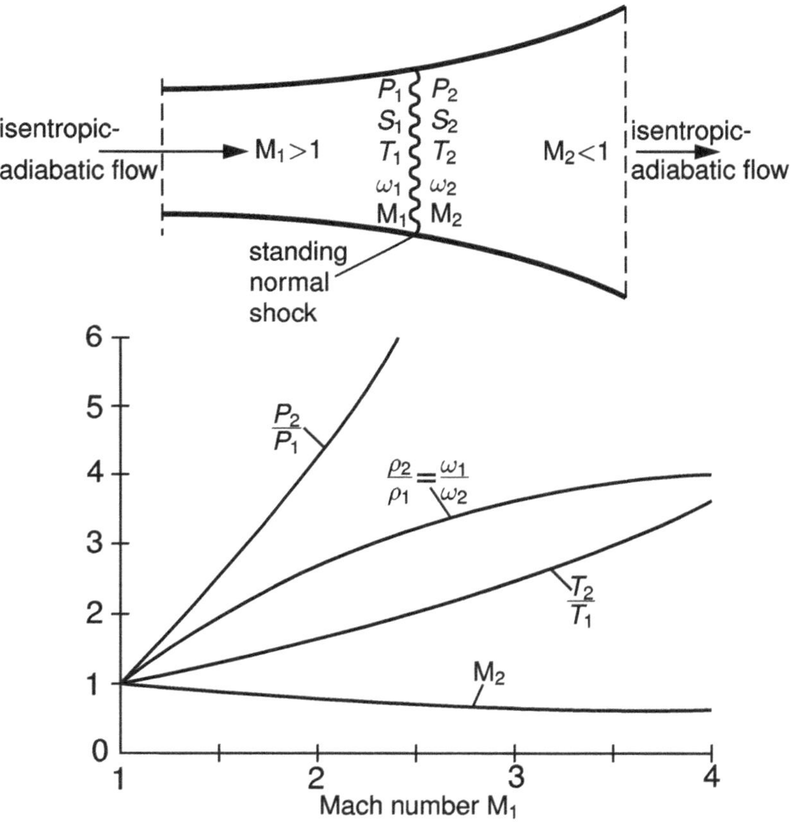

Fig. 11.6: A steady adiabatic gas flow moving in the diverging part of a converging-diverging nozzle with a standing normal gasdynamic shock (the wavy line in the figure). The subscripts 1 and 2 denote the properties of the flowing gas 'just before' and 'just after' the shock, respectively. The parts of the flow before the shock and after the shock can often be considered adiabatic-isentropic flows if there are no work and heat transfers between the nozzle and its surroundings. Passing of the flow through the shock causes a significant increase of pressure P, entropy S, temperature T, and density ρ of the flowing gas as well as a decrease of the flow's speed ω. The shock transforms the supersonic flow into subsonic flow.

converging-diverging nozzle, can be written as (see expressions (11.17)–(11.19) and notation in Fig. 11.6)

$$\rho_1\omega_1 = \rho_2\omega_2 \qquad \text{the continuity equation,} \qquad (11.64)$$

$$h_1 + \frac{\omega_1^2}{2} = h_2 + \frac{\omega_2^2}{2} = h_\mathrm{o} \qquad \text{the energy equation,} \qquad (11.65)$$

and

$$P_1 + \rho_1\omega_1^2 = P_2 + \rho_2\omega_2^2 \qquad \text{the momentum equation,} \qquad (11.66)$$

where subscripts 1 and 2 denote the conditions of the flowing gas *just before* and *just after* the shock, respectively, and where we assumed that the cross sections of the flow channel just before and just after the shock are practically the same ($A_1 \simeq A_2$) because the width of the shock is usually very small. As before, ρ, P, ω, and h are the mass density, pressure, speed, and specific enthalpy of the flow, respectively, and h_o is the flow specific stagnation enthalpy which represents the specific total (kinetic plus thermal) energy of the flowing matter – see expression (11.18). (The changes of the potential (gravitational) energy of the matter flowing horizontally are usually negligible.) This total energy is constant during the flows studied here because they exchange neither heat nor work with the environment.

The thermodynamic properties of the discussed flows *just before* the shock (state 1) and *just after* the shock (state 2) can be given as (see the previous sections)

$$P_{1,2} = \rho_{1,2}RT_{1,2}, \qquad (11.67)$$

$$h_2 - h_1 = c_P(T_2 - T_1), \qquad (11.68)$$

$$a_{1,2} = \sqrt{\kappa P_{1,2}/\rho_{1,2}}, \qquad (11.69)$$

and

$$M_2^2 = \frac{1 + [(\kappa - 1)/2]M_1^2}{\kappa M_1^2 - (\kappa - 1)/2}, \qquad (11.70)$$

where $M_1 > 1$ and $M_2 < 1$. Thus, the Mach number just behind the standing normal shock in a quasi-one-dimensional steady flow of a calorically perfect ideal gas is a function of only the Mach number of the flow just before the shock. (When $M_1 = M_2 = 1$, the resulting infinitely weak shock is sometimes called the *Mach wave*.) However, when M_1 becomes progressively greater than one, the standing normal shock becomes stronger (that is, the changes of pressure, temperature, and density of the flow across the shock become larger), and M_2 becomes progressively smaller than one. Indeed, expressions (11.64)–(11.70) lead to the following relationships 'connecting' the flow properties just after the standing normal shock with the corresponding properties of the flow just before the shock:

$$\frac{\rho_2}{\rho_1} = \frac{\omega_1}{\omega_2} = \frac{(\kappa + 1)M_1^2}{2 + (\kappa - 1)M_1^2}, \tag{11.71}$$

$$\frac{P_2}{P_1} = 1 + \frac{2\kappa}{\kappa + 1}(M_1^2 - 1), \tag{11.72}$$

and

$$\frac{T_2}{T_1} = \frac{h_2}{h_1} = \left[1 + \frac{2\kappa}{\kappa + 1}(M_1^2 - 1)\right]\left[\frac{2 + (\kappa - 1)M_1^2}{(\kappa + 1)M_1^2}\right]. \tag{11.73}$$

One can see that M_2 and ratios ρ_2/ρ_1, P_2/P_1, and T_2/T_1 for a standing normal shock in a steady adiabatic-isentropic flow of calorically perfect ideal gas are functions of M_1 only. In a similar flow of thermally perfect ideal gas (where heat capacities depend on temperature), the changes of the flow properties across the shock depend on both M_1 and T_1, whereas in reacting flows the changes also depend on pressure P_1.

Relationships derived in this section often give a reliable picture of the dynamics of steady flows of calorically perfect ideal gases with standing normal shocks when the upstream Mach numbers M_1 are smaller than about three. In flows with the Mach numbers greater than that, the temperature and pressure behind the normal shock are usually too high for the isentropic exponent κ to be constant. Also, in such high-temperature flows, gas dissociation, ionization, and an increase of the rates of chemical reactions can be significant and, therefore, must be considered.

Expressions (11.70)–(11.73) are mathematically valid for $M_1 < 1$ as well as $M_1 > 1$. Therefore, one has to test the validity of the expressions against the Second Law of Thermodynamics to see which of the two solutions has the correct physical meaning. The change of the specific entropy (a function of state) of the considered steady flow of a calorically perfect ideal gas across the standing normal shock can be given as (see expression (7.58))

$$s_2 - s_1 = c_P \ln \frac{T_2}{T_1} - R_g \ln \frac{P_2}{P_1}, \tag{11.74}$$

which together with expressions (11.72) and (11.73) leads to

$$
s_2 - s_1 = c_P \ln \left\{ \left[1 + \frac{2\kappa}{\kappa + 1}(M_1^2 - 1) \right] \left[\frac{2 + (\kappa - 1)M_1^2}{(\kappa + 1)M_1^2} \right] \right\}
$$
$$
- R_g \ln \left[1 + \frac{2\kappa}{\kappa + 1}(M_1^2 - 1) \right]. \tag{11.75}
$$

Therefore, if $M_1 = 1$ then $s_2 - s_1 = 0$, if $M_1 < 1$ then $s_2 - s_1 < 0$, and if $M_1 > 1$ then $s_2 - s_1 > 0$. Since the very narrow shock in steady flow with no heat and work transfers can be considered an isolated system, the Second Law of Thermodynamics requires that $s_2 - s_1 > 0$. Thus, the Mach number M_1 (of the flow just before the standing shock) is greater than one while the Mach number M_2 (of the flow just after the shock) is smaller than one.

Rewriting relationship (11.65) for the stagnation conditions of each of the flow's two parts (the part before the shock and the part after the shock) gives

$$
c_P T_{o,1} + \frac{\omega_{o,1}^2}{2} = c_P T_{o,2} + \frac{\omega_{o,2}^2}{2}, \tag{11.76}
$$

or

$$
T_{o,1} = T_{o,2} = T_o, \tag{11.77}
$$

because $\omega_{o,1} = 0$ and $\omega_{o,2} = 0$ under the stagnation conditions. One can see that the **stagnation temperature of a steady adiabatic-isentropic flow of a calorically perfect ideal gas before the standing normal shock is the same as the stagnation temperature of the flow after the shock. Subsequently, the specific stagnation enthalpies of the flow before the shock and the flow after the shock also are the same** (see expression (11.18)),

$$
h_{o,1} = h_{o,2} = h_o. \tag{11.78}
$$

One should recall again that the relationships of this section were derived using constant specific heats, ideal-gas equations of state, and the assumption that the gas thermal energy, enthalpy, and entropy are properties of state. Therefore, in any calculations based on these relationships, the **flowing gas states marked with subscripts 1, 2, $\star$, and $\circ$ must be some local-equilibrium states**.

Another fact to remember is that **any quasi-one-dimensional and steady flow of a calorically perfect ideal gas with a standing normal shock can**

be studied by dividing the flow into three parts that can be investigated independently: the upstream flow approaching the standing shock, the shock itself, and the downstream flow moving away from the shock. In many applications, the flows before and after the shock can be treated as adiabatic-isentropic processes.

Problem 11.6: A normal gasdynamic shock is standing in a steady flow in a convergent-divergent nozzle. The Mach number and temperature of the flow just before the shock are $M_1 = 1.70$ and $T_1 = 300\,\text{K}$, respectively. The flow is isentropic on each side of the shock, and neither work nor heat are exchanged between the flow and the environment. The flowing gas is an ideal diatomic gas with heat capacities that are constant (calorically perfect ideal gas). The ratio of A_e (the cross section of the nozzle at its exit) to A_* (the cross section of the nozzle's throat) is equal to two. Find: (a) the value of the Mach number just after the shock, (b) the strength of the shock (that is, the ratio of the flow pressure just behind the shock to the pressure just before the shock), (c) the ratio A_e/A_2 (A_2 is the cross section of the nozzle just after the shock), and (d) the stagnation temperatures of the flows on both sides of the shock. Assume that the flowing gas is in a local equilibrium in each considered location.

Solution: The properties of the flow just before the shock (where $M_1 = 1.70$) can be obtained from expressions (11.29)–(11.33) and (11.40),

$$\frac{P_1}{P_{\text{o},1}} = 0.202, \quad \frac{T_1}{T_{\text{o},1}} = 0.633, \quad \frac{\rho_1}{\rho_{\text{o},1}} = 0.319, \quad \frac{A_1}{A_{*,1}} = \frac{A_1}{A_*} = 1.34, \quad (11.79)$$

where, as before, P_1, T_1, and ρ_1 are pressure, temperature, and mass density of the flow just before the shock, respectively, and $P_{\text{o},1}$, $T_{\text{o},1}$, and $\rho_{\text{o},1}$ are stagnation pressure, temperature, and mass density of the flow just before the shock, respectively. One should notice that the value of the critical cross section for the flow before the shock is $A_{*,1} = A_*$ (the flow in the converging part of the nozzle reaches $M_1 = 1$ at the nozzle throat of cross section A_*). However, the cross-section $A_{*,1}$ differs from the critical cross section $A_{*,2}$ that characterizes the after-the-shock part of the flow. This is because there is a discontinuity (the entropy-producing shock) between the two parts. Each of the two parts can be treated in calculations as a separate adiabatic-isentropic flow with its own critical parameters, including its own value of critical cross section. If there was no shock in the nozzle, both adiabatic-isentropic parts would 'merge' smoothly without producing the shock-type discontinuity, and the entire flow consisting of the two parts would have a single critical cross section A_*. In other words, one has in the discussed flow:

$$A_{*,1} \equiv A_* \quad \text{but} \quad A_{*,2} \neq A_{*,1}. \tag{11.80}$$

According to expression (11.70), the Mach number of the flow just behind the shock is $M_2 = 0.64$. Thus, one has for the part of the flow behind the shock (see expressions (11.29)–(11.33), and (11.40))

$$\frac{P_2}{P_{o,2}} = 0.759, \quad \frac{T_2}{T_{o,2}} = 0.924, \quad \frac{\rho_2}{\rho_{o,2}} = 0.821, \quad \frac{A_2}{A_{*,2}} = 1.145, \tag{11.81}$$

and, using expressions (11.71)–(11.73)

$$\frac{P_2}{P_1} = 3.20, \quad \frac{T_2}{T_1} = 1.46, \quad \frac{\rho_2}{\rho_1} = 2.19. \tag{11.82}$$

Thus (see expression (11.79)),

$$\frac{P_{o,2}}{P_{o,1}} = 0.855, \quad T_2 = 1.46 T_1 = 438 \text{ K}, \quad T_{o,1} = \frac{T_1}{0.633} = 473 \text{ K}, \tag{11.83}$$

where $T_{o,1} = T_{o,2}$ (also $h_{o,1} = h_{o,2}$) because the stagnation temperatures (and the specific stagnation enthalpies) of the flows on both sides of the shock are equal when the flow's heat capacities are constant – see expressions (11.77) and (11.78).

Since $A_1/A_{*,1} = 1.34$, $A_2/A_{*,2} = 1.145$, and $A_1 \simeq A_2$ (the widths of gasdynamic shocks are usually very small), one can write

$$\frac{A_{*,1}}{A_{*,2}} = \frac{1.145}{1.34} = 0.86, \tag{11.84}$$

and

$$\frac{A_e}{A_2} = \frac{A_e}{A_{*,1}} \times \frac{A_{*,1}}{A_{*,2}} \times \frac{A_{*,2}}{A_2} = 2 \times 0.86 \times \frac{1}{1.145} = 1.50. \tag{11.85}$$

End of Problem 11.6.

Problem 11.7: A diatomic nitrogen gas is steadily flowing out of a tank, through a converging-diverging nozzle. The gas pressure and temperature in the regions not close to the nozzle entrance are constant and equal to $P_t = 10\,\text{bar}$ and $T_t = 800\,\text{K}$, respectively. (The gas in those regions can be considered as being in thermodynamic equilibrium.) The cross section of the nozzle throat is $A_* = 10^{-3}\,\text{m}^2$, and the cross section of the nozzle exit is $A_e = 2 \times 10^{-3}\,\text{m}^2$. The pressure of the gas at the nozzle exit is $P_e = 5.8\,\text{bar}$ and the speed of the flow at the exit is $\omega_e = 280\,\text{m/s}$. Neither heat nor work are exchanged between the nozzle and its surroundings, and the flow is isentropic on each side of the shock. Assuming that the nitrogen is a calorically perfect ideal gas, find the properties of the flow at the throat and at the exit of the nozzle. Is there a normal shock standing somewhere in the nozzle? If yes, where?

Solution: The individual gas constant for diatomic nitrogen is (see expression (5.5))

$$R_g = \frac{R}{\mu} = \frac{8.314\,\text{kJ/kmol}}{28\,\text{kg/kmol}} = 297\,\text{J/kg·K} \tag{11.86}$$

where R and μ are the universal (molar) gas constant and the molecular mass of the gas, respectively.

The stagnation density of the flow before the possible shock is (see expression (11.67))

$$\rho_{\text{o},t} = \frac{P_{\text{o},t}}{R_g T_{\text{o},t}} = \frac{P_t}{R_g T_t} = 4.21\,\text{kg/m}^3, \tag{11.87}$$

where the stagnation pressure $P_{\text{o},t}$ and the stagnation temperature $T_{\text{o},t}$ of the flow before the shock are taken as the pressure P_t and temperature T_t, respectively, because the gas is practically at rest in the tank's regions away from the nozzle. The stagnation speed of sound in the regions can be given as (see expression (11.7))

$$a_{\text{o},t} = \sqrt{\kappa R_g T_{\text{o},t}} = \sqrt{\kappa R_g T_t} = 576\,\text{m/s}, \tag{11.88}$$

where $\kappa = 1.4$ is the isentropic exponent for processes in ideal diatomic gases. Assuming that the thermodynamic states of the flowing matter at the nozzle throat and at the other locations considered in the calculations are some local-equilibrium states, the critical pressure, temperature, and mass density of the flow at the locations are (see expressions (11.44), and (11.46)–(11.47)):

$$P_* = P_{\text{o},t}\left(\frac{2}{\kappa+1}\right)^{\kappa/(\kappa-1)} = P_t\left(\frac{2}{\kappa+1}\right)^{\kappa/(\kappa-1)} = 5.28\,\text{bar}, \tag{11.89}$$

$$T_* = T_{o,t}\frac{2}{\kappa+1} = T_t\frac{2}{\kappa+1} = 667\,\text{K}, \tag{11.90}$$

and

$$\rho_* = \rho_{o,t}\left(\frac{2}{\kappa+1}\right)^{1/(\kappa-1)} = \rho_t\left(\frac{2}{\kappa+1}\right)^{1/(\kappa-1)} = 2.67\,\text{kg/m}^3. \tag{11.91}$$

The flow critical speed and critical speed of sound are (see expression (11.41)),

$$\omega_* = a_* = \sqrt{\kappa R_g T_*} = 527\,\text{m/s}. \tag{11.92}$$

The mass flow rate in the nozzle's throat (where $M = 1$) and at any other location in the nozzle is the same because the flow under consideration is a steady flow. The rate is (see expression (11.42))

$$\dot{m} = \rho_*\omega_*A_* = 1.41\,\text{kg/s}. \tag{11.93}$$

The stagnation temperature of the part of the flow before the shock is the same as the stagnation temperature of the part of the flow after the shock (see expression (11.77)),

$$T_{o,t} = T_{o,e} = T_t = 800\,\text{K}, \tag{11.94}$$

and the flow temperature at the nozzle exit can be obtained from the equation of state for ideal gas at the exit,

$$T_e = \frac{P_e}{\rho_e R_g} = 759\,\text{K}. \tag{11.95}$$

Thus (see expression (11.29)),

$$\frac{T_e}{T_{o,e}} = \frac{759\,\text{K}}{800\,\text{K}} = 0.948 \rightarrow M_2 = 0.52. \tag{11.96}$$

When $M_2 = 0.52$, then (see expression (11.70))

$$M_1 = 2.05, \tag{11.97}$$

and, according to expressions (11.71)–(11.73), one has

$$\frac{P_2}{P_1} = 4.74, \quad \frac{T_2}{T_1} = 1.72, \quad \frac{\rho_2}{\rho_1} = 2.73. \tag{11.98}$$

Relationship (11.40) leads to

$$\frac{A_1}{A_{*,1}} = \frac{A_1}{A_*} = 1.76 \qquad \text{and} \qquad \frac{A_2}{A_{*,2}} = 1.22, \tag{11.99}$$

and either of these two relationships can be used to find the location of the standing shock because $A_1 \simeq A_2$. Thus,

$$A_1 = 1.76 A_* = 1.76 \times 10^{-3} \, \text{m}^2. \tag{11.100}$$

Indeed, there is a standing normal shock in the diverging part of the nozzle where the nozzle's cross section is 1.76×10^{-3} m^2.

The density of the gas at the exit of the nozzle can be calculated from the flow continuity equation (11.42),

$$\rho_e = \frac{\dot{m}}{\omega_e A_e} = \frac{1.41 \, \text{kg/s}}{(280 \, \text{m/s})(2 \times 10^{-3} \, \text{m}^2)} = 2.57 \, \text{kg/m}^3. \tag{11.101}$$

End of Problem 11.7.

Chapter 12

Fundamentals of Thermochemistry

12.1 Stoichiometric Equations

Consider a thermodynamic system containing a fixed amount of matter which is a *mixture* of several *components* (pure substances) produced by a single chemical reaction. No work is exchanged between the system and the environment. Since the matter of the system is conserved, the number of atoms of a particular kind that are free and bound in the mixture's molecules is constant if ionization of the particles can be ignored. The balance of matter during the reaction can be written in the form of the so-called **stoichiometric equation** of the reaction,

$$\nu_1 C_1 + \nu_2 C_2 + \cdots \rightleftharpoons \nu_l C_l + \nu_m C_m + \cdots, \tag{12.1}$$

where C_k is the kth component participating in the reaction, ν_k is the **molar stoichiometric coefficient** (the number of mols) of the kth component, and the arrows indicate that the reaction can proceed 'forward' as well as 'backward'. The components C_1, C_2, ... on the left-hand side of equation (12.1) are called the **reactants**, and the components C_l, C_m, ... on the right-hand side of the equation are called the **products** of the reaction. The stoichiometric coefficients can be arbitrary numbers, but their signs are chosen in thermochemical calculations as follows: $\nu_k > 0$ for the reaction products and $\nu_k < 0$ for the reactants. For example, the stoichiometric coefficients of a reaction of 1 mol of diatomic hydrogen, 0.5 mol of diatomic oxygen, and 3 mols of diatomic nitrogen,

$$H_2 + \frac{1}{2}O_2 + 3N_2 \rightleftharpoons H_2O + 3N_2, \tag{12.2}$$

are $v_{H_2} = -1$, $v_{O_2} = -1/2$, v_{N_2} (reactant) $= -3$, $v_{H_2O} = +1$, and v_{N_2} (product) $= +3$, and the diatomic nitrogen is called the **inert component** of the reaction since it does not participate chemically in the process. Note that diatomic nitrogen can be a 'non-inert component' in some other reactions such as $N_2 + O_2 \rightleftharpoons 2NO$.

One should notice in equation (12.2) that the total number of mols of all products in reaction (12.1) does not have to be equal to the total number of mols of all reactants. However, the number of atoms of a particular kind (both free and bound in compounds) present on the left-hand side of equation (12.1) must be the same as the number of such atoms (both free and bound in compounds) on the right-hand side of the equation, as required by the Principle of Conservation of Matter.

Multiplying both sides of the equation (12.1) by an arbitrary number does not change the characteristics of the corresponding reaction. Thus, reactions

$$H_2 + \frac{1}{2}O_2 \rightleftharpoons H_2O \qquad \text{and} \qquad 2H_2 + O_2 \rightleftharpoons 2H_2O \qquad (12.3)$$

are equivalent, and either of them can be used to study thermochemical properties of the chemical reaction of diatomic hydrogen and diatomic oxygen.

12.2 Enthalpy and Entropy of Chemical Reactions

Changes of thermodynamic properties of state (the internal energy, enthalpy, entropy, the Helmholtz free energy, and the Gibbs free energy) during chemical reactions are of great importance in a variety of applications. For example, the change of enthalpy during reaction (12.1) can be given as (the subscript r indicates the fact that the property change applies to the *entire* reaction):

$$\Delta_r H = H_{\text{products}} - H_{\text{reactants}}, \qquad (12.4)$$

where H_{products} and $H_{\text{reactants}}$ are the sum of the enthalpies of all the products and the sum of the enthalpies of all the reactants, respectively,

$$H_{\text{products}} = \sum_j v_j \bar{H}_j \qquad \text{and} \qquad H_{\text{reactants}} = \sum_i |v_i| \bar{H}_i, \qquad (12.5)$$

where v_j and $\bar{H}_j$ are the stoichiometric coefficient (a positive number) of the jth product and the molar value of the product's enthalpy, respectively, v_i and

$\bar{H}_i$ are the stoichiometric coefficient (a negative number) of the ith reactant and the molar value of the reactant's enthalpy, respectively (as before, the bar over the symbol H_i indicates that the value of the property is that of *one mol* of the component); the sums $\sum_j$ and $\sum_i$ are taken over all products and all reactants present in the reaction, respectively.

The change of entropy of the discussed reaction is

$$\Delta_r S = S_{\text{products}} - S_{\text{reactants}}, \tag{12.6}$$

where S_{products} and $S_{\text{reactants}}$ are the sum of the entropies of all the products and the sum of the *absolute* entropies of all the reactants, respectively,

$$S_{\text{products}} = \sum_j \nu_j \bar{S}_j \qquad \text{and} \qquad S_{\text{reactants}} = \sum_i |\nu_i| \bar{S}_i, \tag{12.7}$$

and $\bar{S}_j$ and $\bar{S}_i$ are the molar values of the entropies of the jth product and the ith reactant, respectively.

Knowledge of the values of the thermodynamic functions (U, H, S, ...) of reactants and products of chemical reactions is necessary for studying the reactions' thermochemical effects. These values for any component (pure substance) in thermodynamic equilibrium at pressure P and temperature T can be obtained by the methods studied in Chapter 8.

As discussed in Section 8.2.4, the molar enthalpy of the kth component (a reactant or a product) of reaction (12.1) at pressure P and temperature T is (see expression (8.25))

$$\bar{H}_k(P,T) = \Delta_f H_k^\circ(T_\circ) + [\bar{H}_k(P,T) - \bar{H}_k^\circ(T_\circ)], \tag{12.8}$$

where $\Delta_f H_k^\circ(T_\circ)$ is the enthalpy change during the reaction of formation of the kth component (see below), and $[\bar{H}_k(P,T) - \bar{H}_k^\circ(T_\circ)]$ is the component's sensible enthalpy.

Once the molar enthalpy $\bar{H}_k(P,T)$ of every component present in reaction (12.1) is known, the change of enthalpy during the entire reaction is

$$\Delta_r H = \sum_k \nu_k \bar{H}_k(P,T), \tag{12.9}$$

where, as before, ν_k is the stoichiometric coefficient (a positive or a negative number) of the kth component (a product or a reactant) present in the stoichiometric equation for the reaction.

The reader should notice that no discussion of pressure P in relationship (12.9) is needed when a considered component is an ideal gas because enthalpies of ideal gases in thermodynamic equilibria are independent of pressure (see Section 7.5). Thus, relationship (12.9) for ideal gas can be written as

$$\Delta_r H = \sum_k v_k \bar{H}_k (\text{any pressure}, T) = \sum_k v_k \bar{H}_k (P^\circ, T), \qquad (12.10)$$

where the temperature-dependent absolute values of molar enthalpies of ideal gases at standard pressure P° ($\bar{H}_k(P^\circ, T) \equiv \bar{H}_k^\circ(T)$) are available in thermo-chemical databases discussed in Chapter 8.

When a component participating in a chemical reaction is a non-ideal substance, then the enthalpy (see Section 8.2.4) of the component can be found using the *thermodynamic departure factors* discussed in Section 8.3.

The molar entropy of the kth component at pressure P and temperature T is (see expression (8.30))

$$\bar{S}_k(P, T) = \bar{S}_k^\circ(T) + [\bar{S}_k(P, T) - \bar{S}_k^\circ(T)]. \qquad (12.11)$$

where $[\bar{S}_k(P, T) - \bar{S}_k^\circ(T_\circ)]$ is the component's sensible entropy.

Once the molar entropy $\bar{S}_k(P, T)$ of every component of a chemical reaction is known, the change of the molar entropy during the completed reaction can be given as

$$\Delta_r S(P, T) = \sum_k v_k \bar{S}_k(P, T). \qquad (12.12)$$

One should recall that the entropy of the kth component, $\bar{S}_k(P, T)$, depends on both the component pressure and its temperature even when the component is an ideal gas. In the case of an ideal gas, the molar sensible entropy can be given as (see expression (8.32))

$$[\bar{S}_k(P, T) - \bar{S}_k^\circ(T)] = -R \ln(P/P^\circ). \qquad (12.13)$$

When the component under consideration is a non-ideal substance, the entropy of the component can be found using the departure factors discussed in Section 8.3.

The molar enthalpies $\bar{H}_k(P, T)$ and entropies $\bar{S}_k(P, T)$ can be used to calculate the molar Helmholtz free energies $\bar{A}_k(P, T)$, and the molar Gibbs free energies $\bar{G}_k(P,T)$ because (see expressions (3.119)–(3.121))

$$\bar{U}_k(P,T) = \bar{H}_k(P,T) - (P\bar{V})_k, \tag{12.14}$$

$$\bar{A}_k(P,T) = \bar{U}_k(P,T) - T\bar{S}_k(P,T), \tag{12.15}$$

and

$$\bar{G}_k(P,T) = \bar{H}_k(P,T) - T\bar{S}_k(P,T). \tag{12.16}$$

When the mixture is an ideal gas, relationships (12.14)–(12.16) can be written as

$$\bar{U}_k(P,T) = \bar{H}_k(\text{any pressure},T) - RT = \bar{H}_k^\circ(T) - RT, \tag{12.17}$$

$$\begin{aligned}
\bar{A}_k(P,T) &= \bar{U}_k(\text{any pressure},T) - T\bar{S}_k(P,T) \\
&= \bar{U}_k^\circ(T) - T\bar{S}_k(P,T),
\end{aligned} \tag{12.18}$$

and

$$\begin{aligned}
\bar{G}_k(P,T) &= \bar{H}_k(\text{any pressure},T) - T\bar{S}_k(P,T) \\
&= \bar{H}_k^\circ(T) - T\bar{S}_k(P,T).
\end{aligned} \tag{12.19}$$

Problem 12.1: A complete combustion of 1 mol of ethylene with 3 mols of diatomic oxygen has produced 2 mols of carbon dioxide and 2 mols of water vapor,

$$C_2H_4(g) + 3O_2(g) \rightarrow 2CO_2(g) + 2H_2O(g). \tag{12.20}$$

The reactants and products can be considered ideal gases, and the combustion process takes place at constant pressure $P = 0.5\,\text{bar}$ and constant temperature $T = 1000\,\text{K}$. There is no work transfer during the process, and the changes of the kinetic and potential energies of the reaction components can be ignored. Find the change of enthalpy in reaction (12.20).

Solution: Since the enthalpy of ideal gas does not depend on pressure, the calculations of the molar enthalpies of the reaction's components can be done for any value of pressure. We choose this value to be the standard pressure $P^\circ = 1\,\text{bar}$. Subsequently, we can replace pressure P in relationship (12.9) with P° which leads to the following change of enthalpy during reaction (12.20):

$$\Delta_r H = \sum_k \nu_k \bar{H}_k(\text{any pressure},T) = \sum_k \nu_k \bar{H}_k^\circ(T), \tag{12.21}$$

where

$$\bar{H}_k^\circ(T) = \Delta_f H_k^\circ(T_\circ) + [\bar{H}_k^\circ(T) - \bar{H}_k^\circ(T_\circ)]. \tag{12.22}$$

According to the *NIST-JANAF Thermochemical Tables*, the standard-state ($P^\circ = 1$ bar) enthalpies of formation of the components at temperature $T_\circ = 298.15$ K are

$$\Delta_f H^\circ(T_\circ)_{CO_2(g)} = -393.522 \, \text{kJ/mol}, \tag{12.23}$$

$$\Delta_f H^\circ(T_\circ)_{H_2O(g)} = -241.826 \, \text{kJ/mol}, \tag{12.24}$$

$$\Delta_f H^\circ(T_\circ)_{C_2H_4(g)} = +52.467 \, \text{kJ/mol}, \tag{12.25}$$

$$\Delta_f H^\circ(T_\circ)_{O_2(g)} = 0, \tag{12.26}$$

and the molar sensible enthalpies of the components at $T = 1000$ K and $T_\circ = 298.15$ K are

$$[\bar{H}^\circ(T) - \bar{H}^\circ(T_\circ)]_{CO_2(g)} = 33.397 \, \text{kJ/mol}, \tag{12.27}$$

$$[\bar{H}^\circ(T) - \bar{H}^\circ(T_\circ)]_{H_2O(g)} = 26.00 \, \text{kJ/mol}, \tag{12.28}$$

$$[\bar{H}^\circ(T) - \bar{H}^\circ(T_\circ)]_{C_2H_4(g)} = 50.665 \, \text{kJ/mol}, \tag{12.29}$$

$$[\bar{H}^\circ(T) - \bar{H}^\circ(T_\circ)]_{O_2(g)} = 22.703 \, \text{kJ/mol}. \tag{12.30}$$

Taking the above into account, expression (12.21) gives the following enthalpy change in reaction (12.20):

$$\Delta_r H = \sum_k \nu_k \left\{ \Delta_f H_k^\circ(T_\circ) + [\bar{H}^\circ(T) - \bar{H}_k^\circ(T_\circ)] \right\}$$

$$= (2 \, \text{mol})(-393.522 + 33.397) \, \text{kJ/mol}$$
$$+ (2 \, \text{mol})(-241.826 + 26.00) \, \text{kJ/mol}$$
$$+ (-1 \, \text{mol})(52.427 + 50.665) \, \text{kJ/mol}$$
$$+ (-3 \, \text{mol})(0 + 22.703) = -1.323 \, \text{MJ}. \tag{12.31}$$

Thus, the complete burning of one mol of ethylene in the discussed chemical reactor produces 1.323 MJ of heat.

End of Problem 12.1.

Problem 12.2: The stoichiometric equation for complete combustion of 1 mol of liquid propane and 5 mols of diatomic oxygen is

$$C_3H_8(l) + 5O_2(g) \rightarrow 3CO_2(g) + 4H_2O(g), \tag{12.32}$$

while the stoichiometric equation for the complete combustion of the propane in air is

$$C_3H_8(l)+5[O_2(g)+3.76N_2(g)] \rightarrow 3CO_2(g)+4H_2O(g)+18.8N_2(g), \quad (12.33)$$

where the atmospheric ratio of diatomic nitrogen to diatomic oxygen is taken as equal to 3.76.

Consider a steady-flow chemical reactor where every second one mol of liquid propane that enters the chamber at temperature $T_p = 298.15$ K reacts completely with air that enters the reactor at temperature $T_a = 280$ K. The temperature of the combustion products exiting the reactor is $T_e = 1000$ K. The reactants and products are in some states of local equilibrium. Pressures of the reactants entering the chamber and the products leaving the chamber are unknown. There is no work transfer in the chamber, and the changes of the kinetic and potential energies of the reaction components during the flow can be ignored. All components participating in the combustion process can be treated as ideal gases. Find the rate of heat transfer (in J/s) between the chamber and the environment.

Solution: Since the enthalpies of ideal gases are independent of pressure, the molar enthalpy of the kth gas component present in reaction (12.33) can be given as (see relationship (12.8))

$$\bar{H}_k^\circ(T) = \Delta_f H_k^\circ(T_\circ) + [\bar{H}_k^\circ(T) - \bar{H}_k^\circ(T_\circ)], \quad (12.34)$$

where temperatures $T = T_e$ (the gas products) and $T = T_a$ (the gas reactants).

Similarly to many other liquids, the dependence of enthalpy of the liquid propane on pressure is weak (see Sections 4.2.2 and 9.2.1, and expression (9.19)) and, therefore, can be assumed being pressure-independent. Then, the molar enthalpy of the propane at any practical pressure (we conveniently choose $P = P^\circ = 1$ bar) is

$$\bar{H}_{C_3H_8(l)}^\circ(T_p) = \Delta_f H_{C_3H_8(l)}^\circ(T_\circ) + [\bar{H}_{C_3H_8(l)}^\circ(T_p) - \bar{H}_{C_3H_8(l)}^\circ(T_\circ)], \quad (12.35)$$

where the reference temperature is $T_\circ = 298.15$ K, and the enthalpy of formation of the liquid propane can be given as

$$\Delta_f H_{C_3H_8(l)}^\circ(T_\circ) \simeq \Delta_f H_{C_3H_8(g)}^\circ(T_\circ) - \Delta_{(l \rightarrow g)}\bar{H}_{C_3H_8}^\circ(T_\circ), \quad (12.36)$$

where $\Delta_f H^\circ_{C_3H_8(g)}(T_o) = -104.70\,\text{kJ/mol}$ is the formation enthalpy of propane gas, and $\Delta_{(l\to g)}\bar{H}^\circ_{C_3H_8}(T_o) = +14.80\,\text{kJ/mol}$ is the enthalpy (the latent heat) required to vaporize one mol of saturated liquid propane into saturated vapor at standard pressure P° and reference temperature T_o). Thus,

$$\Delta_f H^\circ_{C_3H_8(l)}(T_o) \simeq -104.70\,\text{kJ/mol} - 14.80\,\text{kJ/mol}$$

$$= -119.50\,\text{kJ/mol}, \tag{12.37}$$

and the molar sensible enthalpy of the liquid propane entering the chamber is

$$[\bar{H}^\circ_{C_3H_8(l)}(T_p = T_o) - \bar{H}^\circ_{C_3H_8(l)}(T_o)] = 0. \tag{12.38}$$

According to the *NIST-JANAF Thermochemical Tables*, the enthalpies of formation of the gases of interest here are

$$\Delta_f H^\circ(T_o)_{O_2(g)} \quad = 0, \tag{12.39}$$
$$\Delta_f H^\circ(T_o)_{N_2(g)} \quad = 0, \tag{12.40}$$
$$\Delta_f H^\circ(T_o)_{CO_2(g)} = -393.522\,\text{kJ/mol}, \tag{12.41}$$
$$\Delta_f H^\circ(T_o)_{H_2O(g)} = -241.826\,\text{kJ/mol}, \tag{12.42}$$

and the related molar sensible enthalpies are ($T_e = 1000\,\text{K}$, $T_a = 280\,\text{K}$)

$$[\bar{H}^\circ(T_a) - \bar{H}^\circ(T_o)]_{O_2(g)} \quad = -0.532\,\text{kJ/mol}, \tag{12.43}$$
$$[\bar{H}^\circ(T_a) - \bar{H}^\circ(T_o)]_{N_2(g)} \quad = -0.528\,\text{kJ/mol}, \tag{12.44}$$
$$[\bar{H}^\circ(T_e) - \bar{H}^\circ(T_o)]_{CO_2(g)} = 33.397\,\text{kJ/mol}, \tag{12.45}$$
$$[\bar{H}^\circ(T_e) - \bar{H}^\circ(T_o)]_{H_2O(g)} = 26.00\,\text{kJ/mol}, \tag{12.46}$$
$$[\bar{H}^\circ(T_e) - \bar{H}^\circ(T_o)]_{N_2(g)} \quad = 21.69\,\text{kJ/mol}. \tag{12.47}$$

Taking the above into account, the enthalpy change during a complete reaction (12.33) is

$$\Delta_r H = \sum_k v_k \left\{ \Delta_f H^\circ_k(T_o) + [\bar{H}^\circ_k(T) - \bar{H}^\circ_k(T_o)] \right\}$$

$$= (3\,\text{mol})(-393.522 + 33.397)\,\text{kJ/mol}$$
$$+ (4\,\text{mol})(-241.826 + 26.00)\,\text{kJ/mol}$$
$$+ (18.8\,\text{mol})(0 + 21.69)\,\text{kJ/mol}$$
$$+ (-1\,\text{mol})(-119.50 + 0)\,\text{kJ/mol}$$
$$+ (-5\,\text{mol})(0 - 0.532)\,\text{kJ/mol}$$
$$+ (-18.8\,\text{mol})(0 - 0.528)\,\text{kJ/mol} = -1.403\,\text{MJ}. \tag{12.48}$$

In a steady flow with negligible changes of kinetic and potential energies of the flowing matter and no work transfer, the heat transfer Q is equal to the enthalpy change during the flow. In other words, $Q = \Delta_r H$ – see expressions (3.44). Thus, our combustion process is exothermic, in agreement with the International Convention for the Sign of Energy (see Section (2.8.2)), because $\Delta_r H$ is negative. The reactor produces 1.403 MJ of heat during every second of its operation.

End of Problem 12.2.

12.3   Adiabatic Flame Temperature

Consider some chemical reactants in a local-equilibrium state 1 (of pressure P_1 and temperature T_1) being transformed during a chemical reaction into some products in another local-equilibrium state 2″ (of pressure $P_{2''}$ and temperature $T_{2''}$). The reaction takes place in a steady flow where changes of kinetic and potential energies of the flowing matter are negligible and there is no work transfer. (Recall that in Classical Thermodynamics the changes of state properties can be accurately calculated only when the initial and final states of the studied processes are equilibrium, quasi-equilibrium, or local-equilibrium states.)

If the *entire* heat produced during the reaction 'stays' within the reaction's space, then the rise of the temperature of the products is the highest possible in the process. Then, the change of enthalpy of the reaction is equal to the heat transfer (the process is shown in Fig. 12.1 as transition 1→2″, also see expressions (12.4) and (12.5)),

$$\Delta_r H = H(P_{2''}, T_{2''})_{\text{products}} - H(P_1, T_1)_{\text{reactants}} = Q = 0, \qquad (12.49)$$

or

$$\sum_j v_j \bar{H}_j (P_{2''}, T_{2''}) - \sum_i |v_i| \bar{H}_i (P_1, T_1) = 0, \qquad (12.50)$$

where, as before, v_j (a positive number) and $\bar{H}_j$ are the number of mols of the jth product and its molar enthalpy, respectively; v_i (a negative number) and $\bar{H}_i$ are the number of mols of the ith reactant and its molar enthalpy, respectively. The sums $\sum_j$ and $\sum_i$ are taken over all products and all reactants, respectively.

Relationship (12.50) can be rewritten as

$$C - D = 0, \qquad (12.51)$$

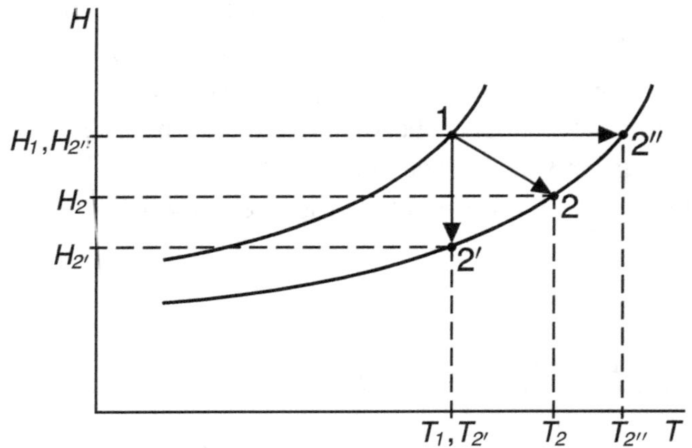

Fig. 12.1: Thermodynamic transitions (marked by the arrows) of reactants in a local-equilibrium state 1 (of pressure P_1 and temperature T_1) to products in a local-equilibrium state 2′ (of pressure $P_{2'}$ and temperature $T_{2'}$), in a local-equilibrium state 2 (of pressure P_2 and temperature T_2), or in a local-equilibrium state 2″ (of pressure $P_{2''}$ and temperature $T_{2''}$). Each of the three transitions is caused by the same chemical reaction but has different constraints: process 1→2′ is an isotherm while process 1→2″ is an adiabat. Symbols H_1, $H_{2'}$, H_2, and $H_{2''}$ denote the enthalpies of states 1, 2′, 2, and 2″, respectively.

where (see expression (12.8))

$$C = \sum_j v_j \left\{ \Delta_f H_j^\circ(T_\circ) + [\bar{H}_j(P_{2''},T_{2''}) - \bar{H}_j^\circ(T_\circ)] \right\}, \qquad (12.52)$$

and

$$D = \sum_i |v_i| \left\{ \Delta_f H_i^\circ(T_\circ) + [\bar{H}_i(P_1,T_1) - \bar{H}_i^\circ(T_\circ)] \right\}. \qquad (12.53)$$

In ideal gases (enthalpy of such gases is pressure-independent), relationships (12.52) and (12.53) become

$$C = \sum_j v_j \left\{ \Delta_f H_j^\circ(T_\circ) + [\bar{H}_j^\circ(T_{2''}) - \bar{H}_j^\circ(T_\circ)] \right\}, \qquad (12.54)$$

and

$$D = \sum_i |v_i| \left\{ \Delta_f H_i^\circ(T_\circ) + [\bar{H}_i^\circ(T_1) - \bar{H}_i^\circ(T_\circ)] \right\}. \qquad (12.55)$$

Solution of equation (12.51) with respect to temperature $T_{2''}$ gives the temperature of the products of the reaction under consideration. This temperature is often called the **adiabatic flame temperature** because it is the maximum temperature that can be reached by the products of the adiabatic reaction.

Problem 12.3: One mol of methane and 2 mols of oxygen (both of temperature $T_1 = 298.15$ K) enter a combustion chamber and burn according to the following chemical reaction:

$$CH_4(g) + 2O_2(g) \rightarrow CO_2(g) + 2H_2O(g). \tag{12.56}$$

The reactants and the products are ideal gases in some local-equilibrium states. Their pressures are unknown but not needed in calculations because enthalpies of ideal gases are pressure-independent. There is neither non-boundary work transfer nor heat transfer between the flow and the environment. The *adiabatic* combustion process is shown in Fig. 12.1 as transition $1 \rightarrow 2''$. Find the adiabatic flame temperature of the combustion products.

Solution: The adiabatic flame temperature of the combustion process can be obtained from relationship (12.51) once the formation enthalpies and the sensible enthalpies of the components of reaction (12.56) are known. These values can be found in the *NIST-JANAF Thermochemical Tables*,

$$\Delta_f H^\circ(T_\circ)_{CH_4(g)} = -74.85 \, \text{kJ/mol}, \tag{12.57}$$

$$\Delta_f H^\circ(T_\circ)_{O_2(g)} = 0, \tag{12.58}$$

$$\Delta_f H^\circ(T_\circ)_{CO_2(g)} = -393.522 \, \text{kJ/mol}, \tag{12.59}$$

$$\Delta_f H^\circ(T_\circ)_{H_2O(g)} = -241.826 \, \text{kJ/mol}, \tag{12.60}$$

and the corresponding molar sensible enthalpies are:

$$[\bar{H}^\circ(T_1 = T_\circ) - \bar{H}^\circ(T_\circ)]_{CH_4(g)} = 0, \tag{12.61}$$

$$[\bar{H}^\circ(T_1 = T_\circ) - \bar{H}^\circ(T_\circ)]_{O_2(g)} = 0, \tag{12.62}$$

$$[\bar{H}^\circ(T_{2''}) - \bar{H}^\circ(T_\circ)]_{CO_2(g)} = ?, \tag{12.63}$$

$$[\bar{H}^\circ(T_{2''}) - \bar{H}^\circ(T_\circ)]_{H_2O(g)} = ? \tag{12.64}$$

Expressions (12.54) and (12.55) can now be written as

$$C = (1 \, \text{mol}) \left\{ -393.522 + [\bar{H}^\circ(T_{2''}) - \bar{H}^\circ(T_\circ)]_{CO_2(g)} \right\} \, \text{kJ/mol}$$
$$+ (2 \, \text{mol}) \left\{ -241.826 + [\bar{H}^\circ(T_{2''}) - \bar{H}^\circ(T_\circ)]_{H_2O(g)} \right\} \, \text{kJ/mol}, \tag{12.65}$$

and

$$D = (1 \text{ mol}) \{-74.850 + 0\} \text{ kJ/mol} + (2 \text{ mol}) \{0 + 0\} \text{ kJ/mol}. \qquad (12.66)$$

Thus, equation (12.51) can be given as

$$(1 \text{ mol}) \left\{ [\bar{H}^\circ(T_{2''}) - \bar{H}^\circ(T_\circ)]_{CO_2(g)} \right\} \text{ kJ/mol}$$
$$+ (2 \text{ mol}) \left\{ [\bar{H}^\circ(T_{2''}) - \bar{H}^\circ(T_\circ)]_{H_2O(g)} \right\} \text{ kJ/mol} = 802.352 \text{ kJ}. \qquad (12.67)$$

We can solve equation (12.67) by analyzing the values of the sensible enthalpies at different values of temperature $T_{2''}$ (see the *NIST-JANAF Thermochemical Tables*). One can see there that the left-hand side of the equation is $(1 \text{ mol})(279.283 \text{ kJ/mol}) + (2 \text{ mol})(242.313 \text{ kJ/mol}) = 800.879 \text{ kJ}$ when $T_{2''} = 5.000$ K. This is close enough to 802.35 kJ to say that the flame temperature of the products leaving our reactor is very close to 5000 K.

The calculated adiabatic flame temperature is the *theoretical* upper limit of the exit temperature of the combustion products. In real systems, the product's temperature is always lower than the theoretical value because no chemical reactor can be perfectly insulated. In addition, dissociation of some (or all) reaction products at high temperature significantly decreases the amount of heat produced inside the reactor because dissociative processes are endothermic.

One should notice that burning of the methane in air, instead of in oxygen, can be represented by the following stoichiometric equation:

$$CH_4(g) + 2O_2(g) + 7.52N_2(g) \rightarrow CO_2(g) + 2H_2O(g) + 7.52N_2(g). \qquad (12.68)$$

Such a combustion process also would lower the adiabatic temperature of the reaction's products because the right-hand side of the reaction has an additional term representing the 'hot' nitrogen (product) of a non-zero enthalpy, while the term representing the 'cold' nitrogen (reactant) on the reaction's left-hand side is zero at temperature $T_1 = T_\circ$.

End of Problem 12.3.

12.4   Chemical Equilibrium in Mixtures

Properties of the mixtures in equilibrium can be effectively studied using the concept of the Helmholtz free energy A (see expression (3.120)) or the Gibbs free energy G (see expression (3.121)). The choice of the free energy depends on the kind of thermodynamic variables used to describe the mixture's properties. If V (volume) and T (temperature) are used for the description, then use of the Helmholtz free energy is usually mathematically more convenient than use of the Gibbs free energy. If P (pressure) and T are used for the description, then use of the Gibbs free energy is usually recommended. The latter pair of thermodynamic variables is used below.

We first consider an equilibrium mixture of several components (pure substances) that are produced by a single chemical reaction. Such a mixture will remain in a complete macroscopic (thermal+mechanical+chemical) equilibrium at (constant) temperature T and (constant) pressure P when the mixture's Gibbs free energy does not change, that is, when

$$\Delta_r G(P, T, v_1, v_2, ...) = \sum_k v_k \mu_k = 0, \tag{12.69}$$

where v_k is the number of mols of the mixture's kth component, the sum is taken over all reactants ($v_k < 0$) and all products ($v_k > 0$) of the reaction, and

$$\mu_k = \left(\frac{\partial \bar{G}_k}{\partial v_k} \right)_{P,T,v_{i \neq k}}, \tag{12.70}$$

is the so-called **chemical potential** of the kth component of the reaction. As before, an overbar is used in the text to denote molar values of the properties of interest. (Note that the chemical potential is, by definition, a molar quantity.) The subscript $P, T, v_{i \neq k}$ of the partial derivative with respect to v_k indicates that the derivative is taken at constant values of P, T, and all mol numbers v_i except one for which $i = k$.

Summarizing the above discussion, one can say that the **conditions for the existence of a complete macroscopic equilibrium in a mixture produced by a single chemical reaction are**

$$T = \text{const}, \tag{12.71}$$

$$P = \text{const}, \tag{12.72}$$

and

$$\sum_k v_k \mu_k = 0. \tag{12.73}$$

Solution of equations (12.71)–(12.73) yields the equilibrium composition of the mixture produced by a single chemical reaction. In many applications, mixtures are produced not by one but by many simultaneous chemical reactions. In such a case, the equilibrium conditions (12.71)–(12.72) are the same as for the single-reaction system, but condition (12.73) is replaced by a set of similar relationships, one for each reaction occurring in the mixture. Subsequently, the equilibrium composition of the multi-reaction mixture can be obtained from solution of the larger set of equations.

12.5   Chemical Activity and Fugacity

The chemical potential of the kth component (a reactant or a product) of a chemical reaction in a gas mixture in equilibrium (or local equilibrium) at pressure P and temperature T can be expressed by the component's standard-state chemical potential μ_k° calculated at the mixture temperature T and the standard-state pressure P° which can be different from the mixture pressure P. (As discussed earlier, a pressure of 1 bar is usually chosen for the standard-state pressure P°.) This can be done by noting that the differential change of the chemical potential μ_k caused by differential changes of the mixture's temperature T and pressure P is (see relationship (3.118))

$$d\mu_k = -\bar{S}_k dT + \bar{V}_k dP, \qquad (12.74)$$

where $\bar{S}_k$ and $\bar{V}_k$ are the molar entropy and molar volume of the kth component, respectively. If the temperature of the mixture remains constant ($dT = 0$), and if the component is an ideal gas, then

$$d\mu_k = \frac{RT}{P} dP. \qquad (12.75)$$

Integrating this relationship from P° to P_k (the partial pressure of the kth component – see Dalton's Law disussed in Section 9.3) at constant temperature T gives the following change of the chemical potential of the kth component of any ideal-gas mixture:

$$\mu_k - \mu_k^\circ = RT \ln(P_k/P^\circ). \qquad (12.76)$$

Relationship (12.76) is valid for ideal gas only. However, Gilbert Lewis (1875–1946) proposed a similar relationship for isothermal change of the chemical potential of *non-ideal mixtures* produced by single chemical reactions. The proposed relationship (later proven to be valid in many thermochemical applications) was

$$\mu_k - \mu_k^\circ = RT \ln a_k, \tag{12.77}$$

where a_k is called **chemical activity** (or just **activity**) of the kth component of the mixture.

The chemical activity of the kth component of a mixture is defined as

$$a_k = f_k / f_k^\circ, \tag{12.78}$$

where f_k is the **fugacity** of the component at temperature T and partial pressure P_k, and f_k° is the fugacity of the component at temperature T and the standard-state pressure P°.

Fugacities of liquids are difficult to calculate, but expression (12.78) can be simplified when the components are gases. The activity of a real gas is

$$a_k = f_k / P^\circ, \tag{12.79}$$

where, using the approach discussed in Section 8.3, the fugacity f_k at temperature T and pressure P_k can be given as

$$f_k \simeq P_k \exp\left[\int_{P_{\text{low}}}^{P_k} \frac{Z_k(P_k, T) - 1}{P_k} dP_k\right], \tag{12.80}$$

where $Z_k(P_k, T)$ is the compressibility factor for the kth component (this factor can be obtained from the Principle of Corresponding States (see Section 5.4)), and P_{low} is some pressure low enough for the component to be considered an ideal gas at temperature T. Thus, the fugacities and chemical activities of the components of real-gas mixtures can be calculated if reliable equations of state and partial pressures of the components are known.

The fugacity of the kth component of an ideal-gas mixture can be taken as the component's partial pressure P_k obtainable from Dalton's Law (see Section 9.3),

$$f_k = P_k = z_k P, \tag{12.81}$$

(z_k is the mol fraction of the kth component) and, according to expression (12.79), the chemical activity of the component is

$$a_k = P_k / P^\circ = z_k P / P^\circ. \tag{12.82}$$

12.6 Equilibrium Constant

On the microscopic (molecular) level, the reactants of the chemical reaction producing a mixture interact with one another creating some products (these are so-called 'forward reactions'). At the same time, the products undergo 'backward reactions' which reproduce the reactants. The mixture remains in a complete (mechanical+thermal+chemical) macroscopic equilibrium when the mixture pressure P and temperature T are constant, and the rates of the 'forward' and 'backward' processes are the same.

The relationship between $\Delta_r H_{\text{forward}}$ (the enthalpy change due to the 'forward reactions') and $\Delta_r H_{\text{backward}}$ (the enthalpy change due to the 'backward reactions') when the mixture is in a complete equilibrium is

$$\Delta_r H_{\text{backward}} = -\Delta_r H_{\text{forward}}, \tag{12.83}$$

and the corresponding changes of the mixture's Gibbs free energy are

$$\Delta_r G_{\text{backward}} = -\Delta_r G_{\text{forward}}. \tag{12.84}$$

According to postulate (12.77), one can write for a mixture produced by a single chemical reaction:

$$\sum_k v_k \mu_k - \sum_k v_k \mu_k^\circ = RT \sum_k v_k \ln a_k, \tag{12.85}$$

where, as before, v_k is the stoichiometric coefficient (a positive or negative number) of the reaction's kth component (reactant or product), and the sums account for contributions of all components of the mixture.

The change of the Gibbs free energy during a single chemical reaction at pressure P and temperature T, and the change of the Gibbs free energy during the reaction at pressure P° and temperature T are, respectively (see expression (12.69)),

$$\Delta_r G = \sum_k v_k \mu_k \quad \text{and} \quad \Delta_r G^\circ = \sum_k v_k \mu_k^\circ. \tag{12.86}$$

Since

$$RT \sum_k v_k \ln a_k = RT \ln \left[\prod_k a_k^{v_k} \right], \tag{12.87}$$

relationship (12.85) can be written as

$$RT \ln \left[\prod_k a_k^{\nu_k} \right] = \Delta_r G - \Delta_r G^\circ, \tag{12.88}$$

where the product $\prod_k$, taken over all components of the reaction that makes the mixture, is sometimes called K_p,

$$K_p = \prod_k a_k^{\nu_k}. \tag{12.89}$$

Since the Gibbs free energy in equilibrium mixture does not change (see expression (12.69)),

$$\Delta_r G = 0, \tag{12.90}$$

one has

$$RT \ln K_r = -\Delta_r G^\circ, \tag{12.91}$$

or

$$K_r = \exp[-\Delta_r G^\circ / RT], \tag{12.92}$$

which is called the **equilibrium constant** of a single chemical reaction (12.1) producing a mixture in macroscopic equilibrium. The values of the constant for various thermochemical systems are available in literature.

Sometimes it is useful to use the concept of the equilibrium constant K_r to relate the rates of the forward and backward reactions in mixtures in complete equilibrium. The relationship is

$$K_r(T)_{\text{backward}} = 1/K_r(T)_{\text{forward}}. \tag{12.93}$$

Thus, in a mixture in equilibrium, the backward reaction is usually 'slow' ('fast') if the corresponding forward reaction is 'fast' ('slow'). The magnitude of the difference between the rates of the forward and backward reactions decides about the chemical composition of the equilibrium mixture produced by the processes. If the forward reaction is fast (slow), then the fraction of the reaction's products in the mixture is high (low) but the fraction of the reaction's reactants is low (high).

Equilibrium constant (12.92) can be used as a measure of the intensity ('speed') of a single chemical reaction that starts as a set of reactants, follows the reaction's stoichiometric equation, and ends as a set of products. This is because K_r is proportional to the amount of the products and reversely proportional to the amount of the reactants – see expressions (12.107) and (12.111). Thus, a reaction can be called 'fast' when its equilibrium constant is much greater than one. Then, most of the particles of the equilibrium mixture are the reaction's products. Similarly, a reaction can be called 'slow' when its equilibrium constant is much smaller than one. In the latter case, most of the particles of the mixture are the reaction's reactants. In other words,

$$\text{fast reaction:} \quad \Delta_r G^\circ \ll 0 \quad \rightarrow \quad K_r \gg 1, \tag{12.94}$$

and

$$\text{slow reaction:} \quad \Delta_r G^\circ \gg 0 \quad \rightarrow \quad K_r \ll 1. \tag{12.95}$$

Such definite statements cannot be made for the reactions where equilibrium constants are close to one. As can be seen in Fig. 12.2, $\Delta_r G^\circ$ for a chemical reaction producing a particular mixture can be negative at one value of temperature T but positive at another value of T.

It should be noticed (see expression (4.75)) that in a single chemical reaction producing an equilibrium mixture at standard-state pressure P° and temperature T, one has

$$\frac{d[\Delta_r G^\circ(T)]}{dT} = -\Delta_r S^\circ(T), \tag{12.96}$$

where $\Delta_r S^\circ$ is the entropy change during the reaction. The definition of the Gibbs free energy gives

$$\Delta_r G^\circ(T) = \Delta_r H^\circ(T) - T\Delta_r S^\circ(T), \tag{12.97}$$

and

$$\begin{aligned}
\frac{d}{dT}\left[\frac{\Delta_r G^\circ(T)}{T}\right] &= \frac{1}{T}\frac{d}{dT}[\Delta_r G^\circ(T)] - \frac{\Delta_r G^\circ(T)}{T^2} \\
&= -\frac{\Delta_r S^\circ(T)}{T} - \frac{\Delta_r G^\circ(T)}{T^2} \\
&= -\frac{\Delta_r H^\circ(T)}{T^2}.
\end{aligned} \tag{12.98}$$

Expressions (12.92) and (12.98) lead to

$$\frac{d}{dT}[\ln K_r] = -\frac{1}{R}\frac{d}{dT}\left[\frac{\Delta_r G^\circ(T)}{T}\right], \tag{12.99}$$

and

$$\frac{d}{dT}\left[\ln K_r\right] = \frac{\Delta_r H^\circ(T)}{RT^2}. \tag{12.100}$$

Relationship (12.100) is called the **Van't Hoff Equation**. Integrating the relationship from the reference temperature $T_\circ$ to the actual temperature T gives

$$\ln\left[\frac{K_r(T)}{K_r(T_\circ)}\right] = \int_{T_\circ}^{T} \frac{\Delta_r H^\circ(T)}{RT^2}\,dT, \tag{12.101}$$

where (see expression (12.9))

$$\Delta_r H^\circ(T) = \sum_k v_k \bar{H}_k^\circ(T), \tag{12.102}$$

and, as before, $\bar{H}_k^\circ(T)$ is the standard-state molar enthalpy of the kth component.

Values of the standard-state molar enthalpies $\bar{H}_k^\circ(T)$ for a large number of components in a broad range of temperatures have been tabulated in thermo-chemical databases discussed in Chapter 8. It is obvious from inspection of the data that $\Delta_r H^\circ(T)$ is usually a weak function of T when T is greater than the reference temperature $T_\circ = 298.15$ K. Therefore, one can assume at such temperatures that

$$\Delta_r H^\circ(T) = \text{const} = \Delta_r H^\circ(T = T_\circ = 298.15\,\text{K}). \tag{12.103}$$

Then, expression (12.101) can be written as

$$\ln\left[\frac{K_r(T)}{K_r(T_\circ)}\right] = \frac{\Delta_r H^\circ(T_\circ)}{R}\left(\frac{1}{T_\circ} - \frac{1}{T}\right). \tag{12.104}$$

Relationship (12.104) is a linear function of $1/T$, a fact used in drawing Fig. 12.2 where the temperature-dependence for several chemical reactions is shown; note that $K_r = K_p$ – see expression (12.107). One can see that some reactions (for example $C+0.5O_2 \rightarrow CO$ and $N+N \rightarrow N_2$) are fast ($K_r \gg 1$) in the entire practical range of temperature, while some other reactions (such as $0.5H_2+O_2 \rightarrow HO_2$ and $0.5N_2+0.5O_2 \rightarrow NO$) are always slow ($K_r \ll 1$). One can see in the figure that the speeds of some reactions depend (sometimes strongly) on their temperature, and that some of them can be slow at some temperatures but fast at other temperatures. For instance, reaction $C+2H_2O \rightarrow CO_2+2H_2$ is fast at

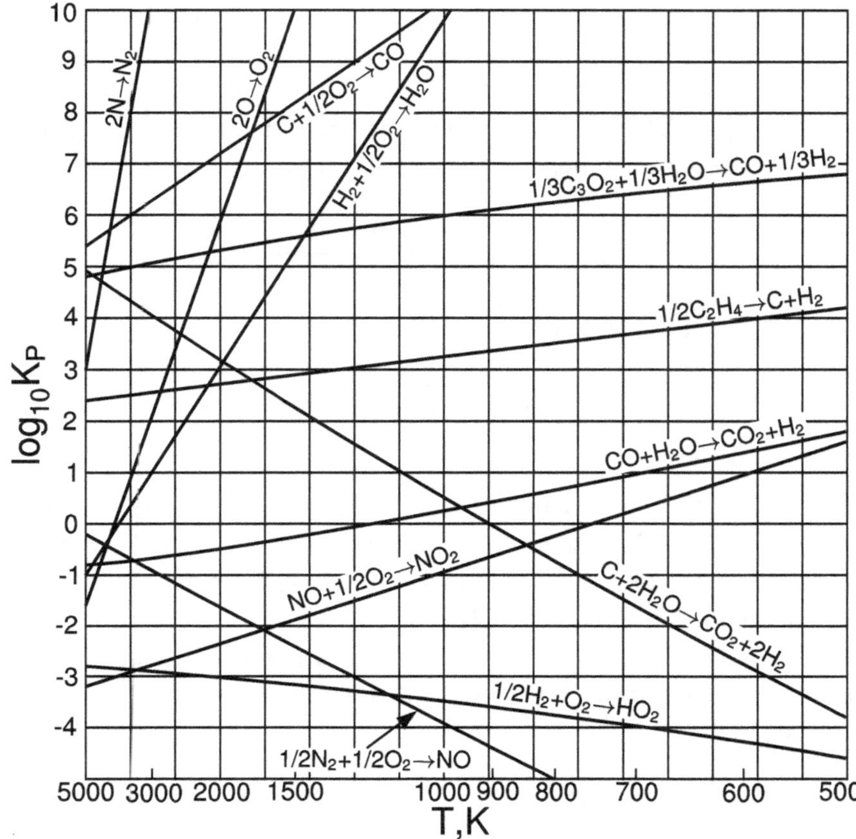

Fig. 12.2: The temperature-dependence of the equilibrium constant K_p ($K_p = K_r$) for several chemical reactions. Some of the reactions are 'fast' ($K_p \gg 1$) while others are 'slow' ($K_p \ll 1$) across a broad range of temperature. The 'speeds' of some of the reactions change significantly with temperature.

high temperatures but slow at low temperatures, but reaction $H_2+0.5O_2\rightarrow H_2O$ is slow at high temperatures but fast at low temperatures.

The thermochemical theory of chemical reactions discussed above assumes that most particles participating in the reactions are in the ground or very low levels of intramolecular energy. However, at higher temperatures, a significant fraction of the particles participating in chemical reactions is often excited to energy levels (rotational, vibrational, electronic) higher than the ground level.

Then, the energy transfers involving the excited particles can have an important impact on the reactions' intensities.

Expression (12.100) and the International Convention for the Sign of Energy tell us that a chemical reaction is exothermic (releasing heat) when the derivative of its constant K_r with respect to temperature is negative, and that the reaction is endothermic (absorbing heat) when the derivative is positive. In other words,

$$\text{exothermic reaction:} \qquad \Delta_r H° < 0 \;\rightarrow\; \frac{d}{dT}\left[\ln K_r\right] < 0, \qquad (12.105)$$

and

$$\text{endothermic reaction:} \qquad \Delta_r H° > 0 \;\rightarrow\; \frac{d}{dT}\left[\ln K_r\right] > 0. \qquad (12.106)$$

The equilibrium constants for chemical reactions defined by their stoichiometric equations (12.1) and taking place in an ideal-gas mixture in macroscopic equilibrium at pressure P and temperature T can be given as (see expression (12.89)):

$$K_r = K_p = \prod_k \left(\frac{P_k}{P°}\right)^{\nu_k} = \left(\frac{P_k}{P°}\right)^{\sum_k \nu_k}, \qquad (12.107)$$

where, as before, P_k is the partial pressure of the kth component of the mixture, and the constant K_p is sometimes called the **partial pressure equilibrium constant of ideal-gas mixture**.

We considered above only mixtures produced by a *single* chemical reaction. However, macroscopic equilibria of mixtures are usually sustained by many reactions. Therefore, the number of equilibrium constants (one for each reaction that contributes to producing the mixture) required to study such mixtures can be significant. This and auxiliary requirements, such as the conservation of matter (that is, conservation of the number of atoms, both free and bound in molecules) and the conservation of energy, allow one to develop computational procedures able to calculate the chemical compositions of multi-reaction mixtures in equilibrium. These procedures are available as commercial software packages – see Problem 12.5.

Problem 12.4: Calculate particle density of NO molecules in unpolluted air in equilibrium at $P = 1$ bar and $T = 298.15$ K, assuming that the molecules are produced by the following chemical reaction:

$$\frac{1}{2}N_2(g) + \frac{1}{2}O_2(g) = NO(g). \qquad (12.108)$$

Solution: The stoichiometric coefficients of chemical reaction (12.108) are $\nu_{N_2} = -1/2$, $\nu_{O_2} = -1/2$, and $\nu_{NO} = +1$. Since the pressure of the air is not high, one can assume that the air is an ideal-gas mixture.

One should notice that reaction (12.108) is just the reaction of formation of $NO(g)$ since the reactants $N_2(g)$ and $O_2(g)$ are chemically stable elements at pressure $P^\circ = 1$ bar and temperature $T = 298.15$ K (see Section 8.2.3). Thus, according to the *NIST-JANAF Thermochemical Tables*, the change of the Gibbs free energy during the reaction is

$$\Delta_r G^\circ (T = 298.15 \, \text{K}) = \Delta_f G^\circ (T = 298.15 \, \text{K})$$
$$= 86.57 \times 10^3 \, \text{J/mol}, \tag{12.109}$$

and, according to expression (12.92), the equilibrium constant for the reaction is

$$K_r = \exp\left[-\frac{86.57 \times 10^3 \, \text{J/mol}}{(8.314 \, \text{J/mol·K})(298.15 \, \text{K})}\right] = 6.804 \times 10^{-16}. \tag{12.110}$$

Subsequently, expression (12.107) leads to

$$K_p = \frac{(P_{NO}/P^\circ)^1}{(P_{N_2}/P^\circ)^{1/2}(P_{O_2}/P^\circ)^{1/2}} = \frac{P_{NO}/(10^5 \, \text{Pa})}{(0.78)^{1/2}(0.21)^{1/2}} = K_r, \tag{12.111}$$

where we used Dalton's Law for ideal-gas mixtures (see Section 9.3) to calculate the partial pressures P_{N_2} and P_{O_2} (the fractions of nitrogen and oxygen in the air are close to 78% and 21%, respectively).

Solution of equation (12.111) gives $P_{NO} = 2.75 \times 10^{-16}$ bar. This and the equation of state for ideal gas give the particle density of NO molecules,

$$n_{NO} = \frac{P_{NO}}{kT} = \frac{2.75 \times 10^{-11} \, \text{Pa}}{(1.381 \times 10^{-23} \, \text{J/K})(298.15 \, \text{K})}$$
$$= 6.69 \times 10^9 \, \text{particles/m}^3, \tag{12.112}$$

while the density of particles in the air at pressure $P = 1$ bar and temperature $T = 298.15$ K is about 3×10^{25} particles/m^3 (see Problem 5.3). This is many orders of magnitude more than the particle density of nitrogen monoxide. Even though some insignificant amount of nitrogen monoxide is produced in the air by chemical reactions other than reaction (12.108), it can be said that only a very small amount of nitrogen monoxide gas is present in unpolluted air. This was already suggested by the very small value of the reaction's equilibrium constant (12.110).

End of Problem 12.4.

Problem 12.5: Consider two different ideal-gas mixtures in macroscopic equilibria produced by the following reactions:

> Mixture A: (some amounts of reactants SO_2 and H_2) $\rightleftharpoons$ (some amounts of possible products in thermodynamic equilibrium).
> Mixture B: (some amounts of reactants N_2 and O_2) $\rightleftharpoons$ (some amounts of possible products in thermodynamic equilibrium).

The making of mixture A starts with the following mass fractions of the reactants:

$$g_{SO_2} = \frac{M_{SO_2}}{M_{SO_2} + M_{H_2}} = 0.8 \quad \text{and} \quad g_{H_2} = \frac{M_{H_2}}{M_{SO_2} + M_{H_2}} = 0.2, \qquad (12.113)$$

(M_i is the mass of the ith reactant), while the making of mixture B starts with the following mass fractions of the reactants:

$$g_{N_2} = \frac{M_{N_2}}{M_{N_2} + M_{O_2}} = 0.8 \quad \text{and} \quad g_{O_2} = \frac{M_{O_2}}{M_{N_2} + M_{O_2}} = 0.2. \qquad (12.114)$$

Find the chemical compositions of the mixtures (the mass fractions of their components) when they reach states of macroscopic equilibrium at pressure $P = 1$ bar and temperature $T = 1000\,\text{K}$.

Solution: The equilibrium compositions of the mixtures can be found using a commercial software able to analyze equilibrium criteria for the mixtures produced by many chemical reactions. One such software, CEARUN, is available on the NASA Glenn Research Center Web Site (www.grc.nasa.gov/WWW/ CEAWeb/ceaHome.htm). CEARUN requires input of the initial mass (or molar) fractions of the reactants used to produce the mixture of interest but does not require specification of all the reactions that can produce the mixture. (The number of such reactions can be large and the degrees of their importance can be quite different.) The software analyzes properties (pressure P, temperature T, and the minima of the Gibbs free energy) of all mixtures that can possibly be made from the particular amounts of the reactants and selects the mixture composition that meets the equilibrium criteria.

Use of the CEARUN software to study mixture A (with initial mass fractions (12.113)) leads to the following composition of the mixture in equilibrium at $P = 1$ bar and $T = 1000$ K:

$$g_{SO_2} \simeq 0, \quad g_{H_2} \simeq 0, \quad g_{H_2S} = 0.478, \quad g_{H_2O} = 0.506, \qquad (12.115)$$

plus some negligibly small amounts of other components. Thus, the equilibrium mixture A at $P = 1$ bar and $T = 1000$ K practically does not have components $SO_2(g)$ and $H_2(g)$.

Use of the CEARUN software to analyze mixture B (with the initial mass fractions (12.114)) leads to the following composition of the mixture in equilibrium at 1 bar and 1000 K:

$$g_{N_2} = 0.799, \quad g_{O_2} = 0.199, \qquad (12.116)$$

plus negligible amounts of N, O, and NO. Thus, the equilibrium mixture B at $P = 1$ bar and $T = 1000$ K does not have meaningful amounts of components other than $N_2(g)$ and $O_2(g)$; note that this conclusion agrees with the results of Problem 12.4.

End of Problem 12.5.

Chapter 13

Irreversibilities

13.1 Dissipation of Energy

As discussed in Section 2.3, the always-present spontaneous production of natural entropy makes every real process irreversible, that is, a process that cannot *spontaneously* (or by any infinitesimally weak activity) return both the system and the environment to their initial states. (However, such a return can take place in reversible processes which are theoretical idealizations of real processes.)

Production of the 'natural' entropy S_{nat} in real processes is caused by the always-present particle-particle, particle-photon, and photon-photon interactions which often are called 'collisions'. (The photon-photon interactions are usually very weak, and contribution of photon-particle interactions to system entropy is negligible in most thermodynamic applications.) The always-present collisions inside the system always try to bring the system closer to equilibrium (that is, increase the entropy of the system), but the system's total entropy S can also be affected (increased or decreased) by transfers of matter, heat, and radiation (each 'carrying' its own 'partial' entropy) between the system and the environment.

When some work is transferred between the system and its environment during a real process, a very important phenomenon, the so-called **dissipation of energy**, takes place in the system. (Dissipation of energy does not occur in reversible processes.) The dissipated energy does not disappear from the system but is moved by the system from its work-designated part to its heat-designated part. Thus, dissipation of energy does not violate the First Law of Thermodynamics which requires conservation of the system's total energy. Since work is more important than heat to users of many thermodynamic devices, the dissipation effects are often considered negative phenomena. The amount of the energy shifted from the work-designated part to the heat-designated part is called the **lost work** (see Section 13.2). The magnitude of the lost work depends on the kind of changes of the thermodynamic conditions during the process under consideration, on the kind and amount of the substance undergoing the process, on the amount of the internal energy of the system, and on the degree of the irreversibility of the process.

Consider a closed nonflow system that undergoes an elementary transition between two states of equilibrium while exchanging some heat δQ and some (boundary plus non-boundary) work δL with the environment. According to the First Law of Thermodynamics, the change of the system thermal energy during the transition is (see expressions (3.123)–(3.124), and (3.168))

$$dU_{\text{rev}} = \delta Q_{\text{rev}} + \delta L_{\text{rev}} \qquad \text{reversible transition,} \qquad (13.1)$$

$$dU_{\text{irr}} = \delta Q_{\text{irr}} + \delta L_{\text{irr}} \qquad \text{irreversible transition,} \qquad (13.2)$$

while

$$dU_{\text{irr}} = dU_{\text{rev}}, \qquad (13.3)$$

because the thermal energy of the system is a function of state. As before, δY and dY denote inexact and exact differentials of function Y, respectively. Thus,

$$\delta Q_{\text{irr}} + \delta L_{\text{irr}} = \delta Q_{\text{rev}} + \delta L_{\text{rev}}. \qquad (13.4)$$

One should notice that relationship (13.4) does not prevent relationships

$$\delta Q_{\text{irr}} \neq \delta Q_{\text{rev}} \qquad \text{and} \qquad \delta L_{\text{irr}} \neq \delta L_{\text{rev}} \qquad (13.5)$$

from being true in all processes.

The amount of work needed to be done on the system to move it irreversibly between two thermodynamic states is always greater than the amount of work that would be needed to move the system reversibly between the states. In other words,

$$|\delta L_{\text{irr}}| > |\delta L_{\text{rev}}| \qquad \text{work is done on the system.} \qquad (13.6)$$

The amount of work done by the system during its irreversible transition between two states is always smaller than the amount of the work that would be done by the system moving reversibly between the states, that is,

$$|\delta L_{\text{irr}}| < |\delta L_{\text{rev}}| \qquad \text{work is done by the system.} \qquad (13.7)$$

The International Convention for the Sign of Energy (see Fig. 2.3) allows one to combine relationships (13.6) and (13.7) into the following single expression valid for work done by or done on the system undergoing any thermodynamic process:

$$\delta L_{\mathrm{irr}} > \delta L_{\mathrm{rev}}. \tag{13.8}$$

Since the reversible work is the theoretical maximum of the entire work available in a real process, it is often called the **available work** L_{avail},

$$L_{\mathrm{avail}} \equiv L_{\mathrm{rev}}. \tag{13.9}$$

The above discussion suggests the following statement about dissipation of energy in real thermodynamic systems (also see Section 2.3): **in every real (that is, irreversible) process designed to produce some amount of work, a (usually not large) fraction of the system's internal energy is always 'dissipated', that is, 'shifted' from the work-designated part of the energy to its heat-designated part**, despite all the brilliance and best intentions of the designers of the process. The magnitude of the dissipation effect is proportional to the amount of natural entropy produced in the system during the process. When the system reaches a state of equilibrium, the production of the natural entropy vanishes. Thus, **dissipation of energy does not occur in systems in equilibrium and in reversible (quasi-equilibrium) processes** which are sequences of quasi-equilibrium states.

13.2   Available Work and Lost Work

Consider an open thermodynamic system that operates in an irreversible manner while exchanging heat, work, and matter with the environment – see Fig. 13.1. The environment may consist of several 'heat reservoirs' plus a 'surrounding' (see Section 10.1). A heat reservoir can be, for example, a lake or a combustion chamber under thermodynamic conditions qualifying them to be considered as heat reservoirs. One of the reservoirs interacting with the system is the earth's atmosphere which is in equilibrium at pressure P_{o} and temperature T_{o}. (It is often assumed in thermodynamic studies that the values of P_{o} and T_{o} are 1 bar and 298.15 K, respectively.) Such atmosphere is called the **standard atmosphere** or the **Atmosphere**. The Atmosphere is distinguished as a special (separate) part of the environment because its properties are close to common

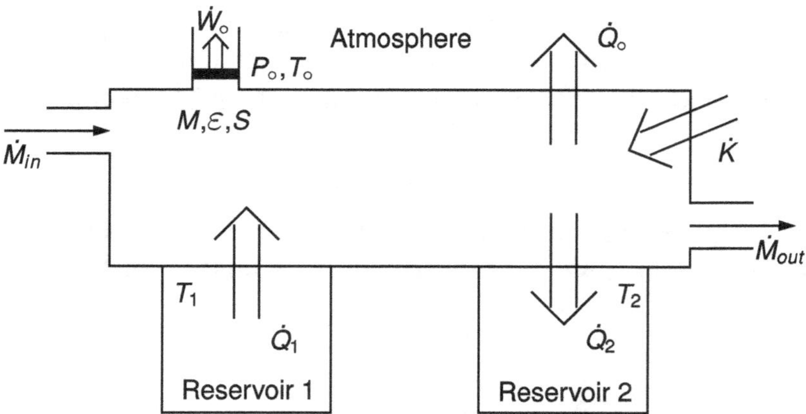

Fig. 13.1: An open thermodynamic system where matter enters the system with mass flow rate $\dot{M}_{in}$ and leaves it with mass flow rate $\dot{M}_{out}$. There are heat transfers between the system and heat reservoirs which are called Atmosphere, Reservoir 1, and Reservoir 2, and the rates of the transfers are $\dot{Q}_0$, $\dot{Q}_1$, and $\dot{Q}_2$, respectively. The work transfers between the system and the Atmosphere are a boundary work (with the transfer rate $\dot{W}_0$) and a non-boundary work (with rate $\dot{K}$); M, $\mathcal{E}$, and S denote the mass of the system matter, its internal energy, and entropy, respectively.

conditions of the earth's atmosphere. Subsequently, the properties are a natural reference point for the studied system and for the other parts of the environment. The heat exchanged between the system and the Atmosphere is Q_0. Since the Atmosphere's temperature T_0 is constant, the Atmosphere's entropy change is Q_0/T_0 – see expression (2.13). The 'surroundings' do not belong to any heat reservoir; they can be, for example, a part of space with a motor that supplies a non-boundary work to the system.

In general, an open system like that shown in Fig. 13.1 can have k_m inlets and j_m outlets, and it can interact with i_m heat reservoirs. The *constant* temperature of the ith heat reservoir is T_i ($i = 1, 2, \ldots, i_m$) while the heat entering or leaving the reservoir is Q_i. Thus, the entropy 'carried' by the heat Q_i changes the reservoir's entropy by Q_i/T_i – see expression (2.13).

The total work of the discussed thermodynamic system is

$$L = W_0 + K, \tag{13.10}$$

where W_0 is the boundary work, and K is the non-boundary work exchanged between the system and the environment. The boundary work is transferred only between the system and the Atmosphere. The non-boundary work is transferred between the system and the surroundings. No work is transferred between the system and the heat reservoirs. (As mentioned earlier, work transfers do not 'carry' entropy.)

Let us study the system defined above using the three Principles of Thermodynamics (the Principle of Conservation of Matter, the Principle of Conservation of Energy, and the Principle of Production of Natural Entropy – see Sections 3.1, 3.2, and 3.3, respectively). This will allow us to estimate the magnitude of the dissipation of energy in the system.

Replacing the theoretical rates of matter, work, and heat by the corresponding practical rates (see Section 3.4), the First Law of Thermodynamics for the system can be written in the following general form (see expression (3.34)):

$$\frac{d\mathcal{E}}{dt} \simeq \sum_{k=1}^{k_m} \dot{M}_k \left(h_k + e_{\mathrm{kin},k} + e_{\mathrm{pot},k} \right) - \sum_{j=1}^{j_m} \dot{M}_j \left(h_j + e_{\mathrm{kin},j} + e_{\mathrm{pot},j} \right)$$

$$+ \dot{L}_{\mathrm{irr}} + \dot{Q}_\mathrm{o} + \sum_{i=1}^{i_m} \dot{Q}_i, \tag{13.11}$$

where $\mathcal{E}$ is the internal energy of the system's matter which must be in some states of local equilibrium at the inlets and outlets because only then the formalism of Classical Thermodynamics can accurately calculate changes of the system's state properties; $\dot{M}_k$, h_k, $e_{\mathrm{kin},k}$, and $e_{\mathrm{pot},k}$ are the mass flow rate, specific enthalpy, specific kinetic energy, and specific potential energy, respectively, of the matter entering the system through its kth inlet; $\dot{M}_j$, h_j, $e_{\mathrm{kin},j}$, and $e_{\mathrm{pot},j}$ are the corresponding properties of the matter leaving the system through its jth outlet. The rate of the net work transferred between the system and the environment is $\dot{L}_{\mathrm{irr}}$, the rate of the heat transfer between the system and the Atmosphere is $\dot{Q}_\mathrm{o}$, and the rate of the heat transfer between the system and the ith reservoir is $\dot{Q}_i$.

The specific kinetic energies of the matter entering the system through the kth inlet and leaving the system through the jth outlet are, respectively,

$$e_{\mathrm{kin},k} = \omega_k^2/2 \qquad \text{and} \qquad e_{\mathrm{kin},j} = \omega_j^2/2, \tag{13.12}$$

and the specific potential energies of the matter entering the system through the kth inlet and leaving the system through the jth outlet are, respectively,

$$e_{\mathrm{pot},k} = gz_k \qquad \text{and} \qquad e_{\mathrm{pot},j} = gz_j, \tag{13.13}$$

where ω_n and z_n are the speed and the elevation, respectively, of the center of mass of the moving matter at the nth inlet/outlet, and $g = 9.81 \text{ m/s}^2$ is the gravitational acceleration.

The Second Law of Thermodynamics for the open system can be written as (see expressions (3.69), (3.70), and (3.71))

$$\frac{dS}{dt} \simeq \sum_{k=1}^{k_m} \dot{M}_k s_k - \sum_{j=1}^{j_m} \dot{M}_j s_j + \dot{Q}_\circ / T_\circ + \sum_{i=1}^{i_m} (\dot{Q}_i / T_i) + \dot{S}_{\text{nat}}, \tag{13.14}$$

where S is the system entropy, s_k and s_j are the specific entropies of the matter entering the system through the kth inlet and leaving it through the jth outlet, respectively, and $\dot{S}_{\text{nat}}$ is the rate at which the natural entropy is produced in the system.

Eliminating $\dot{Q}_\circ$ between expression (13.11) and expression (13.14) leads to the following rate of the net work transfer in the system (note that the rate of the transfer is just the power of the system):

$$\dot{L}_{\text{irr}} = \frac{d}{dt}(\mathcal{E} - T_\circ S) - \sum_{i=1}^{i_m} (1 - T_\circ / T_i)\, \dot{Q}_i$$

$$- \sum_{k=1}^{k_m} \dot{M}_k \left(h_k + e_{\text{kin},k} + e_{\text{pot},k} - T_\circ s_k \right)$$

$$+ \sum_{j=1}^{j_m} \dot{M}_j \left(h_j + e_{\text{kin},j} + e_{\text{pot},j} - T_\circ s_j \right) + T_\circ \dot{S}_{\text{nat}}. \tag{13.15}$$

If the system operates in a reversible way ($\dot{S}_{\text{nat}} = 0$), then relationship (13.15) gives the following *maximum rate* of the net work transfer:

$$\dot{L}_{\text{rev}} = \frac{d}{dt}(\mathcal{E} - T_\circ S) - \sum_{i=1}^{i_m} (1 - T_\circ / T_i)\, \dot{Q}_i$$

$$- \sum_{k=1}^{k_m} \dot{M}_k \left(h_k + e_{\text{kin},k} + e_{\text{pot},k} - T_\circ s_k \right)$$

$$+ \sum_{j=1}^{j_m} \dot{M}_j \left(h_j + e_{\text{kin},j} + e_{\text{pot},j} - T_\circ s_j \right). \tag{13.16}$$

Since $\dot{S}_{nat} > 0$ in all real processes, the irreversibilities cause some dissipation of energy, that is, loss of some work which is called **lost work** L_{lost}. The work is used in thermodynamic studies as a measure of the degree of energy dissipation in real processes. The rate of the lost work is (recall the International Convention for the Sign of Energy)

$$\dot{L}_{lost} = \dot{L}_{irr} - \dot{L}_{rev} = T_{o}\dot{S}_{nat}, \tag{13.17}$$

a relationship known as the **Gouy-Stodola Theorem**.

As already mentioned, expressions (13.15) and (13.16) represent the rates of the *total* (net) work transfers in the discussed system,

$$\dot{L}_{irr} = \dot{W}_{o,irr} + \dot{K}_{irr} \tag{13.18}$$

and

$$\dot{L}_{rev} = \dot{W}_{o,irr} + \dot{K}_{rev}, \tag{13.19}$$

where $\dot{W}_{o,irr}$ and $\dot{W}_{o,rev}$ are rates of the boundary works of the system operating in an irreversible or reversible way, respectively, and $\dot{K}_{irr}$ and $\dot{K}_{rev}$ are rates of the non-boundary work of the system operating in an irreversible or reversible way, respectively.

Since the standard atmosphere's pressure P_{o} and temperature T_{o} are constant, the isobaric-isothermal transfer of boundary work $W_{o,irr}$ is 'seen' by the Atmosphere as a transition with a negligibly small degree of irreversibility – see the discussion leading to expression (2.71). Therefore, the transfer can be treated in calculations as a reversible (quasi-equilibrium) process. In such a process, pressure P on the system's side of the piston in Fig. 13.1 is close to pressure P_{o} on the Atmosphere's side of the piston. Then, one can write

$$\delta W_{o,irr} \simeq \delta W_{o,rev} = -P_{o}dV, \tag{13.20}$$

and the rate of the boundary work transfer between the system and the Atmosphere can be given as

$$\frac{\delta W_{o,irr}}{dt} = \frac{\delta W_{o,rev}}{dt} = -P_{o}\frac{dV}{dt}. \tag{13.21}$$

Relationships (13.15) and (13.16) can now be written as

$$\dot{L}_{\text{irr}} = \frac{d}{dt}(\mathcal{E} + P_\circ V - T_\circ S) - \sum_{i=1}^{i_m}(1 - T_\circ/T_i)\,\dot{Q}_i$$

$$- \sum_{k=1}^{k_m} \dot{M}_k\left(h_k + e_{\text{kin},k} + e_{\text{pot},k} - T_\circ s_k\right)$$

$$+ \sum_{j=1}^{j_m} \dot{M}_j\left(h_j + e_{\text{kin},j} + e_{\text{pot},j} - T_\circ s_j\right) + T_\circ \dot{S}_{\text{nat}}, \tag{13.22}$$

and

$$\dot{L}_{\text{rev}} = \frac{d}{dt}(\mathcal{E} + P_\circ V - T_\circ S) - \sum_{i=1}^{i_m}(1 - T_\circ/T_i)\,\dot{Q}_i$$

$$- \sum_{k=1}^{k_m} \dot{M}_k\left(h_k + e_{\text{kin},k} + e_{\text{pot},k} - T_\circ s_k\right)$$

$$+ \sum_{j=1}^{j_m} \dot{M}_j\left(h_j + e_{\text{kin},j} + e_{\text{pot},j} - T_\circ s_j\right). \tag{13.23}$$

Expressions (13.22) and (13.23) give the rates of actual and available works, respectively, of processes in thermodynamic systems. The work L_{irr} is the **actual work** of the system during its irreversible operation where dissipation of energy always occurs. The work L_{rev} is the maximum theoretically possible **available work** to be produced during the operation. Therefore, the following terminology is often used in the literature of the subject:

$$\text{actual work:} \qquad L_{\text{actual}} = L_{\text{irr}},$$
$$\text{available work:} \qquad L_{\text{avail}} = L_{\text{rev}}. \tag{13.24}$$

Relationships (13.22) and (13.23) also give the rate of the work that is lost in real processes in open systems because of the presence of irreversibilities,

$$\dot{L}_{\text{lost}} = \dot{L}_{\text{irr}} - \dot{L}_{\text{rev}} = \dot{L}_{\text{actual}} - \dot{L}_{\text{avail}} = T_\circ \dot{S}_{\text{nat}}. \tag{13.25}$$

The magnitudes of the available work and actual work depend on the characteristics of the process taking place in the system and on the characteristics of the heat transfers between the heat reservoirs and the system.

13.2.1 Open Systems

As discussed in Section 3.5.2, the total energy, mass, entropy, and volume of the matter inside the control volume of a steady-flow open system do not change with time during the flow. Applying expressions (13.23) and (13.25) to such a system gives

$$\frac{d(\mathcal{E} + P_\circ V - T_\circ S)}{dt} = 0, \tag{13.26}$$

$$\dot{L}_{\text{actual}} = \dot{L}_{\text{avail}} + T_\circ \dot{S}_{\text{nat}}, \tag{13.27}$$

and

$$\dot{L}_{\text{avail}} = - \sum_{i=1}^{i_m} \left(1 - T_\circ / T_i\right) \dot{Q}_i$$

$$- \sum_{k=1}^{k_m} \dot{M}_k \left(h_k + e_{\text{kin},k} + e_{\text{pot},k} - T_\circ s_k\right)$$

$$+ \sum_{j=1}^{j_m} \dot{M}_j \left(h_j + e_{\text{kin},j} + e_{\text{pot},j} - T_\circ s_j\right). \tag{13.28}$$

The actual work L_{actual} is often called the **useful work** or **shaft work**, and the available work L_{avail} is often called the **available useful work** or **available shaft work**. The corresponding rates of the works are called the **useful power**, **shaft power**, and **available power**.

If the steady-flow open system has only one inlet and one outlet (a situation common in applications), then $\dot{M}_{\text{in}} = \dot{M}_{\text{out}} = \dot{M}$, and expression (13.28) becomes

$$\dot{L}_{\text{avail}} = - \dot{M} \left[(h_{\text{in}} - h_{\text{out}}) + (e_{\text{kin,in}} - e_{\text{kin,out}}) + (e_{\text{pot,in}} - e_{\text{pot,out}}) + \right.$$

$$\left. - T_\circ (s_{\text{in}} - s_{\text{out}}) \right] - \sum_{i=1}^{i_m} \left(1 - T_\circ / T_i\right) \dot{Q}_i. \tag{13.29}$$

13.2.2 Closed Systems

In studies of closed systems, expressions (13.22) and (13.23) are integrated over the duration of the process taking place in the system. The process begins in state 1 at time t_1 and ends in state 2 at time t_2. Its actual work is

$$L_{\text{actual}} = L_{\text{avail}} + T_{\text{o}} \Delta S_{\text{nat}}, \tag{13.30}$$

and its available work is (see expression (13.23))

$$L_{\text{avail}} = \Delta \mathcal{E} + P_{\text{o}} \Delta V - T_{\text{o}} \Delta S - \sum_{i=1}^{i_m} (1 - T_{\text{o}}/T_i)\, Q_i, \tag{13.31}$$

where $\Delta V = M(v_2 - v_1)$, M is the mass of the matter locked up inside the system (in general, the matter can be moving inside the system), v_1 and v_2 are the specific volumes of the matter in states 1 and 2, respectively, $\Delta S = M(s_2 - s_1)$ (s_1 and s_2 are the specific entropies of the matter in states 1 and 2, respectively), and $\Delta \mathcal{E}$ is the change of the system internal energy during the process (see expression (3.53)),

$$\Delta \mathcal{E} = M(\varepsilon_2 - \varepsilon_1) = M[(u_2 - u_1) + (\omega_2^2/2 - \omega_1^2/2) + g(z_2 - z_1)], \tag{13.32}$$

where, as before, ε_1, u_1, ω_1, and z_1 are the specific internal energy, specific thermal energy, speed, and elevation, respectively, of the matter at the beginning of the process; ε_2, u_2, ω_2, and z_2 are the corresponding properties of the matter at the end of the process, and g is the gravitational acceleration.

Taking the above into account, expression (13.31) can be written as

$$L_{\text{avail}} = M\left[(u_2 - u_1) + (\omega_2^2/2 - \omega_1^2/2) + g(z_2 - z_1) + P_{\text{o}}(v_2 - v_1) + \right.$$
$$\left. - T_{\text{o}}(s_2 - s_1)\right] - \sum_{i=1}^{i_m} (1 - T_{\text{o}}/T_i)\, Q_i, \tag{13.33}$$

where the terms representing the kinetic and potential energies of the matter become zero in nonflow systems.

One should notice that the amounts of the available and actual works of thermodynamic processes depend on the properties of the initial and final states of the processes and on the properties of the involved heat reservoirs.

13.3 Exergy of Thermodynamic Systems

The available work of most stationary open systems and closed systems can be calculated from expressions (13.29) and (13.33), respectively. Below we study the expressions while introducing the concept of the **dead state** and the related concept of **exergy**.

A system is in the 'dead state' when it is in equilibrium with the standard atmosphere at pressure $P_o = 1$ bar and temperature $T_o = 298.15$ K. The largest amount of work that can be produced by a system in some (initial) thermodynamic state is the available work of the system when it passes from this state to the dead state. The work is called the **exergy** of the initial state.

13.3.1 Open Systems

Consider a steady-flow open system that has only one inlet (the properties of the flowing matter at the inlet are denoted by subscript *in*) and only one outlet (the matter properties at the outlet are denoted by subscript *out*). The properties of the dead state are denoted by subscript o. Since we are using the formalism of Classical Thermodynamics to evaluate thermodynamic functions of state, the matter at the inlet and at the outlet must be in some local-equilibrium states. The power available in such a system is given by relationship (13.29) which can be rewritten as

$$\dot{L}_{\text{avail}} = -\dot{M}(b_{\text{in}} - b_{\text{out}}) - \sum_{i=1}^{i_m} (1 - T_o/T_i)\,\dot{Q}_i, \qquad (13.34)$$

where the term

$$b_{\text{in}} = (h_{\text{in}} - h_o) - T_o(s_{\text{in}} - s_o) + \omega_{\text{in}}^2/2 + gz_{\text{in}}, \qquad (13.35)$$

is called the **specific exergy of the flow at the inlet of the system**, and the term

$$b_{\text{out}} = (h_{\text{out}} - h_o) - T_o(s_{\text{out}} - s_o) + \omega_{\text{out}}^2/2 + gz_{\text{out}}, \qquad (13.36)$$

is called the **specific exergy of the flow at the outlet of the system**.

Every amount of matter (moving or at rest), except the matter in the dead state, has some exergy. Exergy is an additive property, and, as can be seen in expressions (13.35) and (13.36), it is a property of state. The value of the exergy cannot be negative (it is zero when the matter is in its dead state). Also, the entropy of the dead state is always greater than the entropy of every state of the matter on its way towards the dead state.

Relationship (13.34) can be given in the following computationally convenient form:

$$\dot{L}_{\text{avail}} = -\sum_{i=1}^{i_m} (1 - T_\circ/T_i)\, \dot{Q}_i + \dot{M}(b_{\text{out}} - b_{\text{in}})$$

$$= -\sum_{i=1}^{i_m} (1 - T_\circ/T_i)\, \dot{Q}_i + \dot{M}[(h_{\text{out}} - h_{\text{in}})$$

$$- T_\circ(s_{\text{out}} - s_{\text{in}}) + (\omega_{\text{out}}^2/2 - \omega_{\text{in}}^2/2) + g(z_{\text{out}} - z_{\text{in}})], \tag{13.37}$$

and, as before, the system's actual power output is (see expression (13.27))

$$\dot{L}_{\text{actual}} = \dot{L}_{\text{avail}} + T_\circ \dot{S}_{\text{nat}}. \tag{13.38}$$

13.3.2   Closed Systems

When a closed system changes its thermodynamic state from an equilibrium state 1 to an equilibrium state 2, relationship (13.33) can be written as

$$L_{\text{avail}} = -M(a_1 - a_2) - \sum_{i=1}^{i_m} (1 - T_\circ/T_i)\, Q_i, \tag{13.39}$$

where the term

$$a_1 = (u_1 - u_o) + P_\circ(v_1 - v_\circ) - T_\circ(s_1 - s_\circ) + \omega_1^2/2 + g z_1 \tag{13.40}$$

is called the **specific exergy of a closed system in state 1** (as before, subscript o denotes the properties of the dead state), and the term

$$a_2 = (u_2 - u_o) + P_\circ(v_2 - v_\circ) - T_\circ(s_2 - s_\circ) + \omega_2^2/2 + g z_2 \tag{13.41}$$

is called the **specific exergy of the system in state 2**.

Expression (13.39) can now be given in the following form:

$$L_{\text{avail}} = -\sum_{i=1}^{i_m} (1 - T_\circ/T_i)\, Q_i + M(a_2 - a_1)$$

$$= -\sum_{i=1}^{i_m} (1 - T_\circ/T_i)\, Q_i + M[(u_2 - u_1) + P_\circ(v_2 - v_1)$$

$$- T_\circ(s_2 - s_1) + (\omega_2^2 - \omega_1^2)/2 + g(z_2 - z_1)], \tag{13.42}$$

and production of the natural entropy in the system can be obtained from a relationship similar to expression (13.27):

$$L_{\text{actual}} = L_{\text{avail}} + T_\circ \Delta S_{\text{nat}}. \tag{13.43}$$

In closed nonflow systems, the changes of kinetic and potential energies of the matter locked in the systems are negligible. Then, the specific exergies (13.40) and (13.41) can be given as

$$a_1 = (u_1 - u_o) + P_\circ(v_1 - v_\circ) - T_\circ(s_1 - s_\circ), \tag{13.44}$$

and

$$a_2 = (u_2 - u_o) + P_\circ(v_2 - v_\circ) - T_\circ(s_2 - s_\circ). \tag{13.45}$$

13.3.3 The Second-Law Efficiency

Reversibly-operating (no dissipation of energy) energy-processing devices are studied in Section 10.6 under the constraints of the First Law of Thermodynamics (see expressions (10.16)–(10.18)). Therefore, the thermal efficiencies of the devices are called the **first-law efficiencies** η_{I}. In this and the following sections, we study the **second-law efficiencies** of the devices. The latter efficiencies, often denoted in literature as η_{II}, account for *both* conservation of energy and entropic effects (the energy dissipation) using the First Law of Thermodynamics as well as the Second Law of Thermodynamics.

The second-law efficiency of a work-producing device is defined as the ratio of the actual work (power) *produced* by the device during its operation to the work (power) that is *available* to the device during the operation,

$$\eta_{\text{II}} = \frac{\dot{L}_{\text{actual}}}{\dot{L}_{\text{avail}}} \ \text{(open system)} \quad \text{or} \quad \eta_{\text{II}} = \frac{L_{\text{actual}}}{L_{\text{avail}}} \ \text{(closed system)}. \tag{13.46}$$

The second-law efficiency of a work-consuming device is

$$\eta_{\text{II}} = \frac{\dot{L}_{\text{avail}}}{\dot{L}_{\text{actual}}} \ \text{(open system)} \quad \text{or} \quad \eta_{\text{II}} = \frac{L_{\text{avail}}}{L_{\text{actual}}} \ \text{(closed system)}. \tag{13.47}$$

Definitions (13.46) and (13.47) do not apply to devices that neither produce nor consume work (see Problem 13.1).

The discussion among thermodynamicists on what is the best definition of the efficiency of energy-processing devices has not ended yet, but most of them agree that the following definition is acceptable:

$$\eta_{\text{II}} = \frac{\text{total exergy output}}{\text{total exergy input}} = 1 - \frac{\text{exergy destroyed}}{\text{total exergy input}}. \tag{13.48}$$

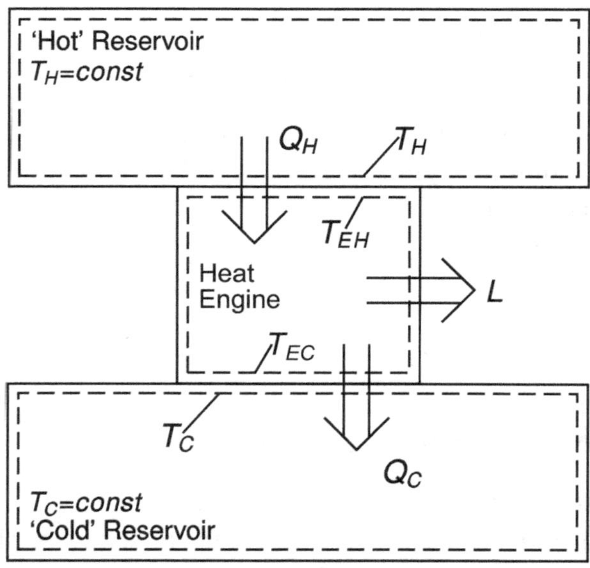

Fig. 13.2: Operation of a heat engine that receives heat Q_H from the 'hot' reservoir, produces work L and rejects heat Q_C to the 'cold' reservoir. Surfaces of the walls through which heat passes have different temperatures ($T_C \neq T_{EC}, T_H \neq T_{EH}$).

13.3.4 Exergy of Thermodynamic Cycles

Consider the power system shown in Fig. 13.2. It consists of a central unit (heat engine) operating as a thermodynamic cycle, a 'hot' reservoir of a (constant) temperature T_H, and a 'cold' reservoir of a (constant) temperature T_C. The engine receives heat Q_H through its heat-receiving port of a variable temperature T_{EH} and rejects heat Q_C through its heat-ejecting port of a variable temperature T_{EC} ($T_H > T_{EH}, T_C < T_{EC}$).

Heat leaves (enters) the ith reservoir ($i = 1$ for the hot reservoir, $i = 2$ for the cold reservoir) at constant temperature $T_{r,i}$ ($T_{r,1} = T_H$ and $T_{r,2} = T_C$) but does not change the reservoir's entropy. Heat enters (leaves) the engine at variable temperatures $T_{e,i}$ ($T_{e,1} = T_{EH}$, and $T_{e,2} = T_{EC}$), generating some natural entropy in the engine. The heat transfers destroy some exergy, and the sum in the relationship (13.39) should be replaced by the following integral:

$$\int \left(1 - \frac{T_\circ}{T_{e,i}}\right) \delta Q_i. \tag{13.49}$$

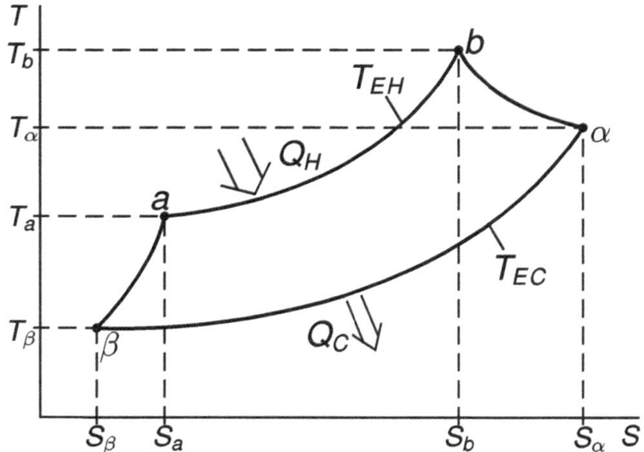

Fig. 13.3: Thermodynamic cycle $a \to b \to \alpha \to \beta$ of a heat engine interacting with two heat reservoirs (see Fig. 13.2). Temperatures T_H (of the 'hot' reservoir) and T_C (of the 'cold' reservoir) are constant, but temperatures T_{EH} (of the engine-side of the wall separating the engine and the 'hot' reservoir) and T_{EC} (of the engine-side of the wall separating the engine and the 'cold' reservoir) change in a limited way ($T_a < T_{EH} < T_b$ and $T_\beta < T_{EC} < T_\alpha$) during the system operations.

When temperatures T_{EH} and T_{EC} change during the engine's operation, the changes of the engine's entropy caused by the non-isothermal heat transfers are, respectively (see expression (2.12)),

$$\Delta S_{H \to E} = \int \frac{\delta Q_H}{T_{EH}}, \tag{13.50}$$

and

$$\Delta S_{E \to C} = \int \frac{\delta Q_C}{T_{EC}}. \tag{13.51}$$

Thus, the available work of the irreversibly-operating engine during one thermodynamic cycle is (see expressions (13.42) and (13.44)–(13.45))

$$L_{\text{avail}} = M(a_2 - a_1) - \int \delta Q_H - \int \delta Q_C + T_\circ \left(\int \frac{\delta Q_H}{T_{EH}} + \int \frac{\delta Q_C}{T_{EC}} \right). \tag{13.52}$$

The thermodynamic cycle of an engine operating in a way similar to that discussed here is shown in Fig. 13.3. One should notice that we have there the following three irreversible processes that produce natural entropy and, therefore, reduce the actual work of the engine: 1) the irreversible absorption of heat Q_H (at $T_{EH} \neq$ const) (during the absorption, the matter in the engine moves from state a to state b), 2) the irreversible ejection of heat Q_C (at $T_{EC} \neq$ const) (during the ejection the matter in the engine moves from state α to state β), and 3) the process inside the engine. The processes 1, 2, and 3 reduce the actual work of the engine by the following amounts, respectively:

$$T_\circ \Delta S_{ab} = T_\circ \int \frac{\delta Q_H}{T_{EH}} = T_\circ M(s_b - s_a), \tag{13.53}$$

$$T_\circ \Delta S_{\alpha\beta} = T_\circ \int \frac{\delta Q_C}{T_{EC}} = T_\circ M(s_\beta - s_\alpha), \tag{13.54}$$

and

$$T_\circ \Delta S_{\text{nat},e}, \tag{13.55}$$

where M is the mass of the matter working in the engine; s_a and s_b are the specific entropies of the matter in states a and b, s_α and s_β are the specific entropies of the matter in states α and β, and $\Delta S_{\text{nat},e}$ is the natural entropy generated inside the engine.

Problem 13.1: The external walls of the heat exchanger shown in Fig. 13.4 and operating in a steady way are insulated well enough to assume that no heat is exchanged between the exchanger and the standard atmosphere (the Atmosphere) in equilibrium at pressure $P_\circ = 1$ bar and temperature $T_\circ = 25\,°C$. A 'hot' stream of superheated water vapor at pressure $P_1 = 7$ bar and temperature $T_1 = 920\,K$ enters the exchanger through inlet 1 and leaves it through outlet 2 where the vapor pressure is $P_2 = 6.8$ bar and its temperature is $T_2 = 690\,K$. The 'cold' stream of air at $P_3 = 1.2$ bar and $T_3 = 400\,K$ enters the exchanger through inlet 3 and leaves it through outlet 4, where the gas pressure is $P_4 = 1.1$ bar. Both of the streams are ideal gases in some local equilibria at all inlets and outlets. The changes of speeds of the flows in both parts of the exchanger are negligible and so are the changes of the flows' potential energies. Enthalpies of ideal gases in local equilibrium depend only on the gases' temperatures – see Section 7.5. The (constant) mass flow rates at the inlets and outlets are $\dot{M}_1 = \dot{M}_2 = 15\,\text{kg/s}$ (the vapor) and $\dot{M}_3 = \dot{M}_4 = 25\,\text{kg/s}$ (the air). Find the temperature of the air at outlet 4, the second-law efficiency of the exchanger, the rate of the exergy destruction, and the amount of the natural entropy produced during the exchanger operation.

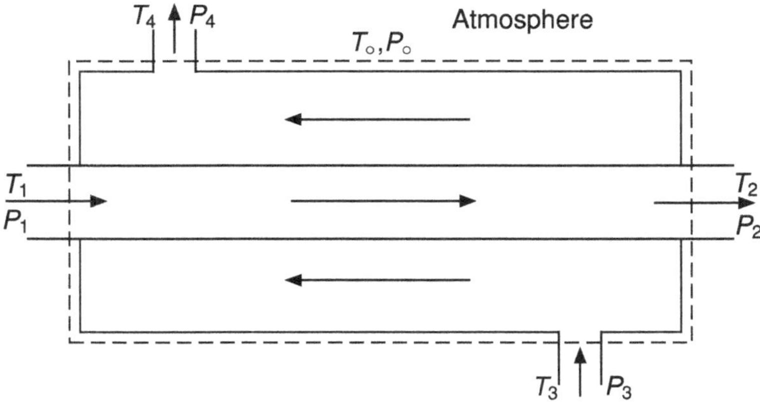

Fig. 13.4: Heat exchanger where 'hot' stream of water vapor enters the device through inlet 1 and leaves it through outlet 2 while 'cold' air enters the device through inlet 3 and leaves it through outlet 4. There is no heat transfer between the exchanger and the Atmosphere.

Solution: The heat exchanger is a steady-flow open system (of control volume marked in Fig. 13.4 with the dashed line) that neither produces nor consumes work, and does not exchange heat with the environment. Since the 'cold' stream gains exergy while the 'hot' stream loses exergy, the second-law efficiency of the system can be given by relationships (13.48) and (13.37) (the convention for the Sign of Exergy is similar to the International Convention for the Sign of Energy),

$$\eta_{II} = \frac{\text{rate of increase of the exergy of the 'cold' stream}}{\text{rate of decrease of the exergy of the 'hot' stream}}$$
$$= \frac{\dot{M}_3(b_4 - b_3)}{-\dot{M}_1(b_2 - b_1)} = \frac{\dot{M}_3[(h_4 - h_3) - T_\circ(s_4 - s_3)]}{\dot{M}_1[(h_1 - h_2) - T_\circ(s_1 - s_2)]}, \qquad (13.56)$$

where we assume that the changes of the kinetic and potential energies of the flowing gases are negligible.

Since we do not know the value of T_4 (the temperature of the 'cold' stream at the outlet 4), we do not know the value of h_4 (the specific enthalpy of the stream at the outlet). Thus, there are two unknowns in relationship (13.56) and another relationship is needed to solve the problem. This relationship is the

First Law of Thermodynamics for steady flows (see expression (3.35)),

$$0 = \dot{M}_1 h_1 + \dot{M}_3 h_3 - \dot{M}_1 h_2 - \dot{M}_3 h_4, \tag{13.57}$$

so that

$$h_4 = \frac{\dot{M}_1 h_1 + \dot{M}_3 h_3 - \dot{M}_1 h_2}{\dot{M}_3}. \tag{13.58}$$

According to the *Tables of Thermodynamic Properties of Water*, the specific enthalpies of the water vapor and the air at the exchanger inlets and outlets where the flowing gases (both treated as ideal gases in states of local equilibrium) are:

$$h_1 = 3980\,\text{kJ/kg}, \qquad h_2 = 3320\,\text{kJ/kg}, \qquad h_3 = 400\,\text{kJ/kg}. \tag{13.59}$$

Thus, according to expression (13.58), one has

$$\begin{aligned}
h_4 &= (\dot{M}_1/\dot{M}_3)(h_1 - h_2) + h_3 \\
&= (15/25)(3980 - 3320)\,\text{kJ/kg} + 400\,\text{kJ/kg} \\
&= 796\,\text{kJ/kg}.
\end{aligned} \tag{13.60}$$

The gases' specific entropies at the heat exchanger's inlets and outlets are

$$s_1 = 8.44\,\text{kJ/kg·K}, \quad s_2 = 7.61\,\text{kJ/kg·K}, \tag{13.61}$$

$$s_3 = 1.99\,\text{kJ/kg·K}, \quad s_4 = 2.67\,\text{kJ/kg·K}, \tag{13.62}$$

where we used expression (8.33) since the specific entropies of ideal gases are functions of both temperature and pressure.

Taking the above into account, the rate of increase of exergy of the 'cold' stream is

$$\begin{aligned}
\Delta \dot{B}_{3 \to 4} &= \dot{M}_3(b_4 - b_3) = \dot{M}_3[(h_4 - h_3) - T_\circ(s_4 - s_3)] \\
&= (25\,\text{kg/s})[(798 - 400) - 298.15(2.67 - 1.99)]\,\text{kJ/kg} \\
&= 4831\,\text{kJ/s},
\end{aligned} \tag{13.63}$$

and the rate of decrease of exergy of the 'hot' stream is

$$\begin{aligned}
\Delta \dot{B}_{1 \to 2} &= \dot{M}_1(b_2 - b_1) = \dot{M}_1[(h_2 - h_1) - T_\circ(s_2 - s_1)] \\
&= (15\,\text{kg/s})[(3320 - 3980) - 298.15(7.61 - 8.44)]\,\text{kJ/kg} \\
&= -6188\,\text{kJ/s}.
\end{aligned} \tag{13.64}$$

Thus, the rate of the total change of exergy in the exchanger is

$$\Delta\dot{B} = \Delta\dot{B}_{1\to2} + \Delta\dot{B}_{3\to4} = (4831 - 6188)\,\text{kJ/s} = -1356\,\text{kJ/s}. \qquad (13.65)$$

Since this rate can be related to the rate $\dot{S}_{\text{nat}}$,

$$|\Delta\dot{B}| = T_\circ\dot{S}_{\text{nat}}, \qquad (13.66)$$

one has

$$\dot{S}_{\text{nat}} = \frac{|\Delta\dot{B}|}{T_\circ} = \frac{1356\,\text{kJ/s}}{298.15\,\text{K}} = 4.55\,\text{kJ/K·s}, \qquad (13.67)$$

and the second-law efficiency of the device is

$$\eta_{\text{II}} = \frac{\Delta\dot{B}_{3\to4}}{-\Delta\dot{B}_{1\to2}} = \frac{4831\,\text{kJ/s}}{6188\,\text{kJ/s}} = 0.78. \qquad (13.68)$$

End of Problem 13.1.

Problem 13.2: A steady stream of superheated water vapor (an ideal gas) enters a turbine at pressure $P_1 = 50\,\text{bar}$, temperature $T_1 = 800\,\text{K}$, and speed $\omega_1 = 180\,\text{m/s}$, and leaves the turbine at pressure $P_2 = 1.1\,\text{bar}$, temperature $T_2 = 380\,\text{K}$, and speed $\omega_2 = 100\,\text{m/s}$ (see Fig. 13.5). The turbine steadily loses Q_r of heat every second. The heat leaves the physical boundary of the turbine at constant temperature $T_r = 600\,\text{K}$ which is the temperature of the outer surface of the turbine. The turbine environment is the Atmosphere. The measured actual useful power (the shaft power) of the turbine per kilogram of the flowing stream is $\dot{L}_{\text{actual}}/\dot{M} = 800\,\text{kJ/kg}$ where $\dot{M}$ is the mass flow rate of the stream driving the turbine. Calculate the rate of the heat leaving the turbine, the turbine's second-law efficiency, and the rate of the natural entropy production in the turbine.

Solution: Since the flow of the matter through the turbine is almost horizontal, the contribution of the changes of the matter potential (gravitational) energy to the energy balance of the discussed open system can beignored.

The Principle of Conservation of Matter for steady flow gives

$$\dot{M}_1 = \dot{M}_2 = \dot{M}, \qquad (13.69)$$

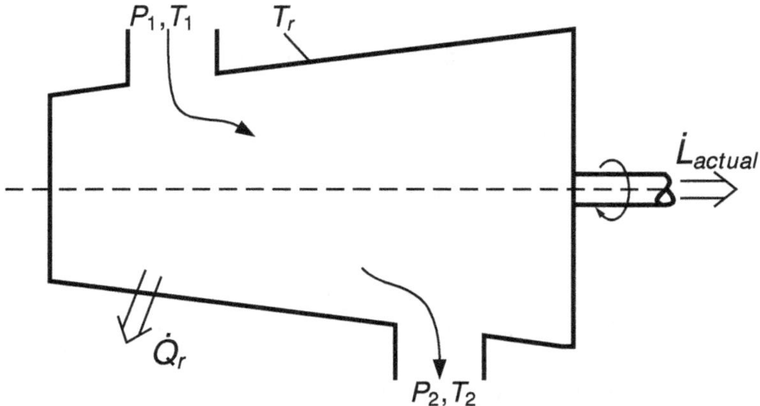

Fig. 13.5: Steady flow of water vapor enters a turbine through inlet 1 and leaves through outlet 2. The rate of heat loss to the environment is $\dot{Q}_r$. The actual useful power $\dot{L}_{actual}$ is transferred to some external device by a rotating shaft. The temperature of the outer surface of the turbine physical boundary is T_r.

where $\dot{M}_1$ and $\dot{M}_2$ are the mass flow rates of the vapor at the system's inlet and outlet, respectively.

The Principle of Conservation of Energy leads to (see expression (3.35))

$$\dot{L}_{actual} + \dot{Q}_r = \dot{M}[(h_2 - h_1) + (\omega_2^2/2 - \omega_1^2/2)]. \qquad (13.70)$$

Relationships (13.69) and (13.70) give the following rate of the heat (per kilogram of the vapor moving through the turbine) that is transferred from the turbine to the surroundings:

$$\frac{\dot{Q}_r}{\dot{M}} = -\frac{\dot{L}_{actual}}{\dot{M}} + (h_2 - h_1) + (\omega_2^2/2 - \omega_1^2/2). \qquad (13.71)$$

According to the *Tables of Thermodynamic Properties of Water*, the specific enthalpies and specific entropies of the superheated vapor in local equilibria at inlet 1 ($T = T_1, P = P_1$) and at outlet 2 ($T = T_2, P = P_2$) are

$$h_1 = 3496 \text{ kJ/kg}, \quad h_2 = 2689 \text{ kJ/kg}, \qquad (13.72)$$

and

$$s_1 = 7.05 \text{ kJ/kg·K}, \quad s_2 = 7.35 \text{ kJ/kg·K}. \qquad (13.73)$$

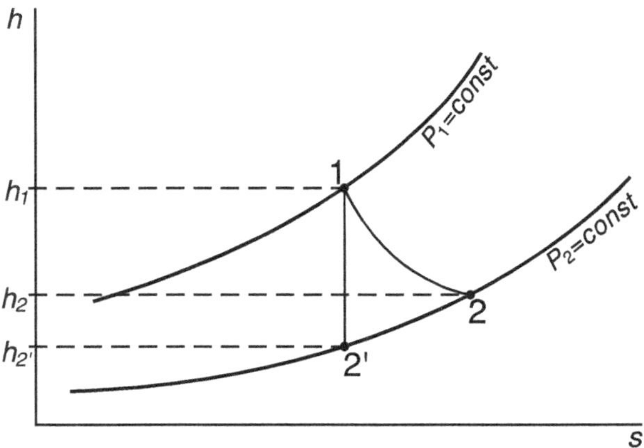

Fig. 13.6: The change of the specific enthalpy of the flow running the turbine in Problem 13.2: $\Delta h = h_{2'} - h_1$ (in isentropic-adiabatic flow) or $\Delta h = h_2 - h_1$ (in real flow where there is some increase of the flow's entropy during the turbine operation).

Thus, according to expression (13.71), the heat leaving the turbine (per one kilogram of the flowing matter) is

$$\frac{\dot{Q}_r}{\dot{M}} = -800\,\frac{\text{kJ}}{\text{kg}} + (2689 - 3496)\,\frac{\text{kJ}}{\text{kg}} + (100^2/2 - 180^2/2)\,\frac{\text{kJ}}{1000\,\text{kg}}$$
$$= -18.2\,\text{kJ/kg}. \tag{13.74}$$

The specific kinetic energies of the vapor entering and leaving the turbine are, respectively,

$$e_{\text{kin},1} = \omega_1^2/2 = (180\,\text{m/s})^2/2 = 16.2\,\text{kJ/kg}, \tag{13.75}$$

and

$$e_{\text{kin},2} = \omega_2^2/2 = (100\,\text{m/s})^2/2 = 5\,\text{kJ/kg}, \tag{13.76}$$

and the maximum work (per kilogram of the flowing vapor) available to the turbine is (see relationship (13.37))

$$\frac{\dot{L}_{\text{avail}}}{\dot{M}} = h_2 - h_1 - T_o(s_2 - s_1) + e_{\text{kin},2} - e_{\text{kin},1} - \left(1 - \frac{T_o}{T_r}\right)\frac{\dot{Q}_r}{\dot{M}}$$

$$= \left[(2689 - 3496)\,\frac{\text{kJ}}{\text{kg}} - (298.15\,\text{K})(7.35 - 7.05)\,\frac{\text{kJ}}{\text{kg}\cdot\text{K}}\right.$$

$$\left. + (5 - 16.2)\,\frac{\text{kJ}}{\text{kg}}\right] - \left(1 - \frac{298.15\,\text{K}}{600\,\text{K}}\right)(-18.2)\,\frac{\text{kJ}}{\text{kg}}$$

$$= -898.49\,\text{kJ/kg}, \tag{13.77}$$

where T_o is the temperature of the Atmosphere.

Taking the above into account, the second-law efficiency of the turbine is

$$\eta_{\text{II}} = \frac{\dot{L}_{\text{actual}}/\dot{M}}{\dot{L}_{\text{avail}}/\dot{M}} = \frac{-800\,\text{kJ/kg}}{-898.49\,\text{kJ/kg}} = 0.89, \tag{13.78}$$

and the amount of the natural entropy (per kilogram of the flowing vapor) generated in the turbine is (see expression (13.38))

$$\frac{\dot{S}_{\text{nat}}}{\dot{M}} = \frac{1}{T_o}\left(\frac{\dot{L}_{\text{actual}}}{\dot{M}} - \frac{\dot{L}_{\text{avail}}}{\dot{M}}\right) = \frac{1}{298.15\,\text{K}}\,[-800 - (-898.49)]\,\frac{\text{kJ}}{\text{kg}}$$

$$= 0.33\,\text{kJ/kg}\cdot\text{K}. \tag{13.79}$$

It should be noticed that the amount of heat leaving the turbine is equal to a small fraction of the useful work produced by the turbine,

$$\frac{(\dot{Q}_r/\dot{M})}{(\dot{L}_{\text{actual}}/\dot{M})} = \frac{-18.2\,\text{kJ/kg}}{-800\,\text{kJ/kg}} = 0.023. \tag{13.80}$$

The ratios of the specific kinetic energies of the stream at the inlet and at the outlet to the corresponding specific enthalpies are, respectively,

$$\frac{e_{\text{kin},1}}{h_1} = \frac{16.2\,\text{kJ/kg}}{3496\,\text{kJ/kg}} = 0.005 \tag{13.81}$$

and

$$\frac{e_{\text{kin},2}}{h_2} = \frac{5\,\text{kJ/kg}}{2689\,\text{kJ/kg}} = 0.002. \tag{13.82}$$

End of Problem 13.2.

13.4 Work Availability in Chemical Reactions

The structures of most (or all) molecules participating in every chemical reaction are broken, and new bonds are formed to create new molecular structures. In other words, an occurrence of a chemical reaction in a thermodynamic system can rearrange the system energy in such a way that less actual work is available than would be in the system if it was a non-reactive system. Therefore, chemical reactions are good examples of irreversible processes.

Consider a steady flow of an ideal gas in an open system where no useful work is transferred between the system and the standard atmosphere of pressure $P_\circ = 1$ bar and temperature $T_\circ = 25\,°C$. The changes of the kinetic and potential energies of the matter flowing steadily through the system are zero. We also assume, in order to simplify the mathematical part of the problem, that there is only one chemical reaction occurring in the flow between the single inlet and the single outlet of the system. The mixture at the inlet consists of the reactants of the chemical reaction occurring in the system. The entering flow is in local equilibrium at pressure P_{inlet} and temperature T_{inlet}. The mixture leaving the system through its outlet consists of the products of the reaction which are in local equilibrium at pressure P_{outlet} and temperature T_{outlet}. The complete chemical reaction taking place in the flow between the system's inlet and outlet follows the reaction's stoichiometric equation. The heat possibly transferred between the system and the standard atmosphere is $Q_\circ$.

According to expression (13.25), the work lost during the reaction in the flow between the system inlet and outlet is

$$L_{lost} = L_{irr} - L_{rev} = L_{actual} - L_{avail} = -L_{avail} \tag{13.83}$$

because we assume that there is no actual (useful) work transferred between the system and the surroundings.

The work available in the steady reactive flow moving between the inlet and the outlet is (see expression (13.34))

$$L_{avail} = B_{outlet} - B_{inlet}, \tag{13.84}$$

where

$$B_{inlet} = \sum_i |\nu_i|[\bar{H}_i(P_i, T_{inlet}) - T_\circ \bar{S}_i(P_i, T_{inlet})], \tag{13.85}$$

and

$$B_{\text{outlet}} = \sum_{j} v_j [\bar{H}_j(P_j, T_{\text{outlet}}) - T_{\text{o}}\bar{S}_j(P_j, T_{\text{outlet}})], \tag{13.86}$$

where, as before, the overbar indicates the *molar* value of the corresponding quantity, P_i is the partial pressure of the ith reactant entering the system, and P_j is the partial pressure of the jth product leaving the system; v_i is the mol number of the ith reactant, and v_j is the mol number of the jth product (the stoichiometric coefficients v_i are negative numbers while the stoichiometric coefficients v_j are positive numbers – see Section 12.1); $\bar{H}_i(P_i, T_{\text{inlet}})$ and $\bar{H}_j(P_j, T_{\text{outlet}})$ are the molar enthalpy of the ith reactant and the molar enthalpy of the jth product, respectively, and $\bar{S}_i(P_i, T_{\text{inlet}})$ and $\bar{S}_j(P_j, T_{\text{outlet}})$ are the corresponding molar entropies.

Problem 13.3: A mixture of 2 mols of diatomic hydrogen and 1 mol of diatomic oxygen enters a combustion chamber through its inlet. The gases react completely during a steady flow between the chamber's inlet and outlet according to the following stoichiometric equation:

$$2H_2(g) + O_2(g) \rightarrow 2H_2O(g). \tag{13.87}$$

As a result, 2 mols of the reaction's single product (gas vapor H_2O) leaves the chamber through the outlet. The changes of the kinetic and potential energies of the components moving steadily through the combustion chamber are negligible. The temperature of the reactants at the inlet and the temperature of the single product at the outlet are the same and equal to the temperature of the standard atmosphere, $T_{\text{inlet}} = T_{\text{outlet}} = T_{\text{o}} = 25\,°C = 298.15\,\text{K}$. The pressure of the mixture (H_2 plus O_2) entering the system is $P_{\text{inlet}} = 1.2\,\text{bar}$, and the pressure of the product leaving the system is $P_{\text{outlet}} = 1\,\text{bar}$. The gases at the inlet and at the outlet are in some states of local-equilibrium. How much work will be lost during reaction (13.87)?

Solution: The temperatures of the gases present in the chamber are not low, and their pressures are not high. Therefore, all components of reaction (13.87) can be treated as ideal gases. As discussed earlier, the enthalpy of ideal gas in equilibrium (or local equilibrium) is pressure-independent, so that molar enthalpies of the components at a given temperature and *any* pressure are the same. We choose the pressure as the standard pressure $P° = 1\,\text{bar}$ because thermochemical properties of standard-state ideal gases in equilibrium (local equilibrium) are available in thermochemical databases. Then, the partial pressures $P_i \equiv [P_{H_2(g)}, P_{O_2(g)}]$ and $P_j \equiv P_{H_2O(g)}$ in relationships (13.85) and (13.86) can be replaced by the standard-state pressure P°. Subsequently, the molar enthalpies of the reactants and the product of reaction (13.87) can be given, respectively, as (see relationship (8.25))

$$\bar{H}_i(P_i, T_{\text{inlet}}) = \bar{H}_i(P^\circ, T_\circ) = \Delta_f H_i^\circ(T_\circ) + [\bar{H}_i(P^\circ, T_\circ) - \bar{H}_i^\circ(T_\circ)]$$
$$= \Delta_f H_i^\circ(T_\circ), \tag{13.88}$$

and

$$\bar{H}_j(P_j, T_{\text{outlet}}) = \bar{H}_j(P^\circ, T_\circ) = \Delta_f H_j^\circ(T_\circ) + [\bar{H}_j(P^\circ, T_\circ) - \bar{H}_j^\circ(T_\circ)]$$
$$= \Delta_f H_j^\circ(T_\circ), \tag{13.89}$$

where $\Delta_f H_i^\circ(T_\circ)$ is the enthalpy of formation of the ith reactant (H_2 or O_2) at temperature $T_{\text{inlet}} = T_\circ = 298.15\,\text{K}$, and $\Delta_f H_j^\circ(T_\circ)$ is the enthalpy of formation of the reaction's product (H_2O) at temperature $T_{\text{outlet}} = T_\circ$.

According to the *NIST-JANAF Thermochemical Tables*, the molar enthalpies of formation of the reactants entering the system are

$$\Delta_f H^\circ(T_\circ)_{H_2(g)} = 0 \qquad \text{and} \qquad \Delta_f H^\circ(T_\circ)_{O_2(g)} = 0, \tag{13.90}$$

and the molar enthalpy of formation of the product leaving the system is

$$\Delta_f H^\circ(T_\circ)_{H_2O(g)} = -241.826\,\text{kJ/mol}. \tag{13.91}$$

The entropies of the reactants in reaction (13.87) depend on the reactants' temperature as well as their partial pressures (see expressions (8.30) and (8.32)),

$$\bar{S}_i(P_i, T_{\text{inlet}}) = \bar{S}_i(P_i, T_\circ) = \bar{S}_i^\circ(T_\circ) + [\bar{S}_i(P_i, T_\circ) - \bar{S}_i^\circ(T_\circ)]$$
$$= \bar{S}_i^\circ(T_\circ) - R\ln(P_i/P^\circ), \tag{13.92}$$

and the entropy of the reaction's product also depends on both temperature and pressure ($P_{\text{outlet}} = P^\circ$, $T_{\text{outlet}} = T_\circ$),

$$\bar{S}_j(P_j, T_{\text{outlet}}) = \bar{S}_j(P_{\text{outlet}}, T_\circ) = \bar{S}_j^\circ(T_\circ) + [\bar{S}_j(P_{\text{outlet}}, T_\circ) - \bar{S}_j^\circ(T_\circ)]$$
$$= \bar{S}_j^\circ(T_\circ), \tag{13.93}$$

where R is the universal gas constant, and, according to Dalton's Law (see Section 9.3), the partial pressures of the reactants are

$$P_i = z_i P_{\text{inlet}}, \tag{13.94}$$

and their molar fractions are (see expression (9.40))

$$z_i = v_i / \sum_i v_i, \tag{13.95}$$

with the sum being taken over all reactants of reaction (13.87). The fractions are

$$z_{\text{H}_2(\text{g})} = 2/3 \qquad \text{and} \qquad z_{\text{O}_2(\text{g})} = 1/3. \tag{13.96}$$

Thus, the partial pressures of the reactants entering the chamber are

$$P_{\text{H}_2(\text{g})} = (2/3)(1.2\,\text{bar}) = 0.8\,\text{bar}, \tag{13.97}$$
$$P_{\text{O}_2(\text{g})} = (1/3)(1.2\,\text{bar}) = 0.4\,\text{bar}, \tag{13.98}$$

and the pressure of the product leaving the chamber is

$$P_{\text{H}_2\text{O}(\text{g})} = P_{\text{outlet}} = 1\,\text{bar}. \tag{13.99}$$

According to the *NIST-JANAF Thermochemical Tables*, the molar entropies of the discussed components in local thermodynamic equilibrium at pressure P° and temperature T_o are

$$\bar{S}^\circ(T_\text{o})_{\text{H}_2(\text{g})} = 130.57\,\text{J/mol·K}, \tag{13.100}$$
$$\bar{S}^\circ(T_\text{o})_{\text{O}_2(\text{g})} = 205.03\,\text{J/mol·K}, \tag{13.101}$$

and

$$\bar{S}^\circ(T_\text{o})_{\text{H}_2\text{O}(\text{g})} = 188.72\,\text{J/mol·K}. \tag{13.102}$$

Taking the above into account, the exergy of the gas entering the system is (see expression (13.85))

$$B_{\text{inlet}} = |v_{\text{H}_2(\text{g})}|B' + |v_{\text{O}_2(\text{g})}|B'', \tag{13.103}$$

where

$$
\begin{aligned}
B' &= \Delta_f H^\circ(T_\text{o})_{\text{H}_2(\text{g})} - T_\text{o}\left[\bar{S}^\circ(T_\text{o})_{\text{H}_2(\text{g})} - R\ln(P_{\text{H}_2(\text{g})}/P^\circ)\right] \\
&= (-298.15\,\text{K})[130.57 - 8.314\ln(0.8/1)]\,\text{J/mol} \\
&= -39{,}470.6\,\text{J/mol}, \tag{13.104}
\end{aligned}
$$

and

$$B'' = \Delta_f H^\circ(T_\circ)_{O_2(g)} - T_\circ \left[\bar{S}^\circ(T_\circ)_{O_2(g)} - R\ln(P_{O_2(g)}/P^\circ)\right]$$
$$= (-298.15\,\mathrm{K})[205.03 - 8.314\ln(0.4/1)]\,\mathrm{J/mol}$$
$$= -63{,}351.8\,\mathrm{J/mol}. \tag{13.105}$$

Thus,

$$B_{\text{inlet}} = -142.34\,\mathrm{kJ}, \tag{13.106}$$

and the exergy of the gas leaving the system is (see expression (13.86))

$$B_{\text{outlet}} = \nu_{H_2O(g)}\left[\Delta_f H^\circ(T_\circ)_{H_2O(g)} - T_\circ\bar{S}^\circ(T_\circ)_{H_2O(g)}\right]$$
$$= (2\,\mathrm{mol})[-241{,}001\,\mathrm{J/mol} - (298.15\,\mathrm{K})(188.72\,\mathrm{J/mol})]$$
$$= -594.53\,\mathrm{kJ}. \tag{13.107}$$

The work available in the chamber during reaction (13.87) is (see expression (13.84))

$$L_{\text{avail}} = B_{\text{outlet}} - B_{\text{inlet}} = [-594.53 - (-142.34)]\,\mathrm{kJ} = -452.17\,\mathrm{kJ}, \tag{13.108}$$

and the work lost during the reaction is (see expression (13.83))

$$L_{\text{lost}} = -L_{\text{avail}} = 452.17\,\mathrm{kJ}. \tag{13.109}$$

One can see that the discussed combustion process has the ability to deliver some work, but this work is entirely lost during the process.

End of Problem 13.3.

Appendix A

Table 8.1: Molar thermodynamic properties of diatomic oxygen (ideal gas) listed in the *NIST-JANAF Thermochemical Tables* (see Chapter 8). Data in the 2nd, 3rd, and 4th columns are in J/K·mol, and data in the 5th, 6th, and 7th columns are in kJ/mol.

$O_2(g)$	$T_o = 298.15$ K, $P^\circ = 0.1$ MPa					
T/K	$\bar{C}_p^\circ$	$\bar{S}^\circ$	$-\dfrac{\bar{G}^\circ - \bar{H}^\circ(T_o)}{T}$	$\bar{H}^\circ - \bar{H}^\circ(T_o)$	$\Delta_f H^\circ$	$\Delta_f G^\circ$
0	0.0	0.0	∞	−8.68	0.	0.
100	29.10	173.30	231.09	−5.77	0.	0.
200	29.12	193.48	207.82	−2.86	0.	0.
298.15	29.37	205.14	205.14	0.	0.	0.
300	29.38	205.32	205.14	0.05	0.	0.
400	30.10	213.87	206.30	3.02	0.	0.
500	31.09	220.69	208.52	6.08	0.	0.
600	32.09	226.45	211.04	9.24	0.	0.
700	32.98	231.46	213.61	12.49	0.	0.
800	33.73	235.92	216.12	15.83	0.	0.
900	34.35	239.93	218.55	19.24	0.	0.
1000	34.87	243.57	220.87	22.70	0.	0.
1100	35.30	246.92	223.09	26.21	0.	0.
1200	35.66	250.01	225.20	29.76	0.	0.
1300	35.98	252.87	227.22	33.34	0.	0.
1400	36.27	255.55	229.15	36.95	0.	0.
1500	36.54	258.06	231.00	40.59	0.	0.

Table 8.2: Molar thermodynamic properties of carbon monoxide (ideal gas) listed in the *NIST-JANAF Thermochemical Tables* (see Chapter 8). Data in the 2nd, 3rd, and 4th columns are in J/K·mol, and data in the 5th, 6th, and 7th columns are in kJ/mol.

CO(g)	$T_\circ = 298.15$ K, $P^\circ = 0.1$ MPa					
$T/$K	$\bar{C}_p^\circ$	$\bar{S}^\circ$	$-\dfrac{\bar{G}^\circ - \bar{H}^\circ(T_\circ)}{T}$	$\bar{H}^\circ - \bar{H}^\circ(T_\circ)$	$\Delta_f H^\circ$	$\Delta_f G^\circ$
0	0.000	0.000	∞	−8.67	−113.80	−113.80
100	29.10	165.85	223.53	−5.76	−112.41	−120.23
200	29.10	186.02	200.31	−2.85	−111.28	−128.52
298.15	29.14	197.65	197.65	0.000	−110.52	−137.16
300	29.14	197.83	197.65	0.05	−110.51	−137.32
400	29.34	206.23	198.79	2.97	−110.10	−146.33
500	29.79	212.83	200.96	5.93	−110.00	−155.41
600	30.44	218.31	203.41	8.94	−110.15	−164.48
700	31.17	223.06	205.89	12.02	−110.46	−173.51
800	31.89	227.27	208.30	15.17	−110.90	−182.49
900	32.57	231.07	210.62	18.40	−111.41	−191.41
1000	33.18	234.53	212.84	21.69	−111.98	−200.27
1100	33.71	237.72	214.96	25.03	−112.58	−209.07
1200	34.17	240.67	216.98	28.43	−113.21	−217.81
1300	34.57	243.43	218.91	31.86	−113.87	−226.50
1400	34.92	246.00	220.76	35.34	−114.54	−235.14
1500	35.21	248.42	222.52	38.85	−115.22	−243.74

Appendix B

Thermodynamic properties of H_2O in saturation states (see Chapter 9). T, P, ρ, v, u, h, and s are the substance temperature, pressure, mass density, specific volume, specific thermal energy, specific enthalpy, and specific entropy, respectively.

Table 9.1. Liquid (quality $x = 0$) saturation states:

| T | P | ρ | v | u | h | s |
K	bar	kg/m^3	m^3/kg	kJ/kg	kJ/kg	kJ/kg·K
273.15	0.0061	999.79	0.00100	−419.06	−419.06	−1.3069
323.15	0.1236	987.99	0.00101	−209.69	−209.67	−0.6029
373.15	1.0145	958.34	0.00104	0.0447	0.1506	0.0004
423.15	4.7629	917.00	0.00109	212.65	213.16	0.5349
473.15	15.553	864.65	0.00115	431.46	433.26	1.0237
523.15	39.768	798.88	0.00125	661.78	666.76	1.4867
573.15	85.891	712.12	0.00140	913.94	926.01	1.9483
623.15	165.31	574.67	0.00174	1223.1	1215.9	2.4716

Table 9.2. Gas (quality $x = 1$) saturation states:

| T | P | ρ | v | u | h | s |
K	bar	kg/m^3	m^3/kg	kJ/kg	kJ/kg	kJ/kg·K
273.15	0.0061	0.0048	205.99	1955.9	2081.9	7.8486
323.15	0.1236	0.0831	12.022	2023.7	2172.2	6.7677
373.15	1.0145	0.5983	1.6712	2087.0	2256.5	6.0471
423.15	4.7629	2.5487	0.3923	2140.0	2326.9	5.5301
473.15	15.553	7.8626	0.1272	2175.2	2373.0	5.1232
523.15	39.768	19.970	0.0501	2182.7	2381.9	4.7651
573.15	85.891	46.176	0.0216	2144.6	2330.6	4.3989
623.15	165.31	113.63	0.0088	1999.0	2144.5	3.9039

Index